AF792308

Active Volcanoes of the World

Series editors
Corrado Cimarelli, München, Germany
Sebastian Müller, Mainz, Germany

For further volumes:
http://www.springer.com/series/10081

Teresa Scolamacchia • José Luis Macías
Editors

Active Volcanoes of Chiapas (Mexico): El Chichón and Tacaná

IAVCEI

Springer

Editors
Teresa Scolamacchia
Earth and Environmental Sciences
Ludwig-Maximilians-Universität
München
Germany

José Luis Macías
Instituto de Geofísica
Universidad Nacional Autónoma de México
(UNAM)
Morelia
Mexico

ISSN 2195-3589 ISSN 2195-7029 (electronic)
Active Volcanoes of the World
ISBN 978-3-642-25889-3 ISBN 978-3-642-25890-9 (eBook)
DOI 10.1007/978-3-642-25890-9

Library of Congress Control Number: 2014960137

Springer Heidelberg New York Dordrecht London
© Springer-Verlag Berlin Heidelberg 2015
This work is subject to copyright. All rights are reserved by the Publisher, whether the whole or part of the material is concerned, specifically the rights of translation, reprinting, reuse of illustrations, recitation, broadcasting, reproduction on microfilms or in any other physical way, and transmission or information storage and retrieval, electronic adaptation, computer software, or by similar or dissimilar methodology now known or hereafter developed.
The use of general descriptive names, registered names, trademarks, service marks, etc. in this publication does not imply, even in the absence of a specific statement, that such names are exempt from the relevant protective laws and regulations and therefore free for general use.
The publisher, the authors and the editors are safe to assume that the advice and information in this book are believed to be true and accurate at the date of publication. Neither the publisher nor the authors or the editors give a warranty, express or implied, with respect to the material contained herein or for any errors or omissions that may have been made.

Printed on acid-free paper

Springer-Verlag GmbH Berlin Heidelberg is part of Springer Science+Business Media
(www.springer.com)

Foreword

El Chichón and Tacaná are both within the limits of the mountainous State of Chiapas in southeastern Mexico and are the only presently active volcanoes in that region. While El Chichón is the youngest of the Quaternary volcanoes forming the Chiapanecan Volcanic Arc, Tacaná is at the border with Guatemala where it is the northernmost active volcano of the Central America Volcanic Arc. The origin of both volcanoes has been traditionally related to the subduction of the oceanic Cocos Plate underneath a complex framework of continental plates.

Chichón volcano, covered by jungle in a remote region was almost unknown until it erupted cataclysmically in March–April 1982 and produced the deadliest volcanic disaster in Mexico's modern history. It was estimated that the eruption caused approximately 2,000 fatalities. Several villages and hamlets were destroyed beyond recognition by pyroclastic flows and pumiceous ash fallout and hundreds of survivors (mostly poor farmers) required urgent help. Many families were relocated to other areas in Mexico or migrated on their own. A few years later, in September 1985, a large magnitude subduction-related earthquake with an epicenter off the Pacific shore of Michoacán shook central Mexico and caused great damage, especially in Mexico City. Hundreds of buildings collapsed burying thousands of people under the rubble. As a result of these tragedies, the awareness of the possible occurrence of future volcanic and seismic disasters on Mexican territory rose dramatically. In both emergencies, the Mexican Army had been ordered to take control of the situation but obviously this simplistic approach was not sufficient and created problems on a political level that had long-lasting effects that had to be dealt with. Public officials, earth scientists, and other professionals that were together with the Army directly involved in the assessment and management of these disasters recognized the urgent need to take more differentiated measures that would mitigate the complexities derived from such types of catastrophes in the future. These efforts crystallized the implementation of a Civil Protection System (coordinated by the federal government), the creation of the Centro Nacional de Prevención de Desastres (CENAPRED), the reform and improvement of existing construction codes, and the strengthening of topic-related curricula in educational institutions (to name just a few among other steps). Although this large endeavor was mostly undertaken and supported by Mexican institutions, technical personnel and financial aid also came from outside of the country (e.g., Japanese International Cooperation Agency, US Geological Survey, etc.).

In the above context, the Instituto de Geofísica at the Universidad Nacional Autónoma de México (UNAM) has also played an important role: New faculty staff devoted to seismology and volcanology have been hired systematically over the past decades; the national seismic network (Servicio Sismológico Nacional) has been modernized and expanded greatly, other types of geophysical instruments have been deployed, chemical and sedimentological laboratories have been developed, etc. Furthermore, numerous research projects focusing on different volcanoes all over the country have been carried out, frequently in conjunction with scientists from other national and foreign institutions. Although numerous scientists have been involved in carrying out more systematic studies at Mexican volcanoes in the recent past, in the particular case of the volcanoes of Chiapas, the personal engagement, endurance, and devotion of my colleagues José Luis Macías (volcanic geology) and Iouri Taran (geochemistry of hydrothermal fluids) have been outstanding and crucial to better understand these volcanoes. Over the past two decades they have guided numerous students (e.g., T. Scolamacchia, L. Capra, J.L. Arce, D. Rouwet, L. Peiffer, all authors or co-authors of different chapters) on expeditions to these remote volcanoes and supervised their respective degree-theses on specific subjects in a certainly difficult terrain. As a result, our knowledge on the subject and the number of publications in specialized journals has increased notably and the present book comes as a natural consequence of all these efforts (not surprisingly, T. Scolamacchia, co-editor of this book is also a former student of J.L. Macías). Furthermore, the geologic studies have culminated in the publication of an updated hazards map that was officially presented to the Governor of Chiapas several months ago. Such a tool is indispensable for adequate land-use planning and for development of future emergency plans. Its final publication occurred more than timely, since fumarolic and hydrothermal activity at Chichón's crater has increased and shown noticeable changes in recent months (Iouri Taran, personal communication). Whether these changes should be envisaged as premonitory phenomena to an impending eruption is not yet clear and requires urgent further investigation. Although the currently existing monitoring program certainly still requires a serious upgrade, and public awareness also needs to rise (especially in the present context), it can objectively be said that, today, the State of Chiapas is in a better position to face the negative effects of a future eruption. Whether the efforts briefly delineated above will suffice to avoid future fatalities is of course uncertain. In any event, the present book represents a milestone in regard to volcanological studies in Chiapas and will serve as a starting point for any scholar interested in pursuing future research.

The first part of the book provides a summary of the geodynamic setting and general tectonic framework of the region (Garduño et al.), as well as the petrology and geochemistry of products erupted from El Chichón and Tacaná volcanoes (Arce et al.). The second part is devoted to Chichón's eruptive history (Scolamacchia and Capra), its hydrothermal system (Peiffer et al), and the analysis of seismic data (Legrand et al.), while the third part focuses on Tacaná's eruptive history (Macías et al.) and hydrothermal system (Rouwet et al.). The fourth part addresses the risk assessment and mitigation (De la Cruz and Tilling), and an outlook into the future (Espíndola). Of these, the chapter, written by Servando de la Cruz and Robert Tilling (both true pioneers of Chichón studies and witnesses of the 1982 calamity), is noteworthy because it provides, 30 years in retrospect, some thoughts in regard to the possible causes of the 1982 disaster and presents a list of lessons learned providing valuable suggestions for the future.

The natural beauty of these volcanoes also deserves mention. In this context, a plan to submit a proposal to designate El Chichón as a "Geopark Site" under the umbrella of UNESCO's natural heritage program is currently being developed (Silvia Ramos, personal communication). Such a step might serve multiple purposes besides preserving the environment (which certainly is an important task). In addition, the victims of the 1982 eruption could be honored in no better way than by preventing further repopulation of hazardous areas, while at the same time helping to provide a sustainable living for the nearby populace by attracting more tourists.

Finally, although principally aimed at reaching the volcanological specialist, the present book might of course also serve as a valuable source of information to the curious layman. Interested tourists wanting to travel beyond the beaten paths leading to famous Maya temples might be able to better understand the gorgeous volcanic landscape and its hidden history after reading this book.

Agua Blanca, Michoacán, January 2014

Claus Siebe
Research Volcanologist and Professor
Instituto de Geofísica
Universidad Nacional Autónoma de México
Ciudad Universitaria
Coyoacán, México, D.F. Mexico

Acknowledgments

We would like to thank all the authors of the chapters of this book, who spent several months walking, sampling, and camping around these beautiful volcanoes en Chiapas, and later analyzing and interpreting the data presented here. Some of them witnessed the catastrophic eruption of El Chichón in 1982, and managed the crisis in 1985–1986 at Tacaná. Thanks for sharing your experience with the general public. We are deeply grateful to all scientists, who spent their already busy time to provide us with timely reviews, and fruitful comments, to the drafts of all chapters:

Jenny Barclay (University of East Anglia, UK), Richard J. Brown (Durham University, UK), Jon Blundy (University of Bristol, UK), Marco Calo (Berkeley Seismological Laboratory, USA), Williams Evans (US Geological Survey), Elisa Fitz Diaz (UNAM, Mexico), Simon Hughes (Terra Dat, UK), Ulrich Kueppers (Ludwig Maximilian University, Germany), Christina Neal (Volcano Science Center, U.S. Geological Survey), Karoly Németh (Massey University, New Zeland), Chris Newhall (Nanyang Technological University, Singapore), Takeshi Ohba (Tokai University, Japan), Peter Schaaf (UNAM, Mexico), Wendy Stovall (Volcano Science Center, U.S. Geological Survey), Roberto Sulpizio (Universitá degli Studi di Bari, Italy), Franco Tassi (Universitá degli Studi di Firenze, Italy), Frank Tepley III (Oregon State University, USA), Robert I. Tilling (US Geological Survey-Emeritus), Johan Varekamp (Wesleyan University, USA), Sebastian Wiesmaier (Ludwig Maximilians University, Germany).

The research summarized in several chapters of this book has been funded by Dirección General de Asuntos del Personal Académico de la Universidad Nacional Autónoma de México (DGAPA-UNAM) to Servando De la Cruz Reyna (#IN106312), Yuri Tarán (#IN10913), Denis Legrand (IACOD programs #IA100911, and #IB101112), José Luis Macías (DGAPA, #IN1142063), José Luis Arce (# IN1032053). The Consejo Nacional de Ciencia y Tecnología (CONACyT) provided funding to José Luis Macías (#47226). Teresa Scolamacchia was funded by a Deutsche Forschunggemeinschaft (DFG) fellowship (SC-152/2-1). Dmitri Rouwet, Loic Peiffer, and Yuri Taran were supported by funding of VUELCO for the period May 2012–2014 (EC FP 7, #282759).

Special thanks go to Felix Sánchez and his family who helped several of us during fieldwork at El Chichón, Dr. Silva Ramos H., Director of the Centro de Monitoreo Sísmico y Volcánico (CMVS) of Chiapas, for her sustained encouragement, support, and hospitality over the decades, and Guillermo Cisneros who prepared several figures of this book.

We hope that the information contained in this book will inspire further research to understand the behavior of these volcanoes, and similar ones around the world.

Teresa Scolamacchia
José Luis Macías

Contents

1 Geodynamic Setting and Pre-volcanic Geology of Active Volcanism in Chiapas

V.H. Garduño-Monroy, J.L. Macías, and R.S. Molina Garza

Abstract

El Chichón and Tacaná, two active volcanoes in the southeastern Mexico Chiapas state, illustrate the complexity of the magmatic systems in the Middle America Trench. Tacaná represents the northwestern end of the Central American Volcanic Arc, whilst El Chichón is the northernmost volcano of the Modern Chiapanecan Volcanic Arc. Tacaná was built on Mesozoic metamorphic and Tertiary intrusive rocks (35–13 Ma) and deposits from mid-Pleistocene calderas (1–2 Ma). El Chichón was built on folded Cretaceous to Miocene sedimentary rocks, established ~ 150 km north of the position of a Miocene arc that existed in the southernmost Maya block. Both the Central American Volcanic Arc and the Modern Chiapanecan Arc have been interpreted in terms of supra-subduction zones of magmatism associated with the subduction of the Cocos Plate beneath the Caribbean and North America plates. A well-defined Wadati-Benioff Zone supports such model for the Central American Volcanic Arc. Nevertheless, available data indicate that the subducted Cocos Plate does not reach the mantle region below the Modern Chiapanecan Volcanic Arc. El Chichón and the other volcanoes of the Chiapanecan Volcanic Arc are situated some 200 km above the down dip projection of the slab over a region that, based on seismic tomography, can be interpreted as a slab gap or an anomalous hot region of the mantle.

1.1 Geodynamic Setting

1.1.1 Regional Plate Interactions

Volcanic activity in southern Mexico and Central America is associated with the subduction process. In a general subduction scheme, an oceanic lithospheric plate plunges into the mantle below a lighter continental plate. Fluids released from the subducting plate promote partial melting of the mantle wedge above it, and the magma produced ascends to form volcanic structures at the surface. However, in southern Mexico, or more precisely in the area corresponding today to the states of Chiapas, Veracruz and Oaxaca, both spatial and

V.H. Garduño-Monroy (✉)
Instituto de Investigaciones en Ciencias de la Tierra, Universidad Michoacana de San Nicolás de Hidalgo, Morelia, Mexico
e-mail: vhgardunom@gmail.com

J.L. Macías
Instituto de Geofísica, Unidad Michoacán, Universidad Nacional Autónoma de México, Campus Morelia, 58090 Morelia, Michoacán, Mexico

R.S. Molina Garza
Centro de Geociencias, Universidad Nacional Autónoma de México, Campus Juriquilla, 76230 Querétaro, Mexico

T. Scolamacchia and J.L. Macías (eds.), *Active Volcanoes of Chiapas (Mexico): El Chichón and Tacaná*, Active Volcanoes of the World, DOI 10.1007/978-3-642-25890-9_1, © Springer-Verlag Berlin Heidelberg 2015

temporal variations in volcanism reflect an intricate geodynamic setting. The dominant tectonic process is the subduction of the oceanic Cocos plate below two continental plates: North America and Caribbean. However, many complications arise from the fact that three plates are involved, and the triple junction between them is unstable. As a result, relative plate motions are accommodated by deformation distributed over a broad region; deformation is not restricted to the major plate boundaries. The two overriding plates, Caribbean and North America, record recent uplift, intense deformation, and scattered manifestations of magmatism including Tacaná and El Chichón volcanoes.

In this chapter we show that Neogene to Recent magmatism in southern Mexico and northern Guatemala is controlled by complex interaction among the Cocos, Caribbean and North America plates, and by internal deformation of the overriding continental plates. According to studies carried out in the last decades, southernmost Mexico is further framed in a "diffuse" and unstable trench-trench-transform triple junction setting that set off distinct patterns of faulting, seismicity, and volcanic activity over an area that extends from the Tehuantepec Isthmus in Mexico to central Honduras (Guzman Speziale et al. 1989; Ratschbacher et al. 2009; Authemayou et al. 2011). We offer the interpretation that the angle of the subducting plate, the subducted segment of the Tehuantepec Ridge, the motion of the Chortis block (nuclear Central America) relative to the North America plate, and intra-plate deformation are the primary factors controlling the distribution of subduction-related magmatism between central Mexico and the Central America Volcanic Arc (CAVA).

The absolute motion of the North America plate has been west-southwestward for most of the Cenozoic, whilst the Caribbean plate has remained nearly stationary since chron 18 (38.4 Ma; Müller et al. 1999). The boundary between the North America and the Caribbean plates is marked by the left-lateral Polochic-Motagua fault system (Fig. 1.1a), and its extension offshore into the Cayman Trough (Burkart 1983; Guzmán-Speziale 2001; DeMets 2001; Franco et al. 2012; Authemayou et al. 2011). This fault system extends ~400 km from the Pacific Coast of Chiapas to the west to the Caribbean Sea to the east, where it merges into the Cayman Trough via the Swan Islands fault zone (Rogers and Mann 2007; Pindell et al. 2005). The strike-slip system is composed of three major arcuate, subparallel, left-lateral faults named from north to south: Polochic, Motagua, and Jocotán (Lyon-Caen et al. 2006).

The Caribbean-North America plate boundary in Central America has been a matter of controversy (e.g., Burkart 1983; Authemayou et al. 2011; and reference therein). The correlation of displaced rocks, folds, and river networks, north and south of the Polochic-Motagua Fault System, suggest a sinistral movement of 132 ± 5 km along the Polochic fault, during the Neogene (Dengo 1982; Burkart 1983; Burkart et al. 1987). Present day deformation along the Polochic-Motagua fault system is accommodated mainly along the Motagua fault, which shows the greater seismicity among the faults in the system (Schwartz et al. 1979; Guzmán-Speziale et al. 1989; Franco et al. 2009, 2012). The relative motion between North America and Cocos plates along the Polochic-Motagua fault system is 18–22 mm/year in eastern Guatemala, 14–20 mm/year in central Guatemala, and it is negligible at the Mexico-Guatemala border (Lyon-Caen et al. 2006; Franco et al. 2012). The deformation attained by this differential motion is transferred both north and south of the plate boundary causing internal deformation in the Caribbean plate and the southern tip of the North America plate. According to Franco et al. (2012) south of the Motagua fault, such deformation is accommodated as a ~9 mm/year east-west extension represented by N-S grabens occurring from Honduras to the Mexico-Guatemala border (Guzmán-Speziale 2001). Examples of this process include the Guatemala City Graben (accommodating ~5 mm/year), and the Tacaná Graben (García-Palomo et al. 2006). García-Palomo et al. (2004, 2006) were among the first to propose that intra-plate deformation, such as the N-S graben systems of Central America, may control the process of magma ascent to the surface.

In Chiapas, in contrast, part of the relative plate motion between North America and the Caribbean plate is accommodated by a system of WNW trending left lateral strike-slip faults (Guzmán-Speziale and Meneses-Rocha 2000), which have been active since Late Miocene time. These faults control the location of volcanic activity in the Chiapas Highlands (García-Palomo et al. 2004).

Some authors have proposed that the Polochic fault connects to the west with yet another important structure, the Tonalá fault. This is a WNW fault parallel to the Middle American trench, along the western

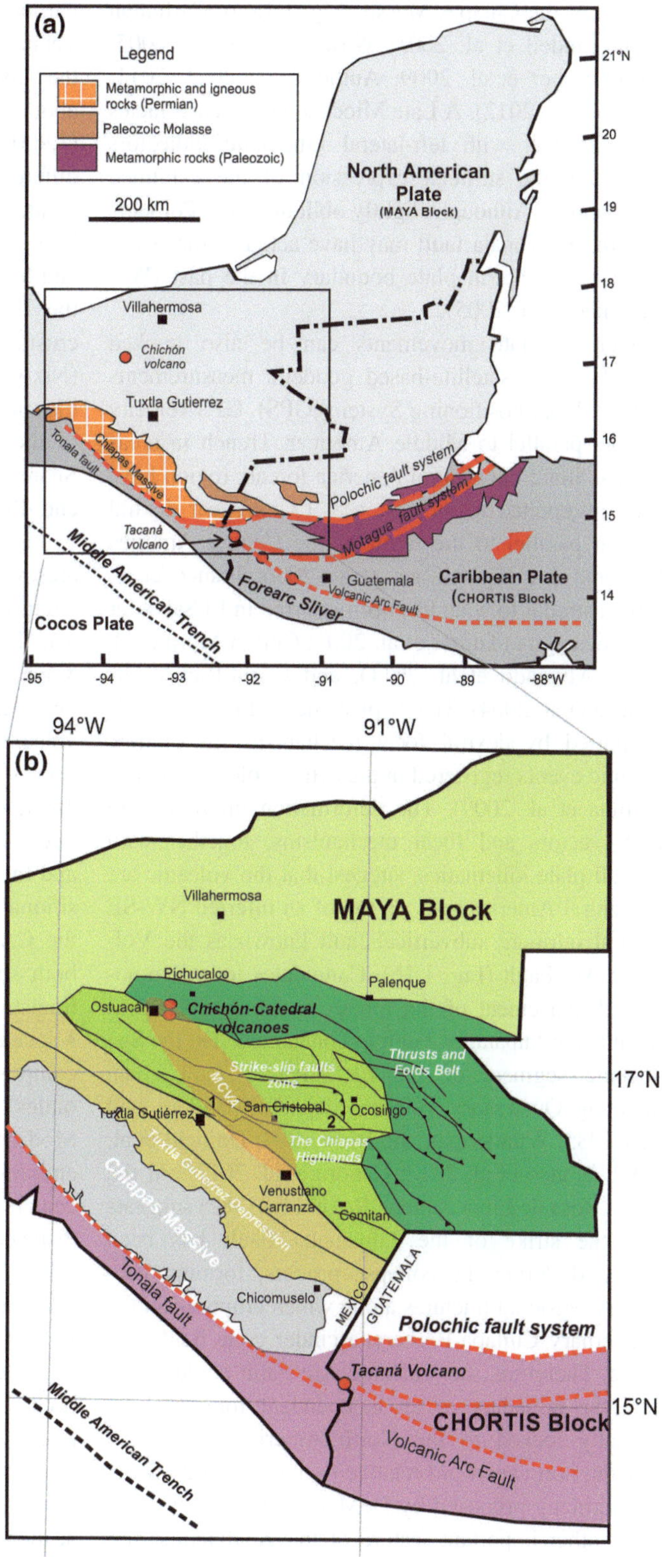

Fig. 1.1 **a** Sketch map showing the tectonic framework of southern Mexico dominated by the presence of three major plates and their boundaries at the Motagua-Polochic fault system and the Middle American Trench. **b** Main geological regional features of Southern Mexico and Guatemala, showing the location of the MCVA, highland fold chains, folds and thrust of the lateral fault zone, Chiapas massif. *1* Chicoasen-Malpaso fault, *2* Ocosingo fault

margin of the Chiapas Massif (Fig. 1.1a, b; Carfantan 1976; Pindell et al. 2005; Wawrzyniec et al. 2005; Ratschbacher et al. 2009; Authemayou et al. 2011; Franco et al. 2012). A Late Miocene greenschist facies mylonite belt with left-lateral kinematic indicators represents the surficial expression of the exhumed Tonalá fault. Although slightly oblique to the Polochic system, the Tonalá fault may have acted as the North America-Caribbean plate boundary in the past (Wawrzyniec et al. 2005).

Current plate movements can be also tracked directly using satellite-based geodetic measurements of the Global Positioning System (GPS). GPS velocity vectors parallel to Middle American Trench in areas located along the Central America forearc region have been interpreted to indicate a ~15 mm/year dextral motion parallel to the trend of the Central America Volcanic Arc (Lyon-Caen et al. 2006; Franco et al. 2012), similar to what has been inferred in El Salvador and Nicaragua (Turner et al. 2007; Correa-Mora et al. 2009; Alvarado et al. 2011), and Costa-Rica (Norabuena et al. 2004). The displacement of the forearc is supported by dextral focal mechanisms in shallow seismic events registered in the active volcanic arc (La Femina et al. 2009). The deformation inferred from GPS vectors and focal mechanisms, together with overall plate kinematics, suggest that the volcanic arc in Central America rests on top of an inferred NW-SE dextral intra-arc subvertical fault known as the Volcanic Arc Fault (Fig. 1.1b). Candidates to accommodate displacement of the forearc in Central America include the Jalpatagua Fault in Guatemala, and the San Vicente segment of the El Salvador fault zone, affecting Quaternary volcanic deposits (Burkart and Self 1985; Wunderman and Rose 1984; Duffield et al. 1992; Franco et al. 2012; Canora et al. 2012). At the Santa Rosa de Lima caldera, Reynolds (1987) suggests that the strike of the Jalpatagua Fault has been deflected during the collapse process, forming subsidiary tensional fractures along which eruptions in the Quaternary Cuilapa-Barbarena cinder cone field took place. Therefore, the Volcanic Arc Fault could delineate the boundary of an independent forearc sliver partially locked to the North American plate and moving northwestward relative to the Caribbean plate, as originally proposed by DeMets (2001).

Another important feature of the geodynamic setting of Chiapas state, and El Chichón volcano in particular, is the Tehuantepec Ridge (Fig. 1.1a). The Tehuantepec Ridge is a narrow, linear feature within the Cocos Plate with a maximum vertical relief of about 2,000 m separating sea floors of different depths (3,900 m to the NW, and 4,800 m to the SE) and different ages. The deepest segment corresponds to the Guatemala basin, to the east of the ridge (Truchan and Larson 1973; Couch and Woodcock 1981; LeFevre and McNally 1985). The morphological expression of the Tehuantepec Ridge separates a 16 Ma oceanic crust, to the west, from a 26 Ma crust to the east (Nixon 1982; Manea and Manea 2006). Furthermore, the subducted Cocos slab dips at low angles of 25–35° to the west of the Tehuantepec Ridge (Pardo and Suárez 1995; Rebollar et al. 1999; Ponce et al. 1992) and 40–45° to the east (Rebollar et al. 1999).

In contrast with typical supra-subduction volcanic arcs, where volcanoes are more or less regularly spaced along the arc, few manifestations of volcanic activity actually exist between the eastern Trans-Mexican Volcanic Belt, in central Mexico, and the CAVA in Guatemala, El Salvador and Honduras. A cluster of volcanic centers characterized by effusive activity occurs to the north of the Tehuantepec Isthmus (i.e., *Macizo de Los Tuxtlas*, or Tuxtlas Massif), and scattered volcanic centers occur between the Tuxtlas Massif and the CAVA. Such regions can be considered transitional between the Trans-Mexican Volcanic Belt and the CAVA itself, being characterized by changes in both subduction geometry and tectonic regime. This transitional region hosts an extinct magmatic arc of Miocene age, represented by plutons of intermediate composition (granodiorites, tonalites and quartzmonzonites) in coastal Chiapas, and further inland it hosts the Modern Chiapanecan Volcanic Arc (MCVA) (Damon and Montesinos 1978). As in the Tuxtlas, volcanic centers in the MCVA are located at a relative large distance (300–350 km) from the trench (Fig. 1.2).

El Chichón represents the youngest of the scattered volcanic centers of the MCVA, which also includes little-studied volcanic structures in *Los Altos de Chiapas* (Chiapas Highlands), such as the Tzontehuitz dome complex, the Apas volcano, and other relatively small volcanic domes (Mora et al. 2007, 2012). The Modern Chiapanecan Volcanic Arc is a rather exceptional arc for several reasons: (a) it is formed by few scattered volcanic structures, (b) a ~200 km gap in magmatism separates the Tuxtlas Massif from the most

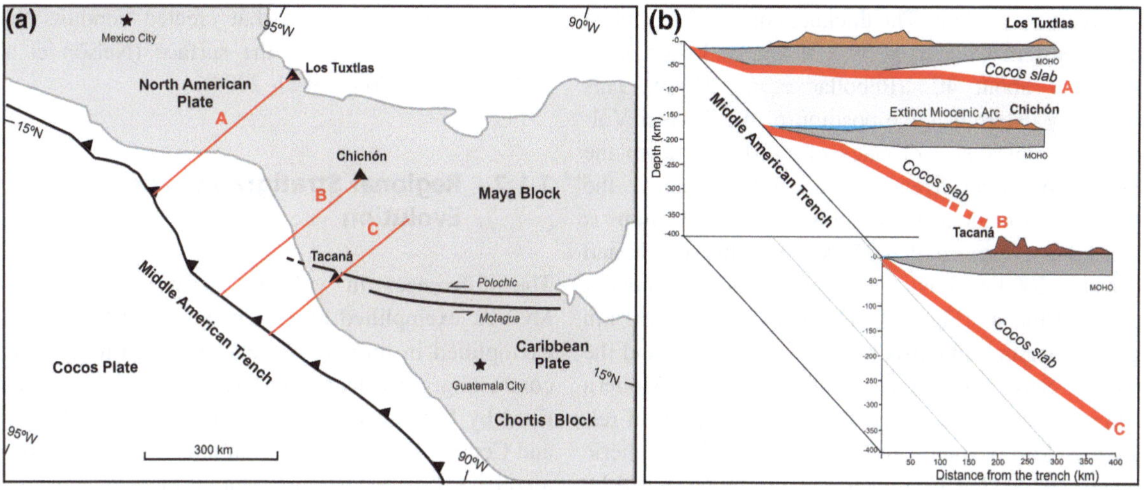

Fig. 1.2 Inferred subduction geometry of the Cocos plate projected below Los Tuxtlas Volcanic Field or Tuxtlas Massif, Veracruz, the Modern Chiapanecan Volcanic Arc (Chichón), and Tacaná volcano

westward volcanic center of the MCVA (i.e. El Chichón volcano), and (c) the volcanic structures of the arc are not aligned parallel to the trench.

The migration of the subduction-related magmatism in Chiapas from its location near the current Pacific coastline during Late Miocene (Fig. 1.1b) toward the present position of the Modern Chipanecan Volcanic Arc in the Chiapas Highlands might be attributed to the tectonic interaction between the Chortis block and southern Mexico. This interaction is explained below.

Generally accepted paleogeographic reconstructions place the Chortis block south of the Mexican continental margin during Paleogene time, south of the Xolapa terrane (Pindell et al. 1988; Schaaf et al. 1995; Rogers et al. 2007; Silva Romo 2008). The Xolapa terrane extends in the region of southern Mexico between Acapulco and Tehuantepec. Tectonic models propose that Chortis was displaced eastward since about 45 Ma along with the Caribbean plate, resulting in the truncation of the Mexican margin together with the uplift and exhumation of mid-crustal plutons and amphibolite facies metamorphic rocks of the Xolapa terrane (Morán et al. 1996). The motion of the Chortis block may explain the migration of magmatism along the southern Mexican Pacific coast (Schaaf et al. 1995), and the presence of mylonites with Cenozoic activity parallel to the continental margin (e.g. Tolson 2007). The position of the Late Miocene Chiapanecan Arc along the Pacific coast of Chiapas, and the inland migration of magmatism during the Pleistocene may be, therefore, partly explained by the same process that acted in the Xolapa area: by the passage of the Chortis block south of Mexico moving in an eastward direction. As the forearc region is tectonically removed, the subduction angle decreases, and the locus of magmatism migrates landward. As for the Xolapa terrane, this process results in uplift and exhumation of mid-crustal rocks.

Although the model and processes described above are widely accepted, alternative paleogeographies, and tectonic scenarios, have been proposed by Keppie and Morán-Zenteno (2005), and by Keppie and Keppie (2012). Keppie and Morán-Zenteno (2005) favored the idea of a Pacific origin for Chortis, whilst Keppie and Keppie (2012) proposed that Chortis originated in the Gulf of Mexico, as the Maya block did. A Gulf of Mexico origin, for example, fails to explain the Cretaceous geology of Chortis such as a long record of subduction magmatism (summarized by Rogers et al. 2007); such model is also difficult to reconcile with the survival of petroleum systems in the southern Gulf of Mexico and Sierra de Chiapas. A Pacific origin, in turn, is inconsistent with the Jurassic to Cretaceous affinity of the stratigraphic records of Chortis and southern Mexico (Silva-Romo 2008). Moreover, both models are in conflict with available paleomagnetic data for the Chortis block (Gose 1985; Molina Garza et al. 2012).

Compared with the CAVA, which follows the general subduction scheme explained in the first paragraphs of this chapter, the MCVA is also relatively anomalous. The thickness of the Cocos plate under eastern Chiapas is 39 ± 4 km, and dips at an angle of about 40° (Rebollar et al. 1999). This geometry determines the position of the Tacaná Volcanic Complex at a distance of ~200 km from the Middle American Trench, and ~100 km above the projected Cocos slab. In contrast, El Chichón is located at a distance of ~350 km from the trench, and lies at ~200 km above the projected Cocos slab under central Chiapas (Fig. 1.2). Furthermore, a ~150 km gap in magmatism exists between the MCVA and the CAVA. Seismic tomography suggests that the Modern Chiapanecan Volcanic Arc lies above a region of relatively low velocity in the upper asthenospheric mantle, which has been interpreted to be produced by the slab detachment (Rogers et al. 2002).

The magmas of MCVA were apparently generated within a complex plate tectonic scenario involving a torn Cocos plate (Tehuantepec fracture zone), fluids released from the subducted slab, and the ascent of hot asthenospheric mantle. Manea and Manea (2006) explain the position of the MCVA in terms of spatial and temporal evolution of the Cocos slab, accompanied by an inflow of hot mantle from NW to SE, and the consequent release of significant amounts of water through deserpentinization of the slab. Combining phase diagrams for sediments, basalt, and peridotite, with a modeled thermal structure of the subduction zone beneath El Chichón, Manea and Manea (2006) proposed that the serpentinized root of the Tehuantepec Ridge dehydrates strongly (~90 %) at depths of ~200 km, which are comparable with those extrapolated for the slab beneath the MCVA (Rebollar et al. 1999).

Alkaline volcanism occurs at El Chichón in the northern part of the MCVA and in some volcanic centers of the Tuxtlas Massif in Veracruz; the alkaline magmas coexist with typical subduction related calc-alkaline lavas. Alkaline magmas have been explained by several authors by the influence of subduction of the Tehuantepec Ridge and the location of the triple junction among internally deforming tectonic plates in the region (Damon and Montesinos 1978; Nixon 1982; Luhr et al. 1984; García-Palomo et al. 2004; Manea et al. 2005; Mora et al. 2007, 2012; Mandujano and Keppie 2009) (Fig. 1.2). Accordingly, magmas produced by a low degree of partial melting of the mantle (and the crust), with fluid components from the slab, reach the surface due to an extensional or transtensional stress field that created conduits that allowed their passage to the surface (Nelson et al. 1995; García-Palomo et al. 2004).

1.1.2 Regional Stratigraphy and Geologic Evolution

The differences in volcanic activity between the MCVA, exemplified by El Chichón, and the CAVA, exemplified in turn by Tacaná may also reflect their contrasting pre-volcanic geologic history. As summarized by Ratschbacher et al (2009), southern Mexico and Central America have been interpreted in terms of a collage of tectonostratigraphic terranes (Campa and Coney 1983). Such interpretation implies the presence of crustal blocks with different basements and overlying successions separated by major faults. The Maya and Chortis blocks (Fig. 1.1a) are those implicated in the evolution of El Chichón and Tacaná volcanoes (Sedlock et al. 1993, and references therein). The Maya block encompasses the Yucatán Peninsula, part of the coastal plain of the Gulf of Mexico (i.e. Tabasco and Veracruz states), northern-central Guatemala, and most of Chiapas state up to the Tehuantepec Isthmus (Fig. 1.1b). The Chortis block includes parts of Guatemala, Honduras, Nicaragua and El Salvador (e.g., Dengo 1985), limited to the north by the Motagua fault zone. The area between the Motagua and Polochic faults has been interpreted as an additional tectonostratigraphic terrane (Ortega et al. 2007). However, the area holds tectonic elements of both Chortis and Maya blocks, and is better understood as a series of tectonic slivers juxtaposed by left lateral strike-slip faults. Tacaná volcano, located between the Polochic fault to the north, and the westward projection of the Motagua fault to the south, is apparently located within the Central America forearc sliver (Franco et al. 2012). On the other hand, El Chichón volcano is located well within the Maya block and overlies a thick succession of sedimentary rocks.

The basement of the Maya block includes the Permian-Triassic plutonic and metamorphic rocks of the Chiapas Massif (Weber et al. 2007), and Lower Paleozoic igneous and metamorphic rocks in Guatemala and Belize (Martens et al. 2010). A deformed, upper Paleozoic sedimentary succession is exposed in southern Chiapas and northern Guatemala (Fig. 1.1a; Clemons et al. 1974). Lower to Middle Jurassic volcanic and sedimentary rocks of continental origin constitute the oldest rocks of the Mesozoic section of the southern part of the Maya block (Godínez et al. 2011); these rocks are overlain by Middle to Upper Jurassic evaporites, found at ~2 km depth in the Malpaso-2 exploration well about 90 km south of El Chichón volcano (Alzaga-Ruiz 1997). Evaporites are present also beneath the volcano (Luhr et al. 1984; Duffield et al. 1984, Chap. 4). The Middle Jurassic rocks record the initial stages of opening of the Gulf of Mexico.

Upper Jurassic and Cretaceous strata (Quezada-Muñetón 1983, 1987) were deposited in a mixed siliciclastic-carbonate shelf, recording a marine transgression in a subsiding passive margin (i.e. San Ricardo Formation and Zacatera Group). Lower to Upper Cretaceous strata were deposited in a carbonate platform during a long episode of tectonic stability in the region (Sierra Madre and Angostura formations). Cenozoic deposition includes syntectonic siliciclastic rocks of a northward prograding deltaic and submarine fan system that reached the Gulf of Mexico (Berlanga-García 2004; Mandujano-Velasquez 1996; Meneses-Rocha 1987, 2001; Padilla and Sánchez 2007). The Cenozoic stratigraphy, sandstone composition, and provenance may be interpreted in terms of a record of the progressive uplift and deformation of the Chiapas Massif and Sierra de Chiapas.

The sedimentary succession of the southern Maya block was affected by the Middle Miocene Chiapanecan orogeny (Sanchez Montes de Oca 1979; García-Molina 1994; Chávez Valois 1997; Martínez Kemp et al. 2006; Meneses-Rocha et al. 1994; Mandujano-Velazquez and Keppie 2009; Witt et al. 2012a). This compressional event influenced the region between easternmost Veracruz (Los Tuxtlas), and western Guatemala, developing a fold and thrust belt detached in Middle-Upper Jurassic evaporites.

From south to north, Meneses-Rocha (2001) recognized four distinct tectonic domains in the Chiapas orogen: the uplifted Chiapas Massif, the Sierra Monocline, the Strike-slip Fault Province, and the Reverse Fault Province. The general orientation of the compressional structures in the foldbelt is WNW in the west, and NW in the east (Fig. 1.1b; Chavez Valois 1997). Upper Miocene to Pleistocene strata buried the foldbelt along the coastal plain in the Macuspana and Comalcalco basins, north of El Chichón (Fig. 1.1a), and offshore in the Campeche sound. Anticline and syncline structures forming the Sierra Chiapas foldbelt are continuous for tens of kilometers, accumulating some 55 km of shortening (García-Molina 1994). According to Witt et al. (2012b) the foldbelt is associated with a NW directed gravitational collapse that occurred between 16 and 10 Ma, as supported by regional unconformities of that age, major movement of salt in the subsurface, and northward progradation of a sedimentary wedge. Structures related to the Chiapanecan orogeny are present in the area of El Chichón volcano.

The Strike-slip Fault Province of the Sierra de Chiapas includes seven major left-lateral faults between 16°N and 17°N, with a nearly East-West trend, which cut previously formed folds (Fig. 1.1b). The faults extend for about 350 km between the Cerro Nanchital in Veracruz, and Altamirano city in eastern Chiapas (Fig. 1.1b). Meneses-Rocha (2001) estimated left-lateral displacements around 4–5 km in each of the faults of the eastern part of the Strike-slip Fault Province, and approximately 1–16 km in faults of its central part; the total shear displacement is about 70 km. Indirect estimates by Witt et al. (2012b) indicate an accumulated displacement between 30 and 43 km during the last 6 Ma in the Strike-slip Fault province. Pull-apart basins with thick Upper Miocene to Pleistocene successions developed within this Province, exemplified by the Ixtapa graben (Meneses-Rocha 2001). All the volcanic structures of the MCVA are located within this province. El Chichón volcano in northwestern Chiapas as well as the Tzontehuitz, Huitepec, and Navenchauc volcanic centers in the Chiapas Highlands erupted within the central part of the Strike-slip Fault Province, and they form the highest elevations in central Chiapas (Damon and Montesinos 1978; Mora et al. 2007, 2012).

1.2 El Chichón and the Modern Chiapas Volcanic Arc

El Chichón was considered by Damon and Montesinos (1978) to be the youngest and most northward volcano of the Modern Chiapanecan Volcanic Arc. The MCVA consists of a series of volcanic domes between the villages of Ostuacán and Venustiano Carranza, in the Los Altos region of central Chiapas (Fig. 1.1b) (Mora et al. 2007, 2012). Volcanic products of the MCVA are dispersed over an area of approximately 4,900 km^2 (Mora et al. 2012), and the arc lays well north of the continuation of the Motagua-Polochic fault system into the Maya block. (Fig. 1.1b). The pre-volcanic basement in the Strike-slip Fault Province consists of Lower to Upper Cretaceous platform limestones of the Chiapas Highlands (Sierra Madre and Angostura Formations), and of Cenozoic siliciclastic rocks (Soyaló and Bosque Formations), which are exposed along tight WNW trending folds formed at the end of the Middle Miocene Chiapanecan orogeny (Fig. 1.1b). The Chiapas Highlands are transected by several sinistral faults (e.g. Chicoasen, Malpaso, and Simojovel faults) that affect Cretaceous to Lower-Middle Miocene rocks (Meneses Rocha 2001, Fig. 1.1b).

1.2.1 Previous Studies

Exploration for hydrocarbons during the 20th century contributed to depict the stratigraphic succession from Paleozoic to Tertiary rocks in Chiapas (Gibson 1936; García-Tijerina 1950; Sánchez-Montes de Oca 1979). The structural studies carried out in the area by Sánchez-Montes de Oca (1979) located possible traps and fracture features related to hydrocarbon deposits. Damon and Montesinos (1978) dated domes and volcanic edifices around the city of San Cristobal de las Casas (Fig. 1.1b) and further north, including El Chichón volcano area. In the 70s, research focused on geothermal exploration around El Chichón edifice, producing the first geological map of the volcano (Canul and Rocha 1981). Canul and Rocha (1981) warned of the high volcanic risk if a geothermal field was to be developed, a warning that turned out to be prophetic (see Chap. 3). Both geological and volcanological investigations increased after the April 1982 eruption of the volcano in an attempt to define the volcanic substrata, the evolution of magmatism through time, and the volcanic hazards.

The tectonic setting of El Chichón volcano during the early eighties was poorly understood due to the lack of studies, but much has been learned from a score of research papers published during the last decade. Rebollar et al. (1999), Guzmán-Speziale and Meneses-Rocha (2000), and Manea and Manea (2005) studied regional geophysical aspects, the crustal structure, the geometry of the slab, and the overall tectonic setting of the area. García-Molina (1994), Meneses Rocha (2001), and García-Palomo et al. (2004) focused on the regional foldbelt structure, and the local structural aspects. A preliminary model for the structural evolution of the volcano was proposed by García-Palomo et al. (2004).

1.2.2 Local Geology

The geological map of El Chichón produced by the Comisión Federal de Electricidad (Canul and Rocha 1981), proposed eight stratigraphic units from Mesozoic to Tertiary. Nevertheless, based on the aerial distribution of the rocks, the local morphology and drainage patterns, Macías et al. (2010) proposed to simplify the pre-volcanic sedimentary stratigraphic column reducing the number of units from 8 to 4 (Fig. 1.3). Their geological map reproduced here shows only three units, considering that the youngest unit (i.e. P-Mtu explained below) has a very small distribution and is not mappable. In this chapter we adopt this subdivision to describe the geology of the volcano.

1.2.2.1 Evaporites and Limestones (lK-evls)

Based on the stratigraphy of the Caimba-12 well, drilled by PEMEX in the decade of 70s a few kilometers SE of the volcano, Canul and Rocha (1981) described two thick sedimentary units. The lowermost unit consisted of a succession of evaporites interbedded with dolomitic limestones and bentonitic beds. The top of this unit, found at a depth of 2,595 m, has a minimum thickness of 1,000 m, and was assigned to the Lower Cretaceous based on its microfossil content (*Colomiella recta, Colomiella Mexicana,* and *Nannoconus* sp., known from the Upper Tamaulipas Formation; López-Ramos 1979). The evaporites and carbonate rocks that constitute this unit, suggest that deposition took place in shallow marine and marginal marine environments.

The Lower Cretaceous evaporites are overlain by light brown to gray massive dolomitic limestones and evaporites with a total thickness of 1,500 m. The dolomitic limestones were petrographically classified as reef calcarenites, breccias, and reef structures by PEMEX (Canul and Rocha 1981), suggesting a reef-rimmed platform depositional environment, including the fore-reef region. Centimetric strata of clayey limestones with chert nodules are present in the uppermost portion of the sequence, indicating deposition took place in a more basinal low-energy environment. This sequence of carbonate rocks crops out to the south of the volcano, and has an estimated thickness of 2,500 m (Fig. 1.3). According to Canul and Rocha (1981), PEMEX concluded that these carbonate rocks are Lower to Upper Cretaceous in age.

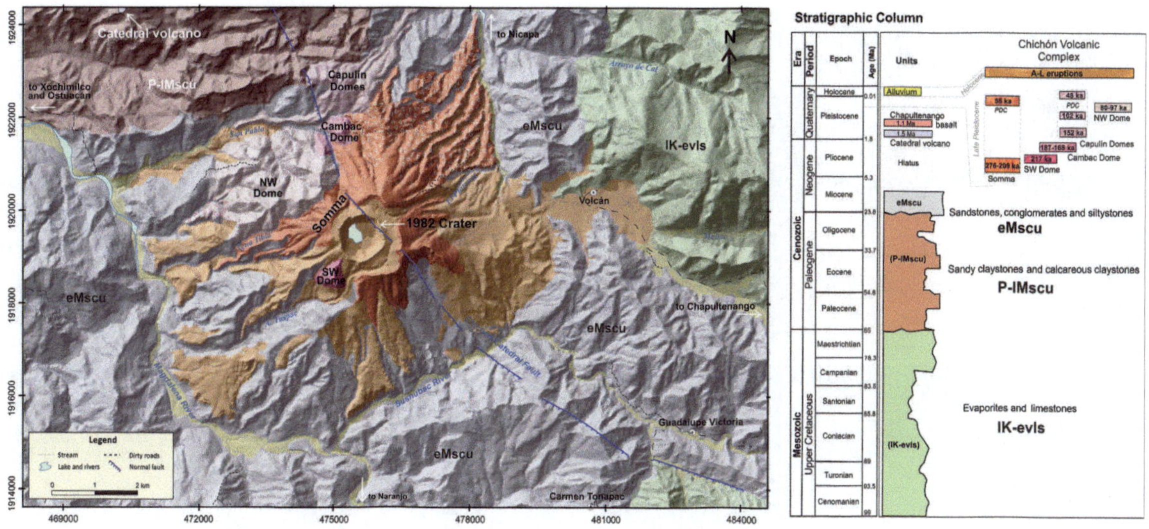

Fig. 1.3 Geological map of the El Chichón Volcanic Complex (after Macías et al. 2010). The geologic map was derived from an interpretation of aerial photographs (scale 1:50,000), a digital elevation model, field reconnaissance, stratigraphic correlation of 47 sections, dating by the $^{40}Ar/^{39}Ar$ method and paleontological dating. The stratigraphic column consists of basement and volcanic units of El Chichón. In the basement we distinguish three major sedimentary units of the early Cretaceous-early Miocene. The cartographic parameters of the map are: UTM zone 15 with the datum and ellipsoid WGS84

Similar carbonate rocks (400 m thick) were found in the Unión-2 well, SW of Francisco León village (Canul and Rocha 1981). The limestones form extensive outcrops to the east of the volcano around the villages of Ocotepec and Ixhuatán as a series of NW trending folds, and in two major thrust faults, cut by younger strike-slip and normal faults (Fig. 1.4a). Carbonate rocks crop out also at Cerro Pelón west of the volcano, and in the Primavera anticline to the East (Fig. 1.3).

1.2.2.2 Sandstones and Calcareous Claystones (P-IMscu)

This unit has the widest distribution of all pre-volcanic units, occupying all the valleys around the volcano, with an inferred minimum thickness of 2,000 m. From the base to the top, it consists of dark-gray claystones with limestone beds, which are gradually replaced by thinly bedded claystones and lenses or beds of sandstones (Fig. 1.4b). This succession may be correlated in time to the Formations Méndez and Nanchital of the Sierra de Chiapas. Numerous cm-thick sandstone beds alternating with lenticular layers of reef limestone crop out along the Susnubac River south of the volcano (Fig. 1.3). The presence of corals in the limestones indicates the deposition in shallow water. Limestone beds with corals, nummulites, and benthic microfauna of Late Paleocene to Lower Miocene age crop out along Arroyo de Cal river, both southeast and south of the volcano. To the NW of the volcanic edifice this unit consists of more calcareous facies. Most of the geothermal and non-geothermal springs that surround the volcano occur at the contact between the terrigenous facies of the P-IMscu unit and the volcanic products of El Chichón (Chap. 4). Regionally, the Cenozoic terrigenous sequence on top of the Upper Cretaceous carbonate sequence suggests a significant change in depositional environment, from a stable marine platform with reefs (i.e. Upper Cretaceous) to a depositional regime characterized by high subsidence in deeper waters, as attested by siliciclastic turbidites (Paleocene-Miocene). The sedimentary record of the area around the crater suggests however, an earlier deposition interval in neritic environments, and a later deposition in shallow marine environments. Such record suggests that differential subsidence occurred in the El Chichón region, with less subsidence southeast of the volcanic edifice, bringing the local preservation

Fig. 1.4 Main sedimentary units exposed around El Chichón volcano. **a** Limestone beds of middle to late Cretaceous of unit lK-evls, **b** sandstones and calcareous claystones (P-lMscu), **c** sandstones and claystones (eMscu) widely exposed around El Chichón volcano, and **d** unit of Sandstones and Conglomerates (P-Mtu)

of shallow marine and coastal depositional settings, similar to what observed in other areas of Sierra de Chiapas (Meneses-Rocha 2001).

1.2.2.3 Sandstones and Claystones (eMscu)

The sedimentary unit overlying unit P-Imscu consists of thick, mica-rich, light-brown sandstone and conglomerate beds, with leaves, stem remains, and charcoal, alternated to dark-gray siltstone beds. Prominent morphological expressions of the sandstone beds are recognizable in fold hinges north of El Chichón. This unit has a maximum thickness of 300 m. The presence of Index foraminifera suggests an Early Miocene age (Canul and Rocha 1981, Figs. 1.3 and 1.4c). Catedral volcano was built on top of this terrigenous unit. Based on the lithological description these rocks correspond to the Depósito Formation (García Tijerina 1950).

1.2.2.4 Sandstones and Conglomerates (P-Mtu)

This unit crops out to the northeast of the volcano, and is composed of limestones, sandstones and reddish conglomerates that overlie all Paleocene-Miocene units through an angular unconformity (Fig. 1.4d). P-Mtu contains Middle-Upper Miocene microfauna. Its distribution is restricted to small areas, some of which have been interpreted as pull-apart basins (Meneses-Rocha 1985, 2001).

1.2.3 Morphostructural Analysis

El Chichón volcano is located in the Sierra de Chiapas physiographic province, where limestones and terrigenous rocks (sandstones to claystones) form E-W to NW-SE oriented Sierras (Fig. 1.5a, b), crossed by E-W strike-slip faults (Meneses-Rocha 2001). The eastern portion of Sierra de Chiapas is occupied by the Reverse Fault Province, which consists of elongated, tightly folded mountains striking NW, separated by narrow valleys. These mountains decrease gradually in elevation from about 2,000 to 500 m to the northeast, near Palenque. The Sierras have a curvature convex to the northeast (Fig. 1.1b). The strike-slip faults in the western portion of Sierra Chiapas imprint a valley-and-ridge topography in the landscape, as they often

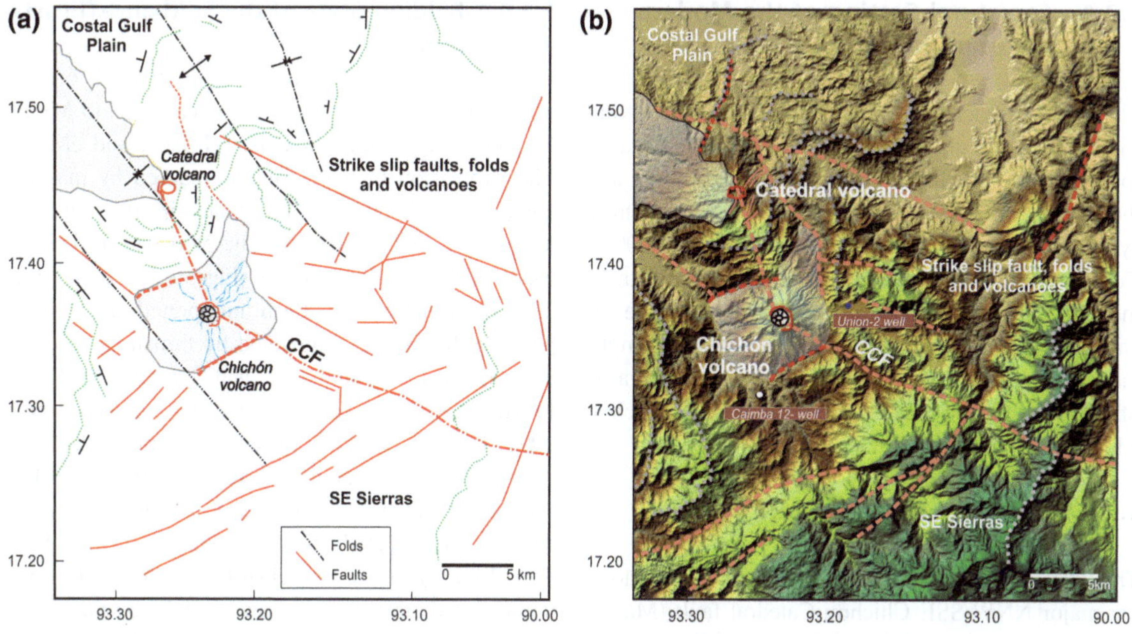

Fig. 1.5 a Morphostructural map of Chichón area displaying NW-SE folds and major faults. **b** Digital elevation model with the main morphostructural features

juxtapose rocks with different resistance to weathering. This area includes large "block-shaped" mountains separated by narrow valleys, which correspond to up-thrown/down-thrown blocks bounded by strike-slip faults. Based on terrane modeling, structural characters, and field observations, the area of El Chichón volcano can be divided into three main sectors:

1.2.3.1 Southeastern Sierras

This area is located south of El Chichón volcano. It consists of 100–500 m-high mountains composed of Lower Cretaceous carbonate rocks and Cenozoic siliciclastic rocks. These low-elevation Sierras show both dendritic and parallel drainage systems. The parallel drainages have a NE-SW orientation associated to normal faults with NW dips. An E-W sinistral strike-slip fault along the Guadalupe Victoria River transects these Sierras (García-Palomo et al. 2004).

1.2.3.2 El Chichón and Catedral Volcanoes and the Strike-Slip Faults Area

This area consists of horizontal to gently plunging NW-trending folds, associated in places with WNW-ESE strike-slip faults (Fig. 1.5a, b). It is characterized by variations in altitude ranging from 100 to 600 m asl. The most competent terrigenous facies (i.e. sandstones and conglomerates) form inverted fold topography, with anticlines occupying topographic highs, and synclines topographic lows.

García-Palomo et al. (2004) and Meneses-Rocha (2001) recognized that the fold axes were displaced by left lateral strike-slip faults during the Late Miocene. These faults determine the orientation of the valleys of major rivers, such as Caimba and Guadalupe Victoria rivers (Fig. 1.3). El Chichón and Catedral volcanoes occur in this area. El Chichón lies inside an isocline core fold affected by NW- and NE-striking faults and fractures (Fig. 1.5a, b).

Catedral volcano lies to the NE of El Chichón. Catedral is an eroded edifice, which collapsed to the east leaving an amphitheater filled by pyroclastic and volcaniclastic deposits of unknown age (Chap. 3). The collapse scar is affected by NE-striking faults and fractures.

1.2.3.3 Lowlands and the Gulf Coastal Plain

This area starts about 70 km north of El Chichón, and extends toward the Gulf of México, with a maximum altitude of 10 m above sea level (Fig. 1.4d), and consists of terrigenous deposits of Pleistocene-Holocene age derived from the Cretaceous-Miocene sedimentary rocks of the fold and thrust belt.

1.2.4 Structural Setting of the Modern Chiapanecan Volcanic Arc

The MCVA is located in a transpressive tectonic setting superimposed on folds and thrust faults of the Chiapas foldbelt. The anticlines in the foldbelt are narrow, and their northern flanks often overthrust the intervening synclines (Chavez Valois 1997; Meneses Rocha 2001). Common fold geometries include box-shaped detachment folds, and fault-propagation folds that produce NE and SW verging asymmetric anticlines of large amplitude. Normal faults displace the fold-axes and the traces of thrust faults. The main structures around El Chichón edifice are described below.

1.2.4.1 Analysis of Left-Lateral Faults and Folds

The volcanism around El Chichón edifice occurs along the major NNW-SSE Chichón-Catedral fault (Macías et al. 2006, 2010); or Nicapa Fault according to Islas Tenorio et al. 2005), which cuts a syncline formed by Upper Miocene terrigenous rocks (Figs. 1.4a, 1.5a, and 1.6a, b). According to Macías et al. (2010), the Chichón-Catedral fault would control most volcanic and geothermal activity on the northern walls of El Chichón crater (Figs. 1.3, and 1.6). This fault has a clear morphological expression in Upper Cretaceous limestones and terrigenous rocks, which are offset by N45°E trending normal faults south of El Chichón (i.e. Chapultenango Fault System of García-Palomo et al. (2004). The earthquake focal mechanisms in NW Chiapas (Guzmán-Speziale and Meneses-Rocha 2000) suggest a stress regime related to left-lateral strike-slip faults and, according to García-Palomo et al. (2004), El Chichón volcano lies in a transtensional setting. The main fault of this system is the San Juan Fault (García-Palomo et al. 2004), named Santa Fe Fault by the Servicio Geológico Mexicano (Islas Tenorio et al. 2005).

We propose that faults around El Chichón can be fitted into a traditional Riedel model with a main E-W left strike-slip fault (San Juan Fault) in which a NNW-SSE (i.e. Catedral-Chichón fault), and a ENE-WSW (i.e. Chapultenango Fault System) structures can be interpreted as R′, and R faults, respectively. In this model, the R′ and R faults affect the Miocene sedimentary cover, producing also NW-SE trending folds by a NE-SW oriented $\sigma 1$ associated to the eastward motion of the Chortis Block.

1.2.4.2 Folding Style of the Fold-and-Thrust Belt

The main folds at El Chichón have NW-SE orientations, and are open folds in the Cretaceous to Middle Miocene rocks (units lK-evls, P-lMscu, and eMscu). The expression of these folds is marked by massive, competent, Cretaceous limestones in the anticlinal hinges and incompetent Middle Miocene terrigenous rocks in the synclinal hinges (Figs. 1.5, 1.6 and 1.7). This deformation has been interpreted as a single phase of folding, accompanied by thrusting, as documented in the Caimba-12 and Union-2 wells by PEMEX (Angeles-Aquino et al. 1994; Chavez Valois 1997; Jenette et al. 2003; Martinez Kemp et al. 2006).

According to Canul and Rocha (1981) PEMEX interpreted the structure of Cretaceous rocks as being due to knee-shaped folds verging to the northeast (Fig. 1.6a, b). However, subsurface PEMEX data also documented thrusted narrow anticlines and large synclines that support our Riedel model (Fig. 1.7). The Cenozoic terrigenous rocks have two principal sets of fractures (N15-30E and N50-65E), with a few fractures oriented parallel to the NW-trending folds (Fig. 1.7).

1.3 Tacaná Volcano

Tacaná volcano is the largest structure of the Tacaná Volcanic Complex (TVC) (Macías et al. 2000); which represent the westernmost active volcanoes of the Central American Volcanic Arc (Fig. 1.1a, b). The Complex consists of four NE-oriented volcanic structures (Chap. 6).

1.3.1 Previous Studies

Bergeat (1894), Sapper (1896, 1899), Böse (1902) and Waibel (1933) provided the first reports about the geology and volcanology of Tacaná volcano. In 1949, a phreatic explosion produced a fumarolic field below the volcano's summit (Mulleried 1951), and Tacaná's activity attracted the attention of Mexico′s national power company (Comisión Federal de Electricidad, CFE), who produced the first geological map, and explored the geothermal energy potential of the volcano (De la Cruz and Hernández 1985; Saucedo and Esquivias

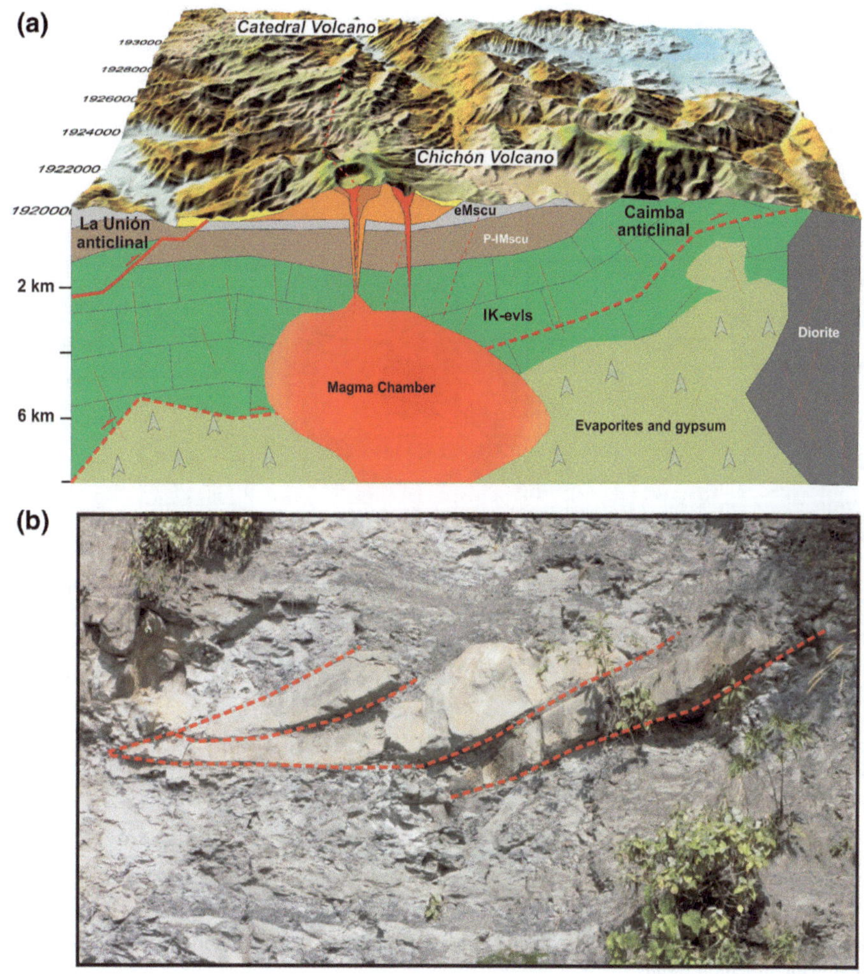

Fig. 1.6 **a** Block diagram showing folding style, ramps, and faults, **b** affecting the Cretaceous and Tertiary sedimentary units below El Chichón volcano. The depth of the inferred magma chamber is also shown

1988). In December 1985, seismic activity in the Tacaná area increased, peaking on May 7, 1986 with a phreatic explosion (De la Cruz-Reyna et al. 1989, see Chap. 6).

1.3.2 Local Geology

The Tacaná volcano is located at the easternmost extreme of the Chiapas Massif. The local geology underlying the volcano consists of a crystalline basement of metamorphic rocks intruded by igneous rocks, which were generated during two main phases during the Late Eocene-Early Oligocene and the Early-Middle Miocene (García-Palomo et al. 2006) (Fig. 1.8). Pliocene to Pleistocene deposits resulted from the activity of three calderas (San Rafael-Chanjale-Sibinal), and rest unconformably on older units. The Tacaná Volcanic Complex overlie these caldera deposits.

1.3.2.1 Mesozoic Metamorphic Basement (Mb)

Metamorphic rocks are exposed along the San Rafael and Coatán rivers (Fig. 1.9a), and consist of relatively small exposures of alternated schists and gneisses. The schists, light to dark green in color, consist of green hornblende-rich bands alternated with white plagioclase-quartz centimeter-thick bands. The general strike of schistosity and foliation is N60°W, dipping about 70°NE. K-Ar dating of hornblende yielded ages of 142 ± 5 Ma (Early Cretaceous) (Mujica 1987). This is

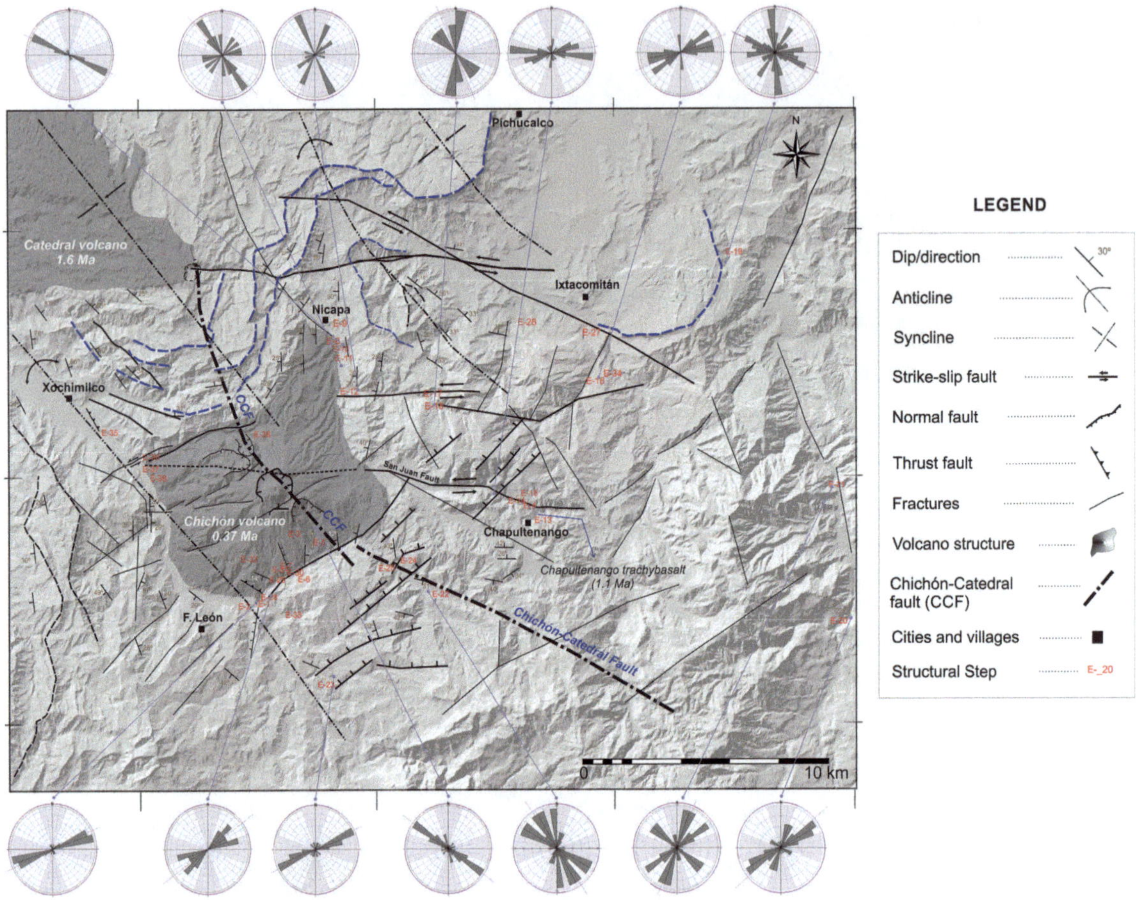

Fig. 1.7 Structural map of the Chichón volcano highlighting the Chichón-Catedral fault and the NW-SE folds axes. The *rose* diagrams show the main structural trends

considered a minimum age for the metamorphic event that formed this unit. Other exposures of metamorphic rocks as gneisses and schists are found to the NW of the volcano.

1.3.2.2 The Cenozoic Coastal Batholith of Chiapas (CB)

The Chiapas Massif was described for the first time by Pantoja Alor et al. (1974), and dated by Damon et al. (1981) as late Paleozoic. This crystalline complex consists mostly of metamorphic and plutonic rocks of Late Permian age ranging between about 245 and 275 Ma (Weber et al. 2005, 2007), and minor Early Paleozoic crystalline rocks, and Cenozoic plutons. The massif is 270 km long and 30 km wide, extending over an area of ~8,000 km^2, from the Isthmus of Tehuantepec to the Chiapas-Guatemala border. Damon and Montesinos (1978) and Mujica (1987) recognized that the massif includes a suite of younger granodiorite, granite, and diorite intrusions, which are mostly exposed along the coastal plain to the south of the Chiapas Massif. Mujica (1987) named these rocks the Coastal Batholith of Chiapas, we refer to the Miocene rocks of this suite as the extinct Miocene Volcanic Arc, as Eocene to Miocene rocks of this suite are not continuous; they were formed in different tectonic settings. Biotite separates from granodiorites in the Tacaná area yielded K-Ar ages of 29–15 Ma (Late Oligocene-Middle Miocene) (Mujica 1987), ages near 13 Ma were reported for intrusions near Arriaga, but ages as young as 3 Ma were obtained for the Toliman Pluton, along the road from Huixtla to Motozintla (Damon and Montesinos 1978) (Fig. 1.1). Based on these ages and new $^{40}Ar/^{39}Ar$ data, García-Palomo et al. (2006) subdivided the crystalline rocks of the Cenozoic Coastal Batholith around Tacaná region into

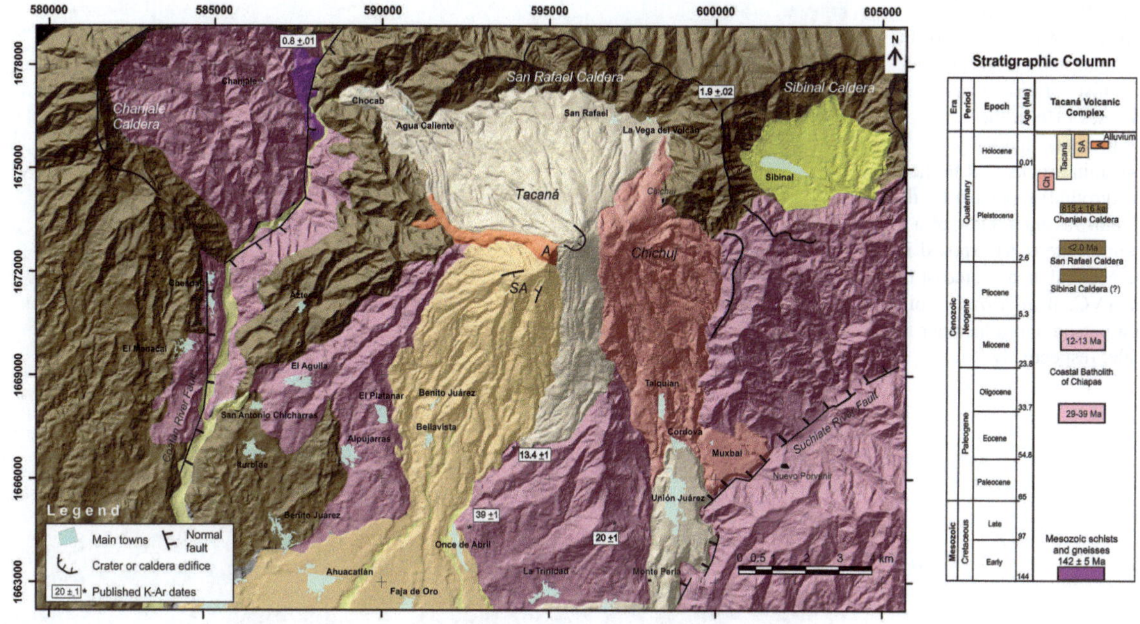

Fig. 1.8 Simplified geologic map of the Tacaná Volcanic Complex and surrounding areas (modified from García-Palomo et al. 2006). It shows the main rocks exposed in the area and available K-Ar and $^{40}Ar/^{39}Ar$ dates

two main groups described below. These episodes of intrusion may be related to the subduction of the Farallon Plate beneath the North American Plate.

Late Eocene-Early Oligocene (Granodiorites, Granites, and Diorites)

Biotite-hornblende granodiorites and hornblende-biotite gneissic diorites crop out near the village of Once de Abril (Fig. 1.8). The granodiorites were dated (K-Ar) at 35 ± 1 Ma, whilst the diorites at 39 ± 1 Ma. A gneissic tonalite that crops out along the Huixtla-Motozintla road (west of the area in Fig. 1.7) yielded a K-Ar of 29 ± 1 Ma (Mujica 1987).

Middle-Late Miocene (Granodiorites and Granites)

Gneissic, biotite-hornblende tonalites and biotite-hornblende granodiorites are widely exposed around the TVC. K-Ar biotite ages of 20 ± 1 Ma were obtained for both a gneissic tonalite at Monte Perla and a granodiorite near Tapachula (south of Tacaná in Fig. 1.7) (Mujica 1987). $^{40}Ar/^{39}Ar$ dating yielded a 13 ± 1 Ma age for a granodioritic stock exposed to the northwest of the TVC (13.3 ± 0.2 Ma) (Fig. 1.9b), and ages of 14 ± 1 and 13 ± 1 Ma for granites that crop out south of San Antonio volcano and in the outskirts of the Sibinal village, respectively (García-Palomo et al. 2006). These ages define the time of activity in the extinct Miocene Volcanic Arc.

1.3.2.3 Late Pliocene-Pleistocene Volcanism

Three calderas, named San Rafael, Chanjale and Sibinal (García-Palomo et al. 2006), and their deposits of Quaternary age overlie the metamorphic and granitic substrate (Fig. 1.8). They are described below.

San Rafael Caldera (<2 Ma)

This caldera has a discontinuous ~9 km wide crater rim (García-Palomo et al. 2006), located to the north and east of the present Tacaná edifice. The caldera sequence consists of a green ignimbrite composed of angular to subrounded dark-gray andesitic lithics, and rounded pumice (53.8 wt% SiO_2) in a matrix of medium-fine ash. A dark-gray andesitic flow (57.94 wt % SiO_2), exposed on the caldera rim, was dated at ~1.9 Ma (García-Palomo et al. 2006). The youngest units of the caldera sequence consist of dark-gray andesitic lava flows, overlain by a massive, matrix-supported, block-and-ash flow deposit, consisting of dark-gray andesitic (55 wt% SiO_2) blocks and minor red, altered, andesitic blocks set in medium-coarse ash.

Fig. 1.9 **a** View from the northeast of an outcrop of the metamorphic basement made of schists of biotite and gneisses (*Mb* in the foreground), and the crystalline rocks (*CB* in the background) separated by the *Coatán River*. **b** View of a granodiorite (*CB*) exposed in the Muxbal area southeast of the TVC. *White arrows* point to a person and a hammer for scale, respectively

Chanjale Caldera (1 Ma)

The Chanjale caldera, west of Tacaná, consists of a 6.5 km-wide crater open toward the Coatán River on its eastern side (García-Palomo et al. 2006). The caldera rim is made of several lava flow units and intercalated debris flows; a gray porphyritic lava flow (58.39 wt% SiO_2) rich in plagioclase and pyroxenes on the caldera flanks yielded a ^{39}Ar-^{40}Ar age of 1 Ma. A white to brown ignimbrite borders the inferred caldera rim. A white to light-yellow pyroclastic flow, with clasts up to 1 m in diameter embedded in a fine ash matrix with mm-size pumice, is exposed on the southern flank of the caldera. The sequence ends with several indurated debris flow deposits (12 m thick), forming a fan on the southern flank of the caldera that probably emplaced during the edifice collapse.

Sibinal Caldera (Unknown Age)

The Sibinal caldera is exposed to the NE of the Tacaná volcano; the town of Sibinal, Guatemala, is located in the center of this structure (García-Palomo et al. 2006). Although no detailed studies of this caldera exist, the deposits reported by García-Palomo et al. (2006) are dark-gray andesitic lava flows with aphanitic texture overlying Cenozoic granitoids at several places around the crater rim. These lava flows are covered by red to yellow hydrothermally altered breccias made of andesitic clasts. On the northern slopes of the caldera gray porphyritic andesites are overlain by pyroclastic fall deposits of the TVC (Chap. 6).

1.3.3 Structural Geology

The Tacaná Volcanic Complex lies on a structural high made up of metamorphic, granitic and volcanic rocks (Fig. 1.8), and is affected by three important fault systems. The oldest system to the west of the volcanic complex consists of fractures and faults striking NW-SE that affect the Mesozoic schist and Cenozoic granitic rocks. A second, NE-striking, fault system is parallel to the TVC. The youngest fault system, with a N-S trend, is superimposed on the others. The NE-SW fault system is the most obvious surficial structure, forming the Tacaná graben that in turn hosts the entire volcanic complex. The graben is 30 km long and 18 km wide, with minimum vertical displacements in the bordering faults of about 600 m. The Chanjale Caldera lies on the horst to the northwest of the graben, whereas the Sibinal caldera is located on the horst to the east (Fig. 1.8). The Coatán and Suchiate rivers follow major NE-striking faults (named after the rivers) that border the Tacaná graben (Fig. 1.8). Activity in the Tacaná graben is Pleistocene as movement on the border faults occurred after the emplacement of the San Rafael (2 Ma) and Chanjale (1 Ma) calderas, but before the edification of the Chichuj volcano—as indicated by the fact that it is not cut by the NE-striking faults. Therefore, it seems that the graben controlled the emplacement and evolution of the TVC; this hypothesis is supported by the N65°E alignment of the Chichuj caldera, the Tacaná volcano, the Las Ardillas dome, and the San Antonio volcano (see Chap. 6). According with this analysis, the region was affected by a NNE stress field with a minimum late Pleistocene principal stress ($\sigma3$) that coincides with the shallow focal mechanism in the region (Lyon-Caen et al. 2006).

1.4 Plate Tectonics and Structural Controls on the Volcanic Evolution of El Chichón and Tacaná Volcanoes

Stratigraphic relationships, supported by radiometric data, structural analyses, as well as geophysical data (such as seismic, GPS, and potential fields) allowed us to better understand the tectonic controls on the emplacement of the Tacaná and El Chichón volcanoes, even if further studies are necessary to assert some hypotheses. According to the best-constrained geophysical models, the Modern Chiapanecan Volcanic Arc and the Central American Volcanic Arc are separated by major discontinuities at crustal and deeper levels (Manea et al. 2005). Both arcs are related to subduction of the Cocos plate, and magma genesis is promoted by dehydration of the slab as it reaches deeper into the mantle. Magmas are likely to follow different evolutionary paths during ascent, being modified by shallow magmatic processes such as contamination-assimilation, fractional crystallization and mixing (Chap. 2). Nevertheless, Tacaná volcano, seems to overlie a well-established mantle wedge, where the magma ascends through a crystalline crust following paths that probably involve a major arc-parallel structure named the Volcanic Arc Fault and N-S normal faults as those in the Tacaná graben. In contrast, the magma ascent at El Chichón and probably at the other volcanic structures of the Modern Chiapanecan Volcanic Arc (MCVA), appear to be controlled by strike-slip faults, fault intersections, and sites of dilation. Pre-volcanic histories of the western CAVA and the MCVB are also different, as inferred considering the geological evolution of southern Mexico from the Early Cretaceous to the Recent.

During Early Cretaceous the northern portion of the State of Chiapas and Tabasco was characterized by sedimentation of evaporites and limestones in a carbonate platform. In contrast, at the same time, one or more episodes of metamorphism affected terrigenous sediments forming gneisses, schists, metavolcanic rocks, and slates around Tacaná. The origin of these metamorphic rocks remains uncertain due to the scarcity of data, and of detailed studies. Similar lithologies and age of metamorphism are observed in the Xolapa Complex, in southwestern Mexico, and can be correlated with metamorphic complexes, delimited by the Polochic and Motagua faults, of similar age in Guatemala (Lyon-Caen et al. 2006; Ratschbacher et al. 2009).

The sedimentary sequence of the southern Maya block was affected by a compressive event during the Late Cretaceous, mostly recorded in rocks of northwestern Guatemala, and again by the Miocenic Chiapanecan orogeny (Sánchez-Montes de Oca 1979; Carfantan 1983; Donnelly et al. 1990). Whilst pre-volcanic rocks at El Chichón area record the effects of these events, as attested by the intense folding and following changes in sedimentation, there is no preserved record of these events in the TVC area.

During most of the Cenozoic, extensive clastic sedimentation occurred in the central and northern parts of the Chiapas State (in the areas of El Chichón volcano and the MCVA), possibly in response to deformation(s) occurring during Late Cretaceous across the Motagua collision zone (Brueckner et al. 2009; Martens et al. 2010), and later due to the interaction between the Chortis block and the southern Chiapas; no record of such depositional systems exists around Tacaná volcano. The deposition of siliciclastic sediments in the northern part of Chiapas continued until Late Miocene, while the intrusion of granodiorites, granites and diorites occurred to the southeast during the Upper Eocene-Lower Oligocene (39–29 Ma) and the Miocene (20–13 Ma).

About 130 km of displacement along the Polochic fault indicates that Eocene-Oligocene rocks of the Tacaná basement formed west of their present position. Additional displacement along the Caribbean-North America plate boundary in the Upper Eocene-Lower Oligocene indicated by the Cayman Trough suggests that the plutons underlying the TVC may have been displaced with the Chortis block from a position west of the Tehuantepec Isthmus (Fig. 1.1a, b).

A second period of intrusive activity occurred during Early to Middle Miocene (20–10 Ma) in southern Chiapas along the Tonalá Fault, probably related to the subduction of the Cocos plate beneath both the Caribbean (Chortis) and North America (Chiapas) plates (García-Palomo et al. 2006). At that time, the TVC was affected by Miocene deformation associated with relative movement of the Chortis and Maya blocks through strike-slip; this deformation was episodically transpressive as suggested locally by reverse faults (Sánchez-Montes de Oca 1979; Carfantan 1983; García-Palomo et al. 2006; Keppie 2012). The eastward motion of Chortis may also explain the presence of mylonites affecting the intrusive rocks of the Coastal Chiapas batholith ($\sim$15 Ma and up to $\sim$5 Ma) along the Tonalá shear zone to the west of Tacaná (Fig. 1.1a; Wawrzyniec et al. 2005; Ratschbacher et al. 2009; Witt et al. 2012a).

A compressive event in the Sierra de Chiapas and the northwestern Chortis block may be related either to: (1) transpression along the Tonalá fault (Wawrzyniec et al. 2005), or transpression transferred north of the Caribbean-North America plate boundary (Witt et al. 2012b); (2) tectonic complexities that resulted from plate reorganization affecting the Cocos and Caribbean plates, which in turn gave rise to stress transferred to southern North America (e.g., ridge jump or ridge collision; Silver et al. 2004); (3) formation of a compressive step-over between the Chamalecón and Motagua faults (Lyon-Caen et al. 2006); or, less likely, (4) to the relative motion between the northern and southern parts of the Maya block (Kim et al. 2011). Relative motion of southern Mexico with respect to North America has also been invoked (Andreani et al. 2008), but such model would be not consistent with the observed northeastward compression in northern Chiapas.

After the Middle Miocene, but before Pliocene, the basement rocks of the extinct Miocene Volcanic Arc were exhumed (Witt et al. 2012a). Exhumation may also be attributed to the lithospheric response to the removal of the forearc region (i.e. shallowing of the subduction angle), although the process of subduction erosion may have played a significant role. Such process consists of the removal of crystalline basement from the base of the crust of the overriding plate by "roughness" in the sinking slab (a process common in all convergent margins), with subsequent isostatic rebound leading to uplift. In the Tacaná area, the rocks of the Coastal Batholith were probably tilted and deeply eroded during the uplift process, creating widespread debris fans in the southeastern portion of the Tapachula area (Macías 2007; Murcia and Macías 2008).

The post-Middle Miocene deformation that formed the Sierra de Chiapas Fold-and-Thrust Belt was also recorded at El Chichón volcano (this study, García-Palomo et al. 2004), and at the Ixtapa graben in central

Chiapas (Meneses-Rocha 2001). The structural analyses of these localities indicate that the orientation of the main principal stress, σ1, was NE-SW. Post-Middle Miocene deformation was transtensional, forming the Ixtapa pull-apart basin. In this and other pull-apart basins, clastic sedimentation continued through Late Miocene and Pliocene. As we mentioned above, transtension has also been invoked as a mechanism to allow magma ascent to the surface in the MCVA. Several authors (Guzmán-Speziale et al. 1989; Guzmán-Speziale and Meneses-Rocha 2001; García-Palomo et al. 2004) indicate that the rocks in northern Chiapas were folded and affected by NW-SE and E-W strike-slip faults, whilst central Chiapas was dominated by strike-slip faults cutting the Cretaceous carbonate platform. These faults may have developed in close association with the long-term evolution of the Tonalá-Polochic-Motagua Fault System (Guzmán-Speziale and Meneses-Rocha 2001).

After the Miocene magmatism in the coastal Chiapas region ceased, volcanism was re-established in Chiapas during the late Pliocene. The inception of volcanism around the TVC began with the formation of the Sibinal and San Rafael Caldera (~2 Ma) (García-Palomo et al. 2006). The formation of the Modern Chiapanecan Volcanic Arc started around 2.1 Ma with the extrusion of the Tzontehuitz dome complex in central Chiapas, and continued sporadically in the Chiapas highlands to form other volcanic structures: (a) the 0.2 Ma Venustiano Carranza dome (Mora et al. 2012); (b) the Catedral volcano (1.6 Ma), and (c) the fissural Chapultenango basalt at 1.1 Ma near El Chichón (García-Palomo et al. 2004; Mora et al. 2012; see Chap. 3).

Around Tacaná, the Chanjale caldera (~0.9 Ma) formed and collapsed to the east, followed by the sedimentation of an extensive/or large volcaniclastic fan to the south that reached the Pacific Ocean (Macías et al. 2000). This was followed by the formation of caldera structures around the TVC that were affected by NE-SW normal faults associated with the present day Tacaná graben, bounded by the Coatán River and Suchiate River faults (García-Palomo et al. 2004). A series of N-S grabens, including the Tacaná graben, are located south of the Polochic-Motagua Fault system. They apparently accommodated an E-W extension that, although insignificant around Tacaná, reaches up to ≈5 mm/year at the Guatemala City graben. According to Lyon-Caen et al. (2006), Tacaná volcano would be located in the northwestern corner of the Caribbean Plate. GPS monitoring indicates that the volcano lies in a compressive stress field that controlled the alignment of volcanic structures and collapses (García-Palomo et al. 2006; Macías et al. (2006, 2010). Such interpretation disagrees with the widespread occurrence of Pleistocene and younger normal faults. El Chichón volcano and the TVC or Tacaná are younger structures formed during Late Pleistocene that continue to be active today. Although no precise ages exist for the onset of volcanism at El Chichón, the oldest age reported is ~0.37 Ma (Layer et al. 2009; Macías et al. 2010; Chap. 3), whereas it is ≥0.22 Ma at the TVC (Chap. 6).

References

Alvarado D, DeMets C, Tikoff B, Hernández D, Wawrzyniec TF, Pullinger C, Mattioli G, Turner HL, Rodríguez M, Correa-Mora F (2011) Forearc motion and deformation between El Salvador and Nicaragua: GPS, seismic, structural, and paleomagnetic observations. Lithosphere 3:3–21. doi:10.1130/L108.1

Alzaga-Ruiz H (1997) Prismas sedimentarios de la Sierra de Chiapas (HST, TST, LST) en función de las velocidades de subsidencia y sedimentación y ambientes sedimentarios. Bol AMGP XLVI(2):1–42

Andreani L, Rangin C, Martínez-Reyes J, Le Roy C, Aranda-García M, Le Pichon X, Peterson Rodriguez R (2008) The Neogene Veracruz fault: evidences for left lateral slip along the southern Mexico block. Bull Soc Géol Fr 179:195–208

Angeles-Aquino FJ, Reyes-Núñez J, Quezada-Muñetón JM, Meneses-Rocha JJ (1994) Tectonic evolution, structural styles, and oil habitat in Campeche sound, Mexico. GCAGS Trans 44:53–62

Authemayou C, Brocard G, Teyssier C, Simon-Labric T, Guttiérrez A, Chiquín EN, Morán S (2011) The Caribbean–North America–Cocos Triple Junction and the dynamics of the Polochic–Motagua fault systems: pull-up and zipper models. Tectonics 30:TC 3010. doi:10.1029/2010TC002814

Bergeat A (1894) Zur kenntnis der jungen eruptivgesteine der Republik Guatemala. Zeitschrift der Deutschen Geologischen Gesellschaft 46:131–157

Berlanga-García JA (2004) Estudio bioestratigráfico con base a macroforaminíferos del Terciario Inferior de la Sierra de Chiapas. Bol Asoc Mex Geól Pet 51:62–70

Böse E (1902) Breve noticia sobre el estado actual del Volcán Tacaná, Chiapas. Memorias y Revista de la Sociedad Científica Antonio Alzate 18:266–270

Brueckner HK, Avé Lallemant HG, Sisson VB, Harlow GE, Hemming SR, Martens U, Tsujimori T, Sorensen SS (2009) Metamorphic reworking of a high pressure–low temperature mélange along the Motagua fault, Guatemala: a record of Neocomian and Maastrichtian transpressional tectonics. Earth Planet Sci Lett 284(2009):228–235

Burkart B (1983) Neogene North American-Caribbean plate boundary across Northern Central America; Offset along the Polochic Fault. Tectonophysics 99:251–270

Burkart B, Self S (1985) Extension and rotation of crustal blocks in northern Central America and effect on the volcanic arc. Geology 13:22–26

Burkart B, Deaton BC, Moreno G (1987) Tectonic wedges and offset Laramide structures along the Polochic fault of Guatemala and Chiapas, Mexico: reaffirmation of large Neogene displacement. Tectonics 6:411–422. doi:10.1029/TC006i004p00411

Campa MF, Coney PJ (1983) Tectono-stratigraphic terranes and mineral resource distributions in México. Can J Earth Sci 20:1040–1051

Canora C, Villamor P, Martínez-Díaz JJ, Berryman KR, Álvarez-Gómez JA, Capote R, Hernández W (2012) Paleoseismic analysis of the San Vicente segment of the El Salvador Fault Zone, El Salvador, Central America. Geol Acta 10:103–123

Canul RF, Rocha VS (1981) Informe geológico de la zona geotérmica de "El Chichónal", Chiapas: Informe 32–81. Unpublished Report of the Geothermal Department of the Comisión Federal de Electricidad, Morelia, Michoacán, México, 30 pp

Carfantan JC (1976) El prolongamiento del sistema de fallas Polochic-Motagua en el sureste de México; una frontera entre dos provincias geológicas. Abstract III Congreso Latino Americano de Geología, Acapulco, México

Carfantan JC (1983) Les ensambles géologiques du Mexique meridional-évolution géodynamique durant le Mésozoique et le Cénozoique. Geofís Int 22:9–37

Chávez Valois VM (1997) Análisis estructural del área Lacantun, porción oriental de la Sierra de Chiapas. Bol AMGP 46:72–92

Clemons RE, Anderson TH, Bohnenberger OH, Burkart B (1974) Stratigraphic nomenclature of recognized Paleozoic and Mesozoic rocks of western Guatemala. Am Assoc Petrol Geol Bull 58:313–320

Correa-Mora F, DeMets C, Alvarado D, Turner HL, Mattioli GS, Hernandez D, Pullinger C, Rodriguez M, Tenorio C (2009) Evidence for weak coupling of the Cocos plate subduction interface and strong coupling of the volcanic arc faults from modeling of GPS data: El Salvador and Nicaragua. Geophys J Int 179:1279–1291

Couch R, Woodcock S (1981) Gravity and structure of the continental margins of southwestern Mexico and northwestern Guatemala. J Geophys Res 86:1829–1840

Damon P, Montesinos E (1978) Late Cenozoic volcanism and metallogenesis over an active Benioff Zone in Chiapas, Mexico. Ariz Geol Soc Digest 11:155–168

Damon PE, Shafiqullah M, Clark KF (1981) Age trends of igneous activity in relation to metallogenesis in the southern Cordillera. Relations of tectonics to ore deposits in the southern Cordillera. Ariz Geol Soc Digest 14:137–154

De la Cruz V, Hernández R (1985) Estudio geológico a semidetalle de la zona geotérmica del Volcán Tacaná, Chiapas. 41/85, Comisión Federal de Electricidad, México

De la Cruz-Reyna S, Armienta MA, Zamora V, Juárez F (1989) Chemical changes in spring waters at Tacaná Volcano, Chiapas, México. J Volcanol Geotherm Res 38:345–353

DeMets C (2001) A new estimate for present-day Cocos-Caribbean plate motion: implications for slip along the Central American volcanic arc. Geophys Res Lett 28:4043–4046. doi:10.1029/2001GL013518

Dengo CA (1982) Structural analysis of the Polochic fault zone in western Guatemala, Central America. Dissertation, Texas A&M University, College Station

Dengo G (1985) Mid America; Tectonic setting for the Pacific margin from southern México to northwestern Colombia. In: Nairn AEM, Stehli FG (eds) The oceanic basins and margins, vol 7a. The Pacific Ocean. Plenum Press, New York, pp 123–180

Donnelly TW, Horne GS, Finch RC, Lopez-Ramos E (1990) Northern Central America; the Maya and Chortis blocks. In: Dengo G, Case JE (eds) The geology of North America, vol H. The Caribbean region. Geological Society of America, Boulder, pp 37–76

Duffield WA, Tilling RI, Canul R (1984) Geology of El Chichón volcano, Chiapas, Mexico. J Volcanol Geotherm Res 20:117–132

Duffield WA, Heiken GH, Wohletz KH, Maasen LW, Dengo G, Mckee EH, Castaneda O (1992) Geology and geothermal potential of the Tecuamburro volcano area, Guatemala. Geothermics 21:425–446. doi:10.1016/0375-6505(92)90001-P

Franco A, Molina E, Lyon-Caen H, Vergne J, Monfret T, Nercessian A, Cortez S, Flores O, Monterosso D, Requenna J (2009) Seismicity and crustal structure of the Polochic-Motagua fault system area (Guatemala). Seismol Res Lett 80:977–984. doi:10.1785/gssrl.80.6.977

Franco A, Lasserre C, Lyon-Caen H, Kostoglodov V, Molina E, Guzman-Speziale M, Monterosso D, Robles V, Figueroa C, Amaya W, Barrier E, Chiquin L, Moran S (2012) Fault kinematics in northern Central America and coupling along the subduction interface of the Cocos Plate, from GPS data in Chiapas (Mexico), Guatemala and El Salvador. Geophys J Int 189:1223–1236. doi:10.1111/j.1365-246X.2012.05390.x

García-Molina, G (1994) Structural evolution of SE Mexico (Chiapas-Tabasco-Campeche) offshore and onshore. Ph.D. thesis, Rice University, 161 pp

García-Palomo A, Macías JL, Espindola JM (2004) Strike-slip faults and K-Alkaline volcanism at El Chichón volcano, southeastern Mexico. J Volcanol Geotherm Res 136:247–268

García-Palomo, A, Macías JL, Arce JL, Mora JC, Hughes S, Saucedo R, Espindola JM. Escobar R, Layer P (2006) Geological evolution of the Tacaná Volcanic Complex, México-Guatemala, In: Rose et al (eds) Natural hazards in Central America. Geological Society of America, Boulder, p 412 (Special Paper)

García-Tijerina N (1950) Bosquejo geológico del istmo de Tehuantepec. Bol Asoc Mex Geól Pet 435–444

Gibson JB (1936) Estratigrafía y tectónica de la zona costera del Golfo entre el 19°34' latitud norte y el Río Cotzacoalcos. Bol Soc Geol Mex, Tomo IX 5:271–288

Godínez A, Lawton TF, Molina Garza RS, Iriondo A, Weber B, Lopez Martinez M (2011) Jurassic volcanic and sedimentary rocks of the La Silla and Todos Santos Formations, Chiapas: record of Nazas arc magmatism and rift-basin formation prior to opening of the Gulf of Mexico. Geosphere 7:121–144

Gose WA (1985) Paleomagnetic results from Honduras and their bearing on Caribbean tectonics. Tectonics 4:565–585

Guzmán-Speziale M (2001) Active seismic deformation in the grabens of northern central America and its relationship to the relative motion of the North American-Caribbean plate boundary. Tectonophysics 337:39–51

Guzmán-Speziale M, Pennington WD, Matumoto T (1989) The triple junction of the North America, Cocos, and Caribbean Plates: seismicity and tectonics. Tectonics 8:981–999

Guzmán-Speziale M, Meneses-Rocha JJ (2000) The Noth America-Caribbean plate boundary best of the Motagua-Polochic fault system: a fault jog in Southeastern Mexico. J South Am Earth Sci 13:459–468

Islas Tenorio JJ, Ramírez García MG, Gómez Áviles A, Moreno Ruiz JP, Wingartz Carranza JA, Mendieta Flores JL (2005) Carta Geológico-Minera Villahermosa E15-8. Servicio Geológico Mexicano, Pachuca

Jennette D, Wawrzyniec T, Fouad KJ, Dunlap DB, Meneses-Rocha J, Grimaldo F, Muñoz R, Barrera D, Williams-Rojas CT, Escamilla-Herrera A (2003) Traps and turbidite reservoir characteristics from a complex and evolving tectonic setting, Veracruz Basin, southeastern Mexico. AAPG Bull 87:1599–1622

Keppie JD, Moran-Zenteno DJ (2005) Tectonic implications of alternative cenozoic reconstructions for Southern Mexico and the Chortis Block. Int Geol Rev 47:473–491

Keppie DF, Keppie JD (2012) An alternative Pangea reconstruction for Middle America with the Chortis Block in the Gulf of Mexico: tectonic implications. Int Geol Rev. doi:10.1080/00206814.2012.676361

Keppie DF (2012) Derivation of the Chortis and Chiapas blocks from the western Gulf of Mexico in the latest Cretaceous–Cenozoic: the Pirate model. Int Geol Rev. doi:10.1080/00206814.2012.676356

Kim YH, Clayton R, Keppie DF (2011) Evidence of a collision between the Yucatan Block and Mexico in the Miocene. Geophys J Int. doi:10.1111/j.1365-246X.2011.05191.x

LaFemina P, Dixon TH, Govers R, Norabuena E, Turner H, Saballos A, Mattioli G, Protti M, Strauch W (2009) Fore-arc motion and Cocos Ridge collision in Central America. Geochem Geophys Geosyst 10:Q05S14. doi:10.1029/2008GC002181

Layer PW, García-Palomo A, Jones D, Macías JL, Arce JL, Mora JC (2009) El Chichón volcanic complex, Chiapas, México: stages of evolution based on field mapping and 40Ar/39Ar geochronology. Geofís Int 48:33–54

LeFevre LV, McNally KC (1985) Stress distribution and subduction of Aseismic Ridges in the Middle America subduction zone. Geophys Res Lett 90(B6):4495–4510

Lopez-Ramos Ernesto (1979) Geologáa de Mexico. México, D. F., t. 3, 453 pp, Edicion Escolar

Luhr JF, Carmichael ISE, Varekamp JC (1984) The 1982 eruptions of El Chichón Volcano, Chiapas, Mexico: mineralogy and petrology of the anhydrite bearing pumices. J Volcanol Geotherm Res 23:69–108

Lyon-Caen H, Barrier E, Lasserre C, Franco A, Arzu I, Chiquin L, Chiquin M, Duquesnoy T, Flores O, Galicia O, Luna J, Molina E, Porras O, Requena J, Robles V, Romero J, Wolf R (2006) Kinematics of the North American Caribbean-Cocos plates in Central America from new GPS measurements across the Polochic-Motagua fault system. Geophys Res Lett 33:1–5

Macías JL (2007) Geology and eruptive history of some active volcanoes of México. In: Alaniz-Álvarez SA, Nieto-Samaniego ÁF (eds) Geology of México: celebrating the centenary of the Geological Society of México, vol 422. Geological Society of America, pp 183–232 (Special Paper). doi: 10.1130/2007.2422(06)

Macías JL, Espíndola JM, García-Palomo A, Scott KM, Hughes S, Mora JC (2000) Late Holocene Peléan style eruption at Tacaná Volcano, Mexico-Guatemala: past, present, and future hazards. Geol Soc Am Bull 112:1234–1249

Macías JL, Arce JL, Garduño-Monroy VH, Rouwet D, Taran Y, Layer P, Jiménez A, Álvarez R (2010) Estudio de prospección geotérmica para evaluar el potencial del volcán Chichonal, Chiapas. Unpublished Report no. 9400047770 IGF-UNA -CFE

Macías JL, Arce JL, García-Palomo A, Mora JC, Layer PW, Espíndola JM (2006) Late-Pleistocene flank collapse triggered by dome growth at Tacaná Volcano, México-Guatemala, and its relationship to the regional stress regime. Bull Volcanol 72:33–53

Mandujano-Velásquez J (1996) Cuatro megasecuencias de evolucion litoestratigráfica en la Sierra de Chiapas. Bol Asoc Mex Geól Pet 44:46–60

Mandujano-Velazquez JJ, Keppie JD (2009) Middle Miocene Chiapas fold and thrust belt of Mexico: a result of collision of the Tehuantepec Transform/Ridge with the Middle America. Geol Soc Lond Spec Publ 327(1):55–69

Manea V, Manea M (2006) Origin of the modern Chiapanecan volcanic arc in southern Mexico inferred from thermal models. GSA Spec Pap 412:27–38

Manea M, Manea VC, Ferrari L, Kostoglodov V, Bandy WL (2005) Tectonic evolution of the Tehuantepec ridge. Earth Planet Sci Lett 238:64–77

Martens U, Weber B, Valencia VA (2010) U/Pb geochronology of Devonian and older Paleozoic beds in the southeastern Maya block, Central America: its affinity with peri-Gondwanan terranes. GSA Bull 122:815–829

Martínez-Kemp HL, González-Posadas JF, Bartok P (2006) Salt involvment in the Jujo-Tecominoacán thrust belt, Chiapas-Tabasco region, south east Basin, Mexico, Gulf Coast. Assoc Geol Soc 55:520–530

Meneses-Rocha JJ (1985) Tectonic evolution of the strike-slip fault province of Chiapas, Mexico. M. A. thesis, University of Texas, Austin

Meneses-Rocha JJ (1987) Marco tectónico y paleogeográfico del Triásico Tardío-Jurásico en el sureste de Mexico. Bol Asoc Mex Geól Pet 39:3–67

Meneses-Rocha JJ (2001) Tectonic evolution of the Ixtapa Graben, an example of a strike-slip basin of southeastern Mexico: implications for regional petroleum systems. In: Bartolini C, Buffler RT, Cantú-Chapa A (eds) The western Gulf of Mexico Basin: tectonics, sedimentary basins, and petroleum systems. AAPG Mem 75:183–216

Meneses-Rocha J, Monroy-Audelo ME, Gomez-Chavarria JC (1994) Bosquejo paleogeografico y tectonico del sur de Mexico durante el Mesozoico. Boletin del la Sociedad Mexicana de Geologos Petroleros 44:18–45

Molina Garza RS, van Hinsbergen DJJ, Rogers RD, Ganerød M, Dekkers MJ (2012) The Padre Miguel Ignimbrite Suite, central Honduras: paleomagnetism, geochronology, and tectonic implications. Tectonophysics 574–575:144–157

Mora JC, Jaimes-Viera MC, Garduño-Monroy VH, Layer PW, Pompa-Mera V, Godinez ML (2007) Geology and geochemistry characteristics of the Chiapanecan volcanic Arc (Central Area), Chiapas Mexico. J Volcanol Geotherm Res 162:43–72

Mora JC, Layer PW, Jaimes-Viera MC (2012) New $^{40}Ar/^{39}Ar$ ages from the Central Part of the Chiapanecan volcanic arc, Chiapas, México. Geofís Int 51(1):39–49

Morán-Zenteno DJ, Corona-Chávez P, Tolson G (1996) Uplift and subduction erosion in southwestern Mexico since the Oligocene: pluton geobarometry constraints. Earth Planet Sci Lett 141:51–65

Mujica R (1987) Estudio petrogenético de las rocas ígneas y metamórficas en el Macizo de Chiapas: México, Instituto Mexicano del Petróleo. Informe interno C-2009:47

Müller RD, Royer JY, Cande SC, Roest WR, Maschenkov S (1999) New constraints on the late cretaceous/tertiary plate tectonic evolution of the Caribbean. In: Mann P (ed) Caribbean basins, sedimentary basins of the World, vol 4. Elsevier Science, London, pp 33–59

Müllerried FKG (1951) La reciente actividad del Volcán de Tacaná, Estado de Chiapas, a fines de 1949 y principios de 1950. Informe del Instituto de Geología de la UNAM, pp 1–28

Murcia H, Macías JL (2008) Registro geológico de inundaciones recurrentes e inundación del 4 de octubre de 2005 en la ciudad de Tapachula, Chiapas, México. Rev Mex Cienc Geol 26:1–17

Nelson SA, Gonzalez-Caver E, Kurtis Kyser T (1995) Constraints on the origin of alkaline and calc-alkaline magmas from the Tuxtla Volcanic Field, Veracruz, Mexico. Contrib Mineral Petrol 122:191–211

Nixon GT (1982) The relationship between Quaternary volcanism in central Mexico and the seismicity and structure of the subducted ocean lithosphere. Geol Soc Am Bull 93:514–523

Norabuena E, Dixon TH, Schwartz S, DeShon H, Newman A, Protti M, Gonzalez V, Dorman L, Flueh ER, Lundgren P, Pollitz F, Sampson D (2004) Geodetic and seismic constraints on some seismogenic zone processes in Costa Rica. J Geophys Res 109:B11403

Ortega-Gutiérrez F, Solari LA, Ortega-Obregón C, Elías-Herrera M, Martens U, Morán-Icál S, Chiquín M, Keppie JD, Rafael Torres De León, Schaaf P (2007) The Maya-Chortis boundary: a tectono stratigraphic approach. Int Geol Rev 49:996–1024

Padilla y Sánchez RJ (2007) Evolución del sureste mexicano desde el Mesozoico al presente en el contexto regional del Golfo de México. Bol Soc Geol Mexico 19–42

Pantoja-Alor J, Fries C Jr, Rincón-Orta C, Silver LT, Solorio-Munguía J (1974) Contribución a la geocronología del Estado de Chiapas. Boletin Asociación Mexicana de Geólogos Petroleros XXVI:205–223

Pardo M, Suárez G (1995) Shape of the subducted Rivera and Cocos plates in southern Mexico: seismic and tectonic implications. J Geophys Res 100:12357–12373

Pindell JL, Cande SC, Pitman WC III, Rowley DB, Dewey JF, Labrecque J, Haxby W (1988) A plate-kinematic framework for models of Caribbean evolution. Tectonophysics 155:121–138

Pindell J, Kennan L, Maresch WV, Stanek K-P, Draper G, Higgs R (2005) Plate-kinematics and crustal dynamics of circum-Caribbean arc continent interactions. In: Avé-Lallemant HG, Sisson VB (eds) Tectonic controls on basin development in Proto-Caribbean Margins. Spec Pap Geol Soc Am 394:7–52

Ponce L, Gaulon R, Suárez G, Lomas E (1992) Geometry and state of stress of the downgoing Cocos plate in the Isthmus of Tehuantepec, Mexico. Geophys Res Let 19:773–776

Quezada-Muñetón JM (1983) Las formaciones San Ricardo y Jerico del Jurásico Medio-Cretácico Inferior en el SE de México. Bol Asoc Mex Geól Pet 35:37–64

Quezada-Muñetón JM (1987) El Cretácico Medio-Superior, y el límite Cretácico Superior-Terciario Inferior en la Sierra de Chiapas. Bol Asoc Mex Geól Pet 39:3–98

Ratschbacher L, Franz L, Min1 M, Bachmann R, Martens U, Stanek K, Stübner K, Nelson BK, Herrmann U, Weber B, López-Martínez, M, Jonckheere R, Sperner B, Tichomirowa M, Mcwilliams MO, Gordon M, Meschede M, Bock P (2009) The North American–Caribbean plate boundary in Mexico- Guatemala-Honduras. In: Martini IP, French HM, Pérez-Alberti A (eds) Ice-marginal and periglacial processes and sediments. Geol Soc Spec Publ 328:219–293. doi:10.1144/SP328.11

Rebollar CJ, Espíndola VH, Uribe A, Mendoza A, Pérez-Vertti A (1999) Distribution of stress and geometry of the Wadati-Benioff zone under Chiapas, Mexico. Geofís Int 38:95–106

Reynolds JH (1987) Timing and sources of neogene and quaternary volcanism in South-Central Guatemala. J Volcanol Geotherm Res 33:9–22

Rogers R, Mann P (2007) Transtensional deformation of the western Caribbean-North America plate boundary. Geol Soc Am Spec Pap 428:37–64

Rogers RD, Kárason H, van der Hilst RD (2002) Epeirogenic uplift above a detached slab in northern Central America. Geology 30:1031–1034

Rogers RD, Mann P, Emmet PA (2007) Tectonic terranes of the Chortis block based on integration of regional aeromagnetic and geologic data. Geol Soc Am Spec Pap 428:65–88

Sánchez-Montes de Oca R (1979) Geología petrolera de la sierra de Chiapas. Bol Asoc Mex Geól Pet 31:67–97

Sapper K (1896) La geografía física y la geografía de la Península de Yucatán. Instituto Geológico de México 3:1–58

Sapper K (1899) Ueper Gebirsbau und Boden des noerdlichen Mittelamerika. Petermanns Geogr Mitt 127:119

Saucedo GR, Esquivias H (1988) Evaluación del riesgo volcánico en el área del Volcán Tacaná, Chiapas. BS thesis, Instituto Politécnico Nacional, Escuela Superior de Ingeniería y Arquitectura, México

Schaaf P, Morán-Zenteno D, Hernández-Bernal MS (1995) Paleogene continental margin truncation in southwestern Mexico: geochronological evidence. Tectonics 14:1229–1350

Schwartz DP, Cluff LS, Donnelly TW (1979) Quaternary faulting along the Caribbean-North American plate boundary in Central America. Tectonophysics 52:431–445

Sedlock RL, Ortega-Gutiérrez F, Speed RC (1993) Tectonostratigraphic terranes and tectonic evolution of Mexico. Geol Soc Am Spec Pap 278:153

Silva-Romo G (2008) Guayape-Papalutla fault system: A continuous Cretaceous structure from southern Mexico to the Chortis block? Tectonic implications. Geology 36:75–78

Silver E, Costa Pisani P, Hutnak M, Fisher A, DeShon H, Taylor B (2004) An 8–10 tectonic event in the Cocos plate offshore Costa Rica: Result of Cocos Ridge Collison. Geophys Res Lett 31. doi:10.1029/2004GL020272

Tolson G (2007) The Chacalapa fault, southern Oaxaca, México. Geol Soc Am Spec Pap 422:343–357. doi:10.1130/2007.2422

Truchan M, Larson RL (1973) Tectonic lineaments on the Cocos plate. Earth Planet Sci Lett 17:426–432. doi:10.1016/0012-821X(73)90211-2

Turner HL, LaFemina P, Saballos A, Mattioli G, Jansma PE, Dixon T (2007) Kinematics of the Nicaraguan forearc from GPS geodesy. Geophys Res Lett 34:L02302. doi:10.1029/2006GL027586

Waibel L (1933) Die Sierra Madre de Chiapas. Mitteil d Geogr Gesell Hamburg 43:13–162

Wawrzyniec T, Molina-Garza RS, Geissman JW, Iriondo A (2005) A newly discovered relic, transcurrent plate boundary—the Tonalá shear zone and paleomagnetic evaluation of the western Maya block, SW Mexico. GSA Abst Prog 37:68

Weber B, Cameron KL, Osorio M, Schaaf P (2005) A Late Permian tectonothermal event in Grenville crust of the southern Maya Terrane; UPb zircon ages from the Chiapas Massif, southeastern Mexico. Int Geol Rev 47:509–529

Weber B, Iriondo A, Premo WR, Hecht L, Schaaf P (2007) New insights into the history and origin of the southern Maya block, SE México: U-Pb–SHRIMP zircon geochronology from metamorphic rocks of the Chiapas massif. Int J Earth Sci 96:253–269

Witt C, Brichau S, Carter A (2012a) New constraints on the origin of the Sierra Madre de Chiapas (south Mexico) from sediment provenance and apatite thermochronometry. Tectonics 31:TC6001. doi:10.1029/2012TC003141

Witt C, Rangin C, Andreani L, Olaez N, Martínez J (2012b) The transpressive left-lateral Sierra Madre de Chiapas and its buried front in the Tabasco plain (southern Mexico). J Geol Soc 169:143–155

Wunderman RL, Rose WI (1984) Amatitlan, an active resurging cauldron 10 km south of Guatemala city. J Geophys Res 89:8525–8539. doi:10.1029/JB089iB10p08525

Petrology and Geochemistry of El Chichón and Tacaná: Two Active, yet Contrasting Mexican Volcanoes

2

José Luis Arce, James Walker, and John Duncan Keppie

Abstract

El Chichón and Tacaná have been widely considered subduction-related volcanoes, although they show differences in mineral assemblage and magma composition. El Chichón emitted potassium- and sulfur-rich trachyandesites and trachybasalts during its eruptive history, whereas Tacaná erupted basalts to dacites with moderate potassium contents, and minor high-Ti magmas. The magmatic evolution in both volcanoes involved similar fractionating assemblages of Fe-Ti oxides, olivine, plagioclase, pyroxenes, amphibole and apatite. Both K_2O/P_2O_5 ratios and isotopic signatures, indicate that the melts of El Chichón and Tacaná experienced significant crustal contamination. Magma genesis for both volcanoes has been related to the northeastward subduction of the Cocos Plate. Even if such origin agrees with the location of Tacaná, situated 100 km above the Cocos Benioff Zone, a subduction origin is at odds with recent tectonic and geophysical data obtained for southern Mexico for El Chichón, located about 400 km from the trench. In this chapter we review the existing petrographic and geochemical data for El Chichón and Tacaná volcanoes, in order to understand their magma genesis and evolution.

Keywords

Tacaná and El Chichón volcanoes · Petrology and geochemistry · Potassium-rich magmas

2.1 Introduction

Volcanism occurs at several tectonic environments, including subduction, rift, and hot spot regions, and magma genesis is controlled mainly by three factors: temperature, pressure, and the presence of volatiles (Schmincke 2004). Petrological and geochemical data, including isotopic ratios, and trace element concentrations, can be fingerprints of the volcano-tectonic environment (Best 2003). Subduction-related magmas are distinctively enriched in large-ion-lithophile elements (LILE), and depleted in high-field-strength elements

J.L. Arce (✉) · J.D. Keppie
Instituto de Geología, Cd. Universitaria Coyoacan, UNAM, 04510 Coyoacan, Mexico
e-mail: jlarce@geologia.unam.mx

J. Walker
Department of Geology and Environmental Geosciences, Northern Illinois University, DeKalb, IL, USA

T. Scolamacchia and J.L. Macías (eds.), *Active Volcanoes of Chiapas (Mexico): El Chichón and Tacaná*, Active Volcanoes of the World, DOI 10.1007/978-3-642-25890-9_2, © Springer-Verlag Berlin Heidelberg 2015

(HFSE) relative to Mid-Ocean Ridge Basalts (MORB) magmas (Tatsumi and Kogiso 1997). These features have been attributed to fluid fluxes deriving from the subducted slab (Saunders et al. 1980), although in tectonically complex subduction zones, magma chemistry may reflect the added tectonic complexity (Gazel et al. 2009) (i.e. the presence of the Tehuantepec Ridge and the Polochic-Motagua fault zone, in southern Mexico).

Both El Chichón and Tacaná volcanoes have been related to the subduction of the Cocos plate beneath the North America plate (Luhr et al. 1984; Mora et al. 2004; Chap. 1). Nevertheless the presence of the Tehuantepec ridge (Fig. 2.1), a prominent lithospheric structure in the Cocos plate, which could extend below El Chichón (Luhr et al. 1984; Manea et al. 2005), complicates the tectonic framework of the area.

Tacaná (4,050 masl) represents the northernmost volcano of the Central American Volcanic Arc (CAVA) (Mercado and Rose 1992; Mora et al. 2004; García-Palomo et al. 2006). El Chichón (1,100 masl) is the youngest structure of the NW-trending Modern Chiapanecan Volcanic Arc (MCVA of Damon and Montesinos 1978), which includes ten volcanoes (Mora et al. 2007) varying in age from 2.2–0.4 Ma (see Chap. 1).

The NW-trending, 150 km long MCVA has been related to the subduction of the Cocos Plate beneath the North American Plate (Damon and Montesinos 1978) and it is located at a distance of approximately 300–400 km from the Middle America Trench (MAT). The Cocos Plate beneath the volcano reaches a depth of about 200 km, with a dip of 40–45° (Rebollar et al. 1999; Manea and Manea 2006).

El Chichón consists of pyroclastic materials, lava domes and other volcanic edifices (see Chap. 3 for stratigraphic details). Both lava domes and pyroclastic material have a uniform composition of alkaline rocks, ranging between K-rich trachyandesitic and trachydacitic rocks (Luhr et al. 1984; McGee et al. 1987; Macías et al. 2003). One of the most important characteristics of the 1982 El Chichón eruption was the great sulfur content of its magma (2.6 wt% SO_3; Luhr et al. 1984). Evidence for such high content was the generation of anhydrite crystals recognized for the first

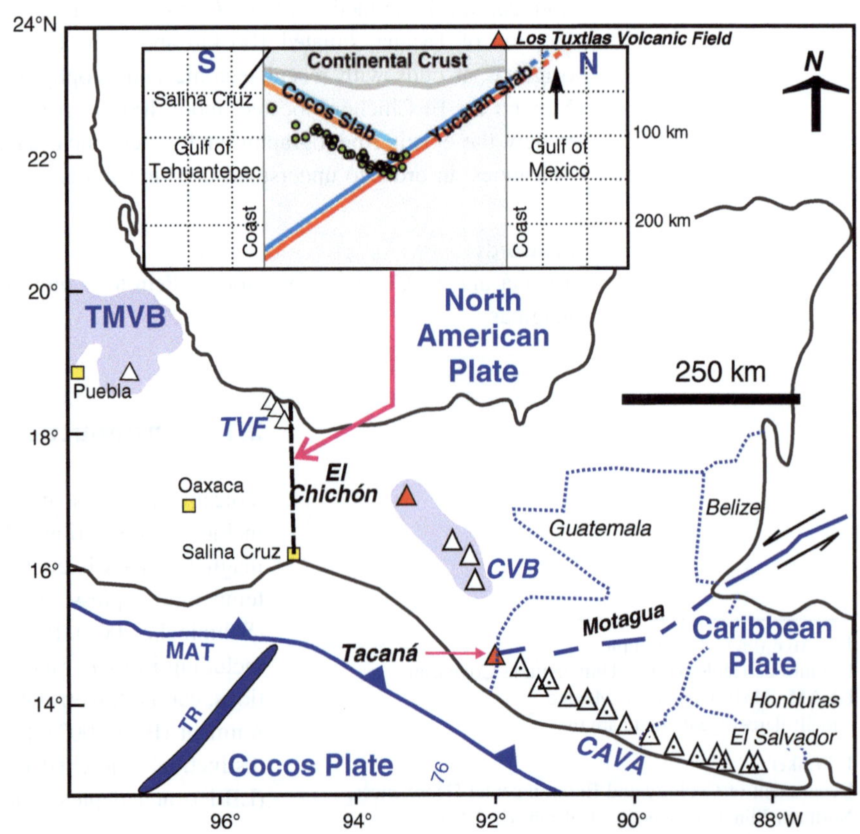

Fig. 2.1 Location map of the El Chichón and Tacaná volcanoes in southern Mexico. *TMVB* Trans-Mexican Volcanic Belt; *CAVA* Central American Volcanic Arc; *CVB* Chiapanecan Volcanic Belt; *MAT* Middle American Trench; *TR* Tehuantepec Ridge. *Inset* is a model showing interaction of the Cocos and Yucatán slabs underneath El Chichón and the Los Tuxtlas Volcanic Field (LTVF) (after Kim et al. 2011)

time in pumice samples of the 1982 eruption. Fresh pumice samples contained up to 2 wt% suhedral to euhedral, anhydrite crystals, in association with hornblende, biotite, and sphene phenocrysts (Luhr et al. 1984). Moreover, the trachyandesites emitted in 1982 were rich in K, Rb, Sr, Th, U, and Cs, compared to other Mexican and Central American volcanoes, fact that was attributed to the relatively large distance from the Middle America Trench (Luhr et al. 1984) (Fig. 2.1). Alkaline rocks in subduction zones have been linked to the presence of fracture zones (DeLong et al. 1975). At El Chichón, such assumption would be consistent with the presence of the Tehuantepec ridge underneath this volcano (Fig. 2.1).

Recent seismic data (Kim et al. 2011) obtained by teleseismic P-to-S converted waves suggest an anomalous southwest-dipping slab in southern Mexico. This interpretation would indicate that the Cocos plate does not directly underlie El Chichón, implying that El Chichón lies just to the south of a projection of the intra-Yucatán subduction zone. Such zone would be a southwest-dipping structure active during the Miocene, which generated a slab descending up to 250 km depth (Kim et al. 2011), truncating the Cocos slab at circa 100 km depth (Fig. 2.1). At a local scale, El Chichón is situated inside a sinistral E-W fault system (San Juan Fault System of García-Palomo et al. 2004; Chap. 1).

Tacaná is a typical stratovolcano with steep slopes that belongs to the Tacaná Volcanic Complex (TVC, see Chaps. 1 and 6). It is composed of calc-alkaline, andesitic to dacitic lava flows, and pyroclastic deposits of similar compositions (Mercado and Rose 1992; Mora et al. 2004; García-Palomo et al. 2006). It lies on the trace of the sinistral Polochic-Motagua Fault zone, which according to Guzman-Speziale et al. (1989) represents the boundary between the North American and Caribbean plates in Guatemala and southern Mexico. The TVC is located ~200 km from the MAT, and approximately 100 km above the subducting Cocos slab which dips about 40° (Rebollar et al. 1999; Syracuse and Abers 2006). The magmatism that generated the CAVA volcanism has been focus of numerous investigations (i.e. Carr et al. 1990; Feigenson and Carr 1993; Carr et al. 2003; Abers et al. 2003; Bolge et al. 2009). Despite the clear relationship of the CAVA with the Cocos plate subduction, large variations in trace elements and isotopic ratios have been reported at regional level (Carr et al. 1990; Bolge et al. 2009), and explained with variations in slab depths, slab inputs, crustal architecture, and tectonic segmentation (Carr et al. 1990; Patino et al. 2000; Bolge et al. 2009).

In this chapter we present the current knowledge of petrological aspects of El Chichón and Tacaná active volcanoes, concerning to their magmatic evolution and present the ideas about the genesis of their magmas, based on published whole-rock chemistry, isotope data, and petrological information.

2.2 Petrography and Geochemistry

2.2.1 El Chichón

Several petrological and geochemical studies have been carried out at El Chichón since the 1982 eruption. Initial investigations indicated the presence of subhedral phenocrysts of anhydrite in fresh ejecta (Luhr et al. 1984). Subsequent petrological and geochemical works reported that not only the 1982 magma, but also the magmas erupted before 1982 were sulfur-rich (Rose et al. 1984; McGee et al. 1987). Rose et al. (1984) analyzed undefined lava dome and pyroclastic flow deposits, outcropping inside the inner 1982 crater walls, consisting of porphyritic trachyandesites, with a crystallinity ranging from 20 to 40 vol% (Rose et al. 1984). Pumice samples of the 1982 eruption (unit A, see Chap. 3) are also porphyritic and vesicular, with crystallinity between 19 and 29 vol% (Luhr et al. 1984) characterized by abundant, and large (0.3–4 mm) phenocrysts of plagioclase and amphibole. Pumice samples from the fallout deposit of Unit B (Chap. 3) show a porphyritic texture (Fig. 2.2a) with up to 20 vol% crystals of plagioclase (6–12 vol %) + amphibole + clinopyroxene + Fe-Ti oxides, set in a groundmass (25–36 vol%) of microlites and glass (Macías et al. 2003). Plagioclase crystals are euhedral to anhedral (Fig. 2.2a, b), mostly showing normal zonation, and have compositions varying from An_{46} to An_{65} in the cores to An_{35} to An_{40} in the rims. The reverse zoning observed in some cases (i.e. An_{41} in the core and An_{47} in the rim) was interpreted as a result of repeated and continuous recharge episodes by relatively mafic magmas (Macías et al. 2003). In Unit B, hornblende represents ~2 vol% with a homogeneous compositions of magnesian hastingsite and minor tschermakite (Macías et al. 2003). Augite and Fe-Ti

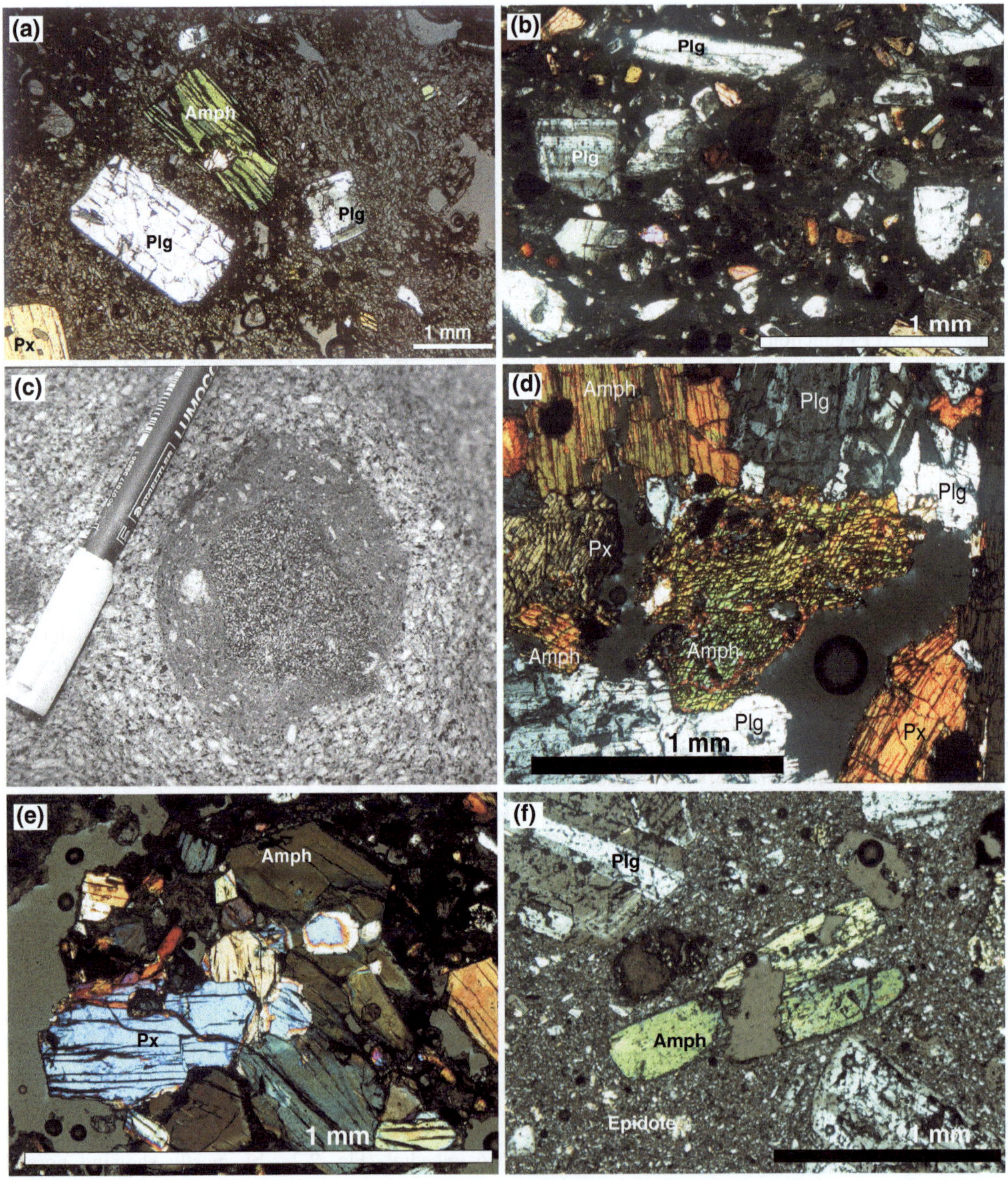

Fig. 2.2 **a** Cross-polarized microphotograph of Unit B pumice sample, showing a porphyritic texture, with phenocrysts of plagioclase (*Plg*), pyroxene (*Px*), amphibole (*Amph*), Fe-Ti oxides (*oxi*) in a groundmass and vesicular matrix (After Macías et al. 2003); **b** Cambac dome sample showing typical porphyritic textures, with abundant plagioclase, pyroxene and amphibole phenocrysts; **c** Mafic enclave in trachyandesitic lava domes (Somma domes) at the head of the Agua Tibia valley. Rounded enclaves commonly show assimilation aureoles, containing abundant plagioclase phenocrysts. The marker is 14 cm long (After Macías et al. 2003); **d** Representative cross-polarized microphotographs of mafic enclaves (CH-ME) found in El Chichón samples, and showing equigranular textures containing plagioclase (*Plg*) pyroxene (*Px*), amphibole (*Amph*), and Fe-Ti oxides (After Arce et al. 2014); microphotographs from **e** Cambac and **f** Capulín dome samples

oxides occur in minor amounts (<1 vol%) (Macías et al. 2003).

Pumice and dense lithic samples from units C, D, E, F, and J (Chap. 3) show similar petrographic characteristics, with a porphyritic texture consisting of phenocrysts and microphenocrysts of plagioclase + amphibole + clinopyroxene ± Fe-Ti oxides (11–42 vol% of crystals). Even if always with euhedral rims, plagioclase crystals show different internal textures, such as sieved and patchy zones, interpreted to have formed during multiple recharge events inside the magma chamber (Andrews et al. 2008). Accessory minerals in every sample include magnetite, apatite, and titanite, whereas biotite is present only in samples of Unit D (1,250 year BP eruption), in which clinopyroxene is absent (Andrews et al. 2008).

Samples from the Capulín, NW, SW and Cambac lava domes, (see Chap. 3 for their location) included numerous mafic enclaves (Espíndola et al. 2000; Macías et al. 2003, 2010a; Arce et al. 2014). The mafic enclaves show an equigranular texture (Fig. 2.2c, d), with a mineral assemblage consisting of amphibole + augite + enstatite + plagioclase + olivine, and minor Fe-Ti oxides (Espíndola et al. 2000; Macías et al. 2003). The lava dome samples are highly crystalline (up to 60 vol%), with a mineral assemblage (in order of abundance) of plagioclase, amphibole, clinopyroxene, ilmenite, titanomagnetite, apatite and sometimes biotite (Fig. 2.2b–f). Apatite is more common as inclusions in plagioclase crystals (Macías et al. 2010a). The presence of anhedral epidote in dome samples matrix was interpreted as a product of hydrothermal activity (Arce et al. 2014).

As mentioned before, primary igneous anhydrite was described for the first time worldwide in the trachyandesitic pumice erupted in 1982 (Luhr et al. 1984), commonly associated with crystals of apatite (Luhr 2008). Given the similar chemistry of the rocks emitted during El Chichón history (see below), magmatic anhydrite was with all probabilities a characteristic feature of its products, even if rarely found in its rocks due to meteoric alteration. In fact Luhr et al. (1984), pointed out that after a single rainy season at El Chichón, the anhydrite in pumice samples was dissolved. Notably, primary anhydrite has been also reported in other volcanoes in the Chiapanecan Volcanic Arc, (i.e. La Lanza and Venustiano Carranza domes located close to San Cristobal de las Casas; Luhr 2008). In summary, all volcanoes of the Chiapanecan Volcanic Arc seem to share similar mineral assemblages and whole-rock chemistry (Mora et al. 2010).

Plagioclase crystals in lava and pumice samples (lava domes and pyroclastic deposits) among different units (i.e. A, B, C, D) show a wide variety of textural features, such as complex compositional zonation, sieve textures, and corroded margins (Fig. 2.2b), all of which have been interpreted as the result of multiple recharge events from deeper and hotter magmas entering in a subvolcanic magma chamber operative throughout the entire history of the volcano (Tepley et al. 2000; Andrews et al. 2008; Macías et al. 2003). Mafic enclaves found in lava domes and pyroclastic deposits (Macías et al. 2003; Layer et al. 2009; Arce et al. 2014) represent another evidence of injections of mafic magmas and mixing as well.

The whole-rock chemistry of El Chichón samples (Luhr et al. 1984; McGee et al. 1987; Espíndola et al. 2000; Macías et al. 2003; Andrews et al. 2008; Layer et al. 2009) has been grouped according to sample ages "or associated structures" into: 1982 eruption, Holocene, Pleistocene, Dome structures, mafic enclaves, and Chapultenango trachybasalt (Arce et al. 2014).

The lava domes and pyroclastic deposits sampled inside the 1982 crater, with trachyandesitic composition (54.5–58 wt% SiO_2) belong to the Holocene group, based on their stratigraphic position and according to Rose et al. (1984) these rocks would be slightly lower in SiO_2, Na_2O, K_2O with respect to the 1982 products. Nevertheless, Layer et al. (2009) did not find a clear temporal correlation with rock composition (Fig. 2.3).

The whole rock pumice samples from 1982 eruption are characterized by similar SiO_2 contents (55–58 wt%) compared to other units of the Holocene, and Pleistocene groups (Espíndola et al. 2000; Andrews et al. 2008). In contrast, samples from the domes (Cambac, Capulin, NW and SW domes) apparently show a broader variation in SiO_2 contents (52–60 wt %) (McGee et al. 1987; Espíndola et al. 2000; Layer et al. 2009; Rose et al. 1984; Arce et al. 2014) (Fig. 2.3). Up today, the few whole-rock chemical data existing for the mafic enclaves show a wide compositional range (44.5–50.5 wt% SiO_2) being mainly trachybasalts (Fig. 2.3a–j) with high contents of K_2O (Macías et al. 2003; Espíndola et al. 2000; Layer et al. 2009; Arce et al. 2014). The most mafic product of the area is the Chapultenango trachybasalt (Espíndola et al. 2000; Layer et al. 2009, see Chap. 3).

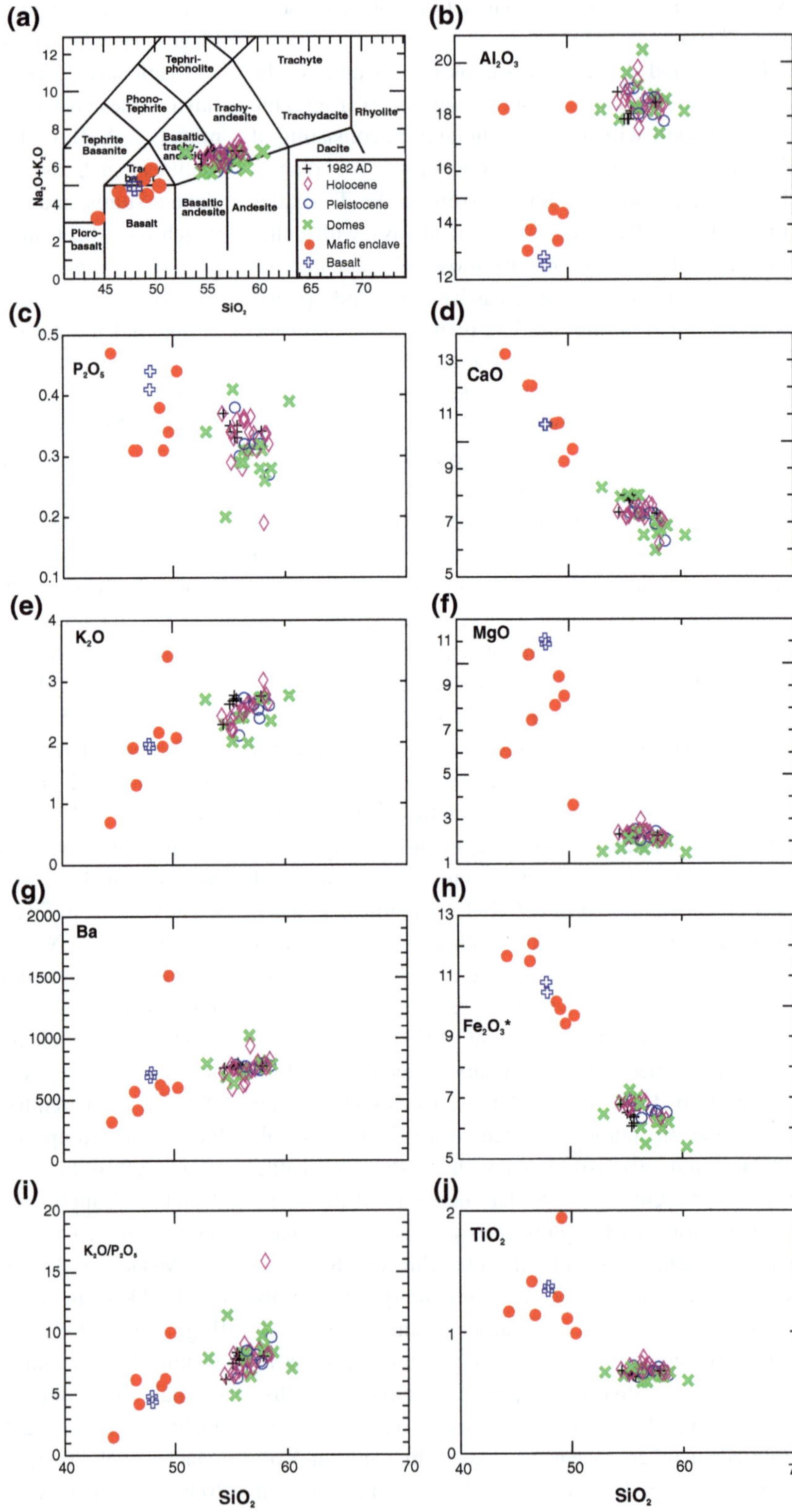

Fig. 2.3 **a** Total alkalies versus silica diagram (Le Bas et al. 1986), of El Chichón samples; **b–j** Harker diagrams. Major elements in wt% (recalculated on anhydrous basis) and trace elements in parts per million (ppm). Data taken from Luhr et al. (1984), Rose et al. (1984), McGee et al. (1987), Espíndola et al. (2000), Macías et al. (2003), Andrews et al. (2008), Layer et al. (2009) and Arce et al. (2014)

In general, all El Chichón samples are K-rich, and some unusually P-rich (Fig. 2.3), particularly some trachybasalts and trachyandesites. Harker diagrams (Fig. 2.3b–j) show that the more mafic compositions (i.e. trachybasalts) could represent the parental compositions for the more evolved trachyandesites erupted at El Chichón. Mafic enclaves are often interpreted as the product of magma mixing (Eichelberger et al. 1976; Stimac and Pierce 1992). The enclaves display a broadly linear array on Harker diagrams, consistent with a mixing relationship for compositions intermediate between parental trachybasalts and more evolved trachyandesites. Some mafic enclaves however, do not lie on simple mixing lines and therefore could indicate a more complex origin (Browne et al. 2006). On an Al_2O_3 versus SiO_2 diagram a notable inflection exists (Fig. 2.3b) explained as indicative of a change from an early plagioclase-poor to a later, plagioclase-rich assemblage (Arce et al. 2014). The steadfast declines in Fe_2O_3 and TiO_2 on Harker plots would attest an early fractionation of either Fe-Ti oxides, or amphibole, and the development of a calc-alkaline trend (Arce et al. 2014; Fig. 2.3h, j). Similarly, decreasing P_2O_5 concentrations during differentiation would

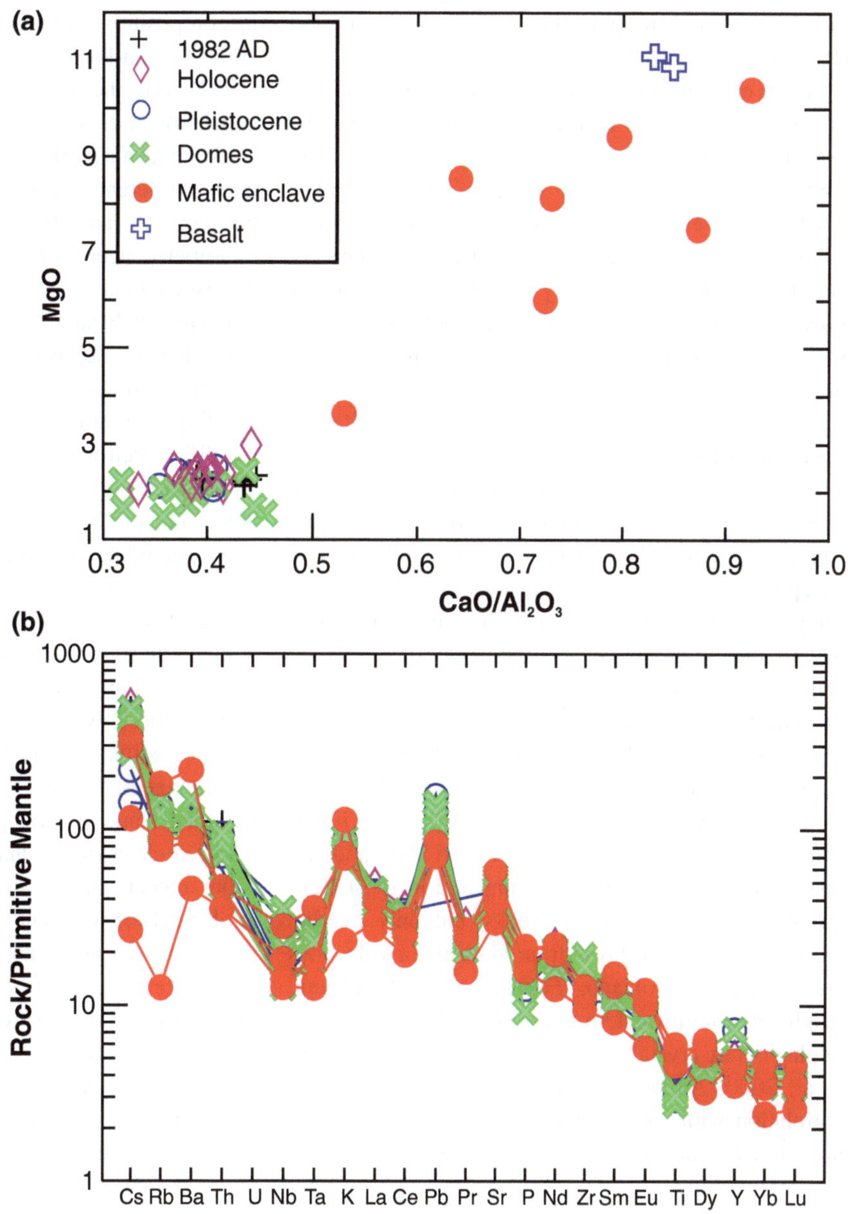

Fig. 2.4 **a** MgO versus CaO/Al_2O_3 ratio for El Chichón samples. All samples were recalculated to 100 % on an anhydrous basis; **b** Spider diagram of selected El Chichón samples, normalized to primitive mantle (Sun and McDonough 1989). Data taken from Arce et al. (2014)

require an early saturation with apatite (Arce et al. 2014; Fig. 2.3c). The decline of the CaO/Al_2O_3 ratio with MgO (Fig. 2.4a) also strongly suggests that early fractionation in the El Chichón magmatic system is dominated by the removal of clinopyroxene (i.e. Walker 1981).

In the Harker diagrams (Fig. 2.3b–j) two geochemical groups are observed with contrasting TiO_2, MgO, Fe_2O_3, Al_2O_3 contents that also have conspicuously different K_2O/P_2O_5 ratios (Fig. 2.3i), likely related by either fractional crystallization, or contamination and magma mixing (Arce et al. 2014). One of the most characteristic features of El Chichón is the production of trachyandesitic anhydrite-bearing magmas, enriched in potassium and sulphur, throughout its active period (Rose et al. 1984; Luhr 2008).

Trace elements in volcanic rocks tend to fractionate into specific minerals and therefore are useful in formulating models for magmatic differentiation and, in some cases, predicting the magma source (Winter 2001). Trace elements in El Chichón volcanic rocks are relatively depleted in Nb, Ta, and Ti, and relatively enriched in LILE, particularly Pb with respect to other trace elements (Fig. 2.4a, b). Such behavior would indicate a subduction related magmatism, involving the subducted slab, either melts or just fluids from it.

Both trachybasalts (i.e. Chapultenango basalt) and trachyandesites (dome samples, and pumice) show similar trace element patterns. The mafic rocks have generally lower concentrations of most incompatible elements (i.e. Ti, Y, Yb, Lu) (Fig. 2.4b), with the notable exceptions of P, as pointed out above. Some high-Ti samples (mafic enclaves) do not show enrichments in other high field strength elements (i.e. Zr, Nb, Ta, Th, U) (Fig. 2.4b) as would have been expected in mafic melts.

Only few $^{87}Sr/^{86}Sr$ and $^{143}Nd/^{144}Nd$ isotopic ratios have been published (Macías et al. 2003; Andrews et al. 2008) for El Chichón volcanic rocks (Fig. 2.5). Values of $^{87}Sr/^{86}Sr$ range from 0.70406 to 0.70426 and $^{143}Nd/^{144}Nd$ from 0.51273 to 0.51279 and are distinctively more radiogenic compared to most Central American Volcanic Arc rocks, but only overlap those ratios of some lavas from Guatemala volcanoes (i.e. Tajumulco volcano, Tacaná's nearest volcanic neighbor in northwestern Guatemala). $^{87}Sr/^{86}Sr$ ratios in plagioclase crystals from Holocene deposits (Tepley et al. 2000; Andrews et al. 2008) show a complex correlation with An contents. Such results have been interpreted with the occurrence of multiple recharge events of a hot magma with low $^{87}Sr/^{86}Sr$ ratios, and high Sr contents, that eventually homogenize with a cooler magma to produce a hybrid magma. The assimilation of sedimentary rocks with high $^{87}Sr/^{86}Sr$ ratios, by El Chichón magmas has been also proposed to explain the complex isotopic zonation in plagioclase crystals (Tepley et al. 2000; Andrews et al. 2008).

2.2.2 Tacaná

The Tacaná Volcanic Complex (TVC) consists of four aligned structures Chichuj, Tacaná, Las Ardillas, and San Antonio (see Chap. 6). All erupted products vary in composition from andesite to dacite (Mora et al. 2004; García-Palomo et al. 2006). Mercado and Rose (1992) were the first to describe the petrographic and geochemical characteristics of TVC rocks. They analyzed sixteen samples of lava and pyroclastic material from different deposits, obtaining compositions between 58 and 64 wt% SiO_2. Nevertheless, no radiometric ages were provided. Macías et al. (2000) published data on eight samples from San Antonio volcano; such samples included the Mixcun block-and-ash flow deposit (see Chap. 6), undifferentiated lava flows (including one mafic enclave), and domes yielding a larger compositional range from 50 to 64 wt % SiO_2. Petrographic descriptions combined with whole-rock chemical analyses of sixty samples, belonging to Chichuj, Tacaná, and San Antonio edifices, including a mafic enclave hosted in andesitic lava from San Antonio volcano were published by Mora et al. (2004). The authors obtained similar compositions (56–64 wt% SiO_2) as Macías et al. (2000) for the entire volcanic complex. A geological map of the TVC area, and the individuation of four volcanic edifices of the TVC was published by García-Palomo et al. (2006) (see Chap. 6). These authors included some general petrographic descriptions and whole-rock chemical analysis of Las Ardillas dome. More recent studies are those published by Macías et al. (2010b) who studied a debris avalanche deposit and its petrographic and geochemical characteristics, and Arce et al. (2012) who studied the eruptive dynamics and the petrology of the Sibinal Pumice deposit, produced by a Plinian eruption around 23, 540 year BP (see Chap. 6).

The rocks of the TVC were grouped according to the volcanic structure they belong to, in: Chichuj (volcano),

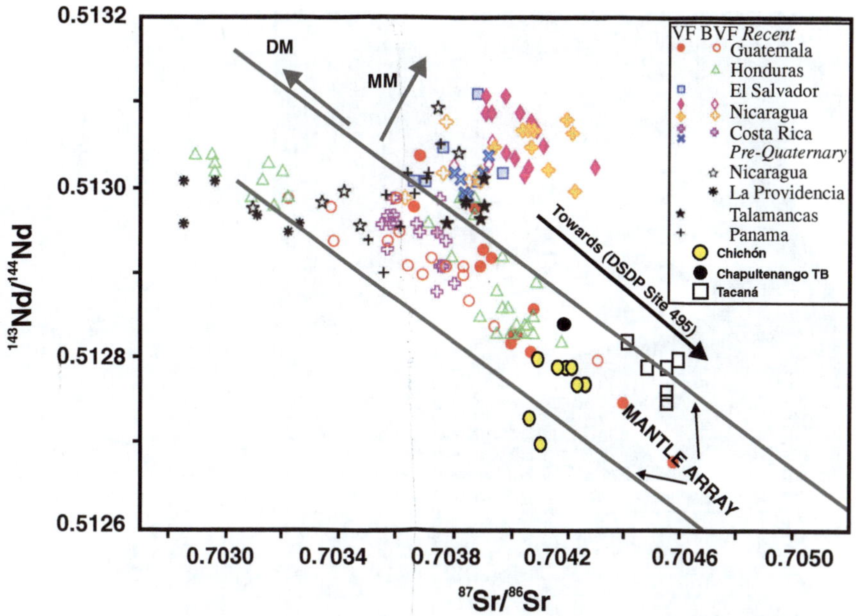

Fig. 2.5 ^{143}Nd/^{144}Nd versus ^{87}Sr/^{86}Sr ratios of El Chichón and Tacaná samples (data from Macías et al. 2003; Mora et al. 2004; Andrews et al. 2008), compared to other Central American Volcanic Arc volcanoes. *DM* depleted mantle; *MM* modified mantle. Isotopic data from CAVA volcanoes were taken from Carr et al. (1990). *DSDP Site 495* refers to hemipelagic sediments from drill Site 495 (von Huene et al. 1980)

Tacaná (volcano), Las Ardillas (dome), and San Antonio (volcano). Each group consisting of lavas and pyroclastic deposits, and mafic enclaves found into lava flows (Mora et al. 2004). Some pumice-rich pyroclastic deposits (i.e. pumice air fall and pyroclastic flow deposits) ejected from the TVC, (i.e. La Vega Pyroclastic flow (LVPF), Sibinal Pumice (SP), and Tacaná Pumice (Arce et al. 2012; Chap. 6) have been grouped separately because their source or vents (i.e. Chichuj, Tacaná, San Antonio volcanic structures) are unknown.

All of TVC rocks display a porphyritic texture, sometimes with seriate and glomeroporphyritic textures (Fig. 2.6) with crystallinity around 32 vol% (Macías et al. 2000; Mora et al. 2004). The mineral assemblage for all TVC structures is, in order of abundance, plagioclase, hornblende, clinopyroxene, orthopyroxene, ± Fe-Ti oxides (Fig. 2.6a–f), set in a glassy and sometimes microlitic matrix. Some lavas from the Tacaná edifice show trachytic textures (oriented plagioclase crystals), commonly developed during the emplacement of lava flows. Mafic enclaves show an intersertal porphyritic texture, but with crystal contents up to 65 vol% (Mora et al. 2004), and a mineral assemblage of plagioclase + hornblende + clinopyroxene + orthopyroxene + olivine, and ± Fe-Ti oxides set in a brown glass matrix (Mora et al. 2004). Olivine phenocrysts were only observed in a basaltic andesite lava flow, located just to the east of Tacana's summit (see Chap. 6). The textures of phenocrysts are quite variable in all of TVC samples, with coexisting subhedral and euhedral phenocrysts (Macías et al. 2000; Mora et al. 2004). Plagioclases exhibit complex compositional zonation and sieve textures (Fig. 2.6a, d, and e). The compositional ranges are similar: anorthite contents in Chichuj samples vary from An_{47} to An_{70} (andesine to labradorite), while San Antonio samples show a range from An_{40} to An_{84}. Plagioclase in mafic enclaves is slightly more anorthitic varying from An_{48} to An_{88} (Macías et al. 2000; Mora et al. 2004). Clinopyroxene and orthopyroxene are ubiquitous in the TVC samples, and have been classified as augite and enstatite. Amphiboles are Ca-rich, and are classified as magnesium-hornblende (Mora et al. 2004). Fe-Ti oxides are represented by titanomagnetite and lesser amounts of ilmenite (Mora et al. 2004).

Pumice samples from three main pyroclastic deposits of the TVC (i.e. La Vega Pyroclastic flow –LVPF-, Sibinal Pumice fall deposit –SP- and Tacaná Pumice fall deposit –TP-, see Chap. 6 for stratigraphic details) display porphyritic and vesicular textures (Fig. 2.6f), with a common mineral assemblage of plagioclase, clinopyroxene, orthopyroxene, Fe-Ti oxides, and minor amphibole (Arce et al. 2012).

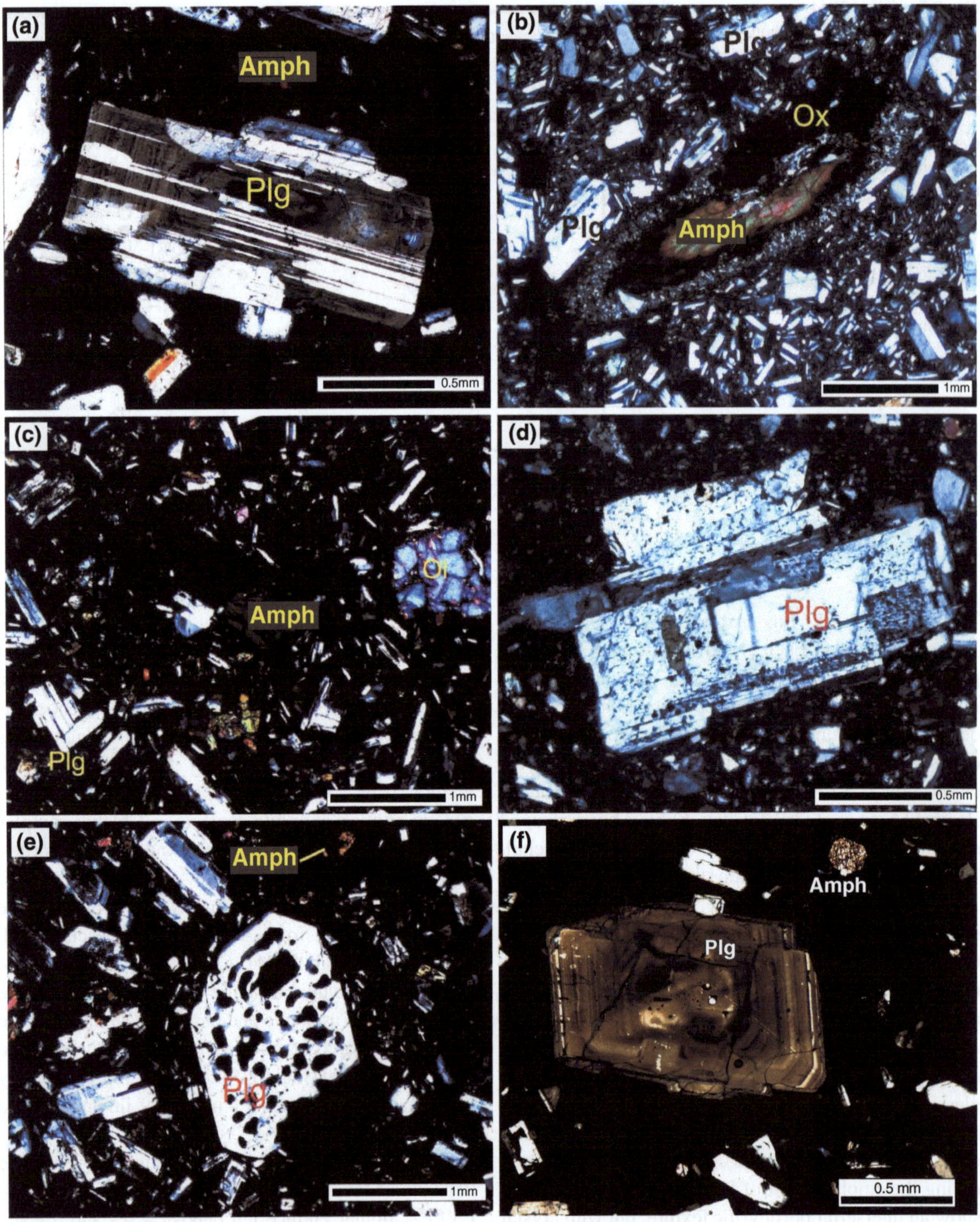

Fig. 2.6 **a–e** Representative cross-polarized microphotographs of Tacaná lava flow samples showing common porphyritic textures, and mineral assemblages (after Arce et al. 2014). *Amph* amphibole; *Plg* plagioclase; *Ol* olivine; *Ox* Fe-Ti oxides. Notice the different textures in plagioclase and amphibole phenocrysts. **f** Cross-polarized microphotograph of the Sibinal Pumice deposit, showing euhedral and zoned plagioclase phenocrysts. *Plg* plagioclase; *Amph* amphibole; *Cpx* clinopyroxene (after Arce et al. 2012)

Plagioclase is the most abundant mineral phase in all pumice samples, often with complex zonation (Mora et al. 2004). Anorthite contents commonly vary from An_{67} to An_{42} for the Sibinal Pumice and Tacaná Pumice deposits (Arce et al. 2008), whereas for the La Vega pyroclastic flow deposit, a wider range is recorded (An_{75}–An_{45}).

Clinopyroxene (diopside) and orthopyroxene (hypersthene) are also abundant in pumice deposits and compositionally homogenous (Arce et al. 2012). They are subhedral to euhedral, and in some samples exhibit reaction rims composed of amphibole and plagioclase. Amphibole phenocrysts are subhedral to anhedral, sometimes with a reaction rim (coronae surrounding the crystals) made of Fe-Ti oxides (Fig. 2.6b) suggesting changes in the ascent velocity during the eruption (Browne and Gardner 2006), or simply oxidation of the amphibole.

Although there are robust whole-rock chemical data sets for the TVC, only few include a complete set of trace elements, for which reason only a reduced number of samples appear in some graphs.

Whole-rock composition for Chichuj samples vary from andesite to dacite (Fig. 2.7a) with 59–63 wt% SiO_2 (Mora et al. 2004); while Tacaná shows a wider compositional range varying from basaltic andesite to dacite with 54–63 wt% SiO_2 (Macías et al. 2000; Mora et al. 2004; Arce et al. 2008), (Fig. 2.7a). One whole-rock chemical analysis for Las Ardillas Dome (García-Palomo et al. 2006) gave an andesitic composition (63 wt% SiO_2). San Antonio volcano has been sampled extensively, and a robust chemical data set has been published (Macías et al. 2000; Mora et al. 2004; Arce et al. 2014). Its products display a wider compositional range from basaltic andesite to dacite (53–64 wt% SiO_2; Fig. 2.7a). The three pumice-rich pyroclastic deposits (SP, TP, and LVPF), represent the most mafic products reported so far for the TVC (Fig. 2.7a), however, the mafic samples commonly show high loss on ignition values (5–6 wt%), suggesting the occurrence of alteration processes (Arce et al. 2008). A couple of mafic enclaves have been found in San Antonio and Tacaná lavas, and they show a compositional range from basalt to basaltic andesite (51–62 wt% SiO_2; Fig. 2.7a) (Macías et al. 2000; Mora et al. 2004).

As a whole, TVC samples exhibit a wide compositional range from basalt to dacite (49–63 wt% SiO_2, Fig. 2.7a), based on the classification scheme of Le Bas et al. (1986), with the majority lying in the andesitic field. Some of the basaltic samples (SP and TP samples) have unusually low contents of CaO, and high in Al_2O_3 (Figs. 2.7b–j and 2.8a), which likely result from a sericitic alteration (Arce et al. 2012). The Tacaná basalts (including mafic enclaves) are generally less rich in MgO than those from El Chichón (i.e. Chapultenango trachybasalt and mafic enclaves) (Fig. 2.7f). The large-ion-lithophile element concentrations of the Tacaná volcanic rocks are also consistently lower than those from El Chichón (Arce et al. 2014). Some compositional trends of the Tacaná samples on the Harker diagrams are similar to those displayed at El Chichón (i.e. P_2O_5, Fe_2O_3, Ba, K_2O, and K_2O/P_2O_5 ratio), possibly indicating analogous fractionating assemblages and conditions (Fig. 2.7b–j). Notably, the TVC rocks extend to higher K_2O/P_2O_5 than those from El Chichón (Fig. 2.7i). Also, four of the Tacaná basalts (corresponding to mafic enclaves) have somewhat higher TiO_2, but, unlike high-Ti samples from El Chichón, show no other compositional distinctions on Harker diagrams (Fig. 2.7b–j). The overall abundances of the highly and moderately incompatible elements are lower in Tacaná volcanic rocks relative to those from El Chichón (Fig. 2.8). The Tacaná rocks show negative Nb, Ta, and Ti anomalies that differ in general from those of El Chichón rocks, with the exception of the strong positive Pb anomaly (Fig. 2.8). A total of six Nd and Sr isotopic ratios for the TVC have been published (Mora et al. 2004). $^{87}Sr/^{86}Sr$ ratios vary from 0.70441 to 0.70459 whereas $^{143}Nd/^{144}Nd$ ranges from 0.51275 to 0.51282. Despite the few data, all values are similar to those obtained for El Chichón but displaced to higher $^{87}Sr/^{86}Sr$ (Fig. 2.5). These results could indicate a more extensive crustal contamination compared to El Chichón samples, however more isotopic analysis from the TVC are needed to confirm this hypothesis.

2.3 Magma Genesis and Evolution

2.3.1 El Chichón

El Chichón is characterized by eruptions of potassium, sulphur, and H_2O-rich magmas (Luhr 2008) that have been referred to as shoshonitic arc-related magmatism. Even if the content in incompatible elements seems to be inconsistent with a simple subduction zone

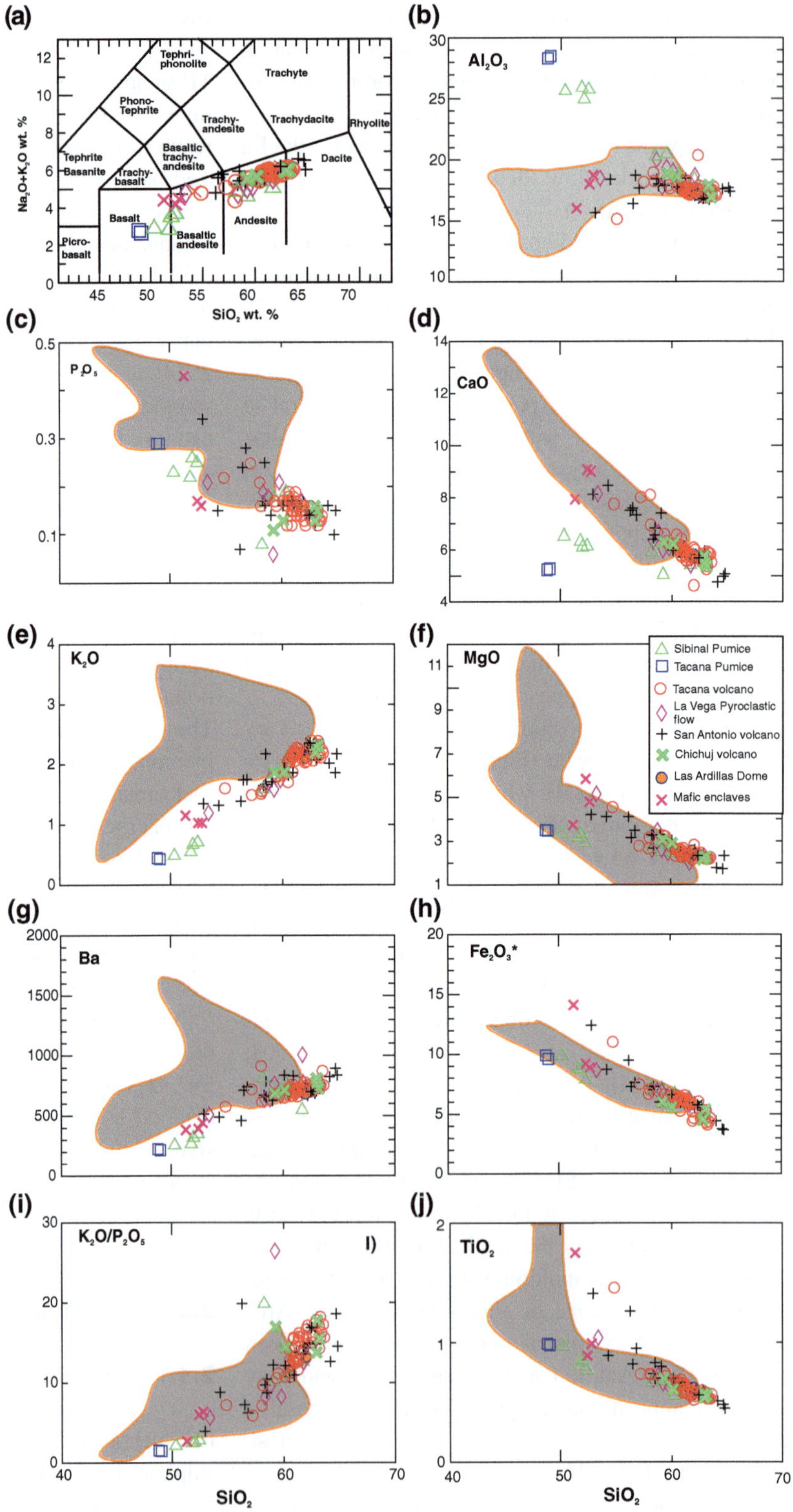

Fig. 2.7 **a** Total alkali versus silica diagram (Le Bas et al. 1986), for Tacaná samples; **b–j** Harker diagrams. Major elements in wt%, and trace elements in ppm. For all samples, major elements were recalculated to 100 % on an anhydrous basis. The *gray* field represents data of El Chichón samples. Data were taken from Macías et al. (2000), Mora et al. (2004), García-Palomo et al. (2006), Macías et al. (2010b), Arce et al. (2012) and Arce et al. (2014)

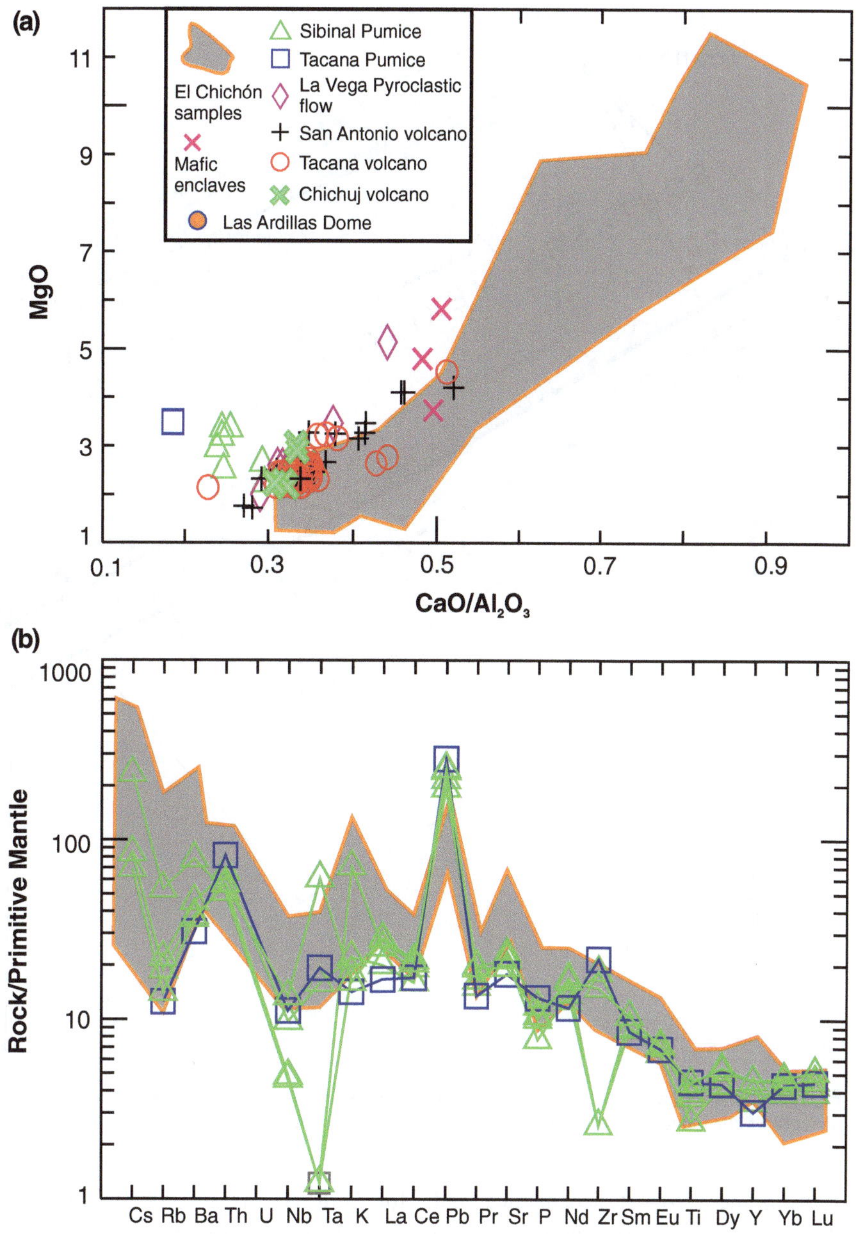

Fig. 2.8 **a** MgO versus CaO/Al_2O_3 ratio for Tacaná samples. All samples were recalculated to 100 % on an anhydrous basis; **b** Spider diagram of selected Tacaná samples normalized to primitive mantle (Sun and McDonough 1989). Note the two samples showing lower values in Ta, Nb, and Zr

signature, the high water contents, and estimated magmatic $\delta^{34}S$ in 1982 eruption products are consistent with a subduction zone provenance (Luhr and Logan 2002; Walker et al. 2003). The strong enrichment in potassium (i.e. Fig. 2.3i) of all products is not well understood. Manea and Manea (2008) attributed the relatively high water contents of El Chichón magmas to the dehydration of a serpentinized oceanic lithosphere, associated with subduction of the Tehuantepec Ridge. This ridge would be highly fractured, and therefore susceptible to deep and pervasive serpentinization.

The geochemistry and mineralogical characteristics of El Chichón rocks, make a difficult task to explain the magma genesis for this volcano. De Ignacio et al. (2003) proposed the presence of adakite-like (high Sr/Y ratios of about 204) that would be produced by melting of the subducted Cocos plate, however only two samples have low Y values (one sample from Unit B and another one for Unit D; see Chap. 6). Although

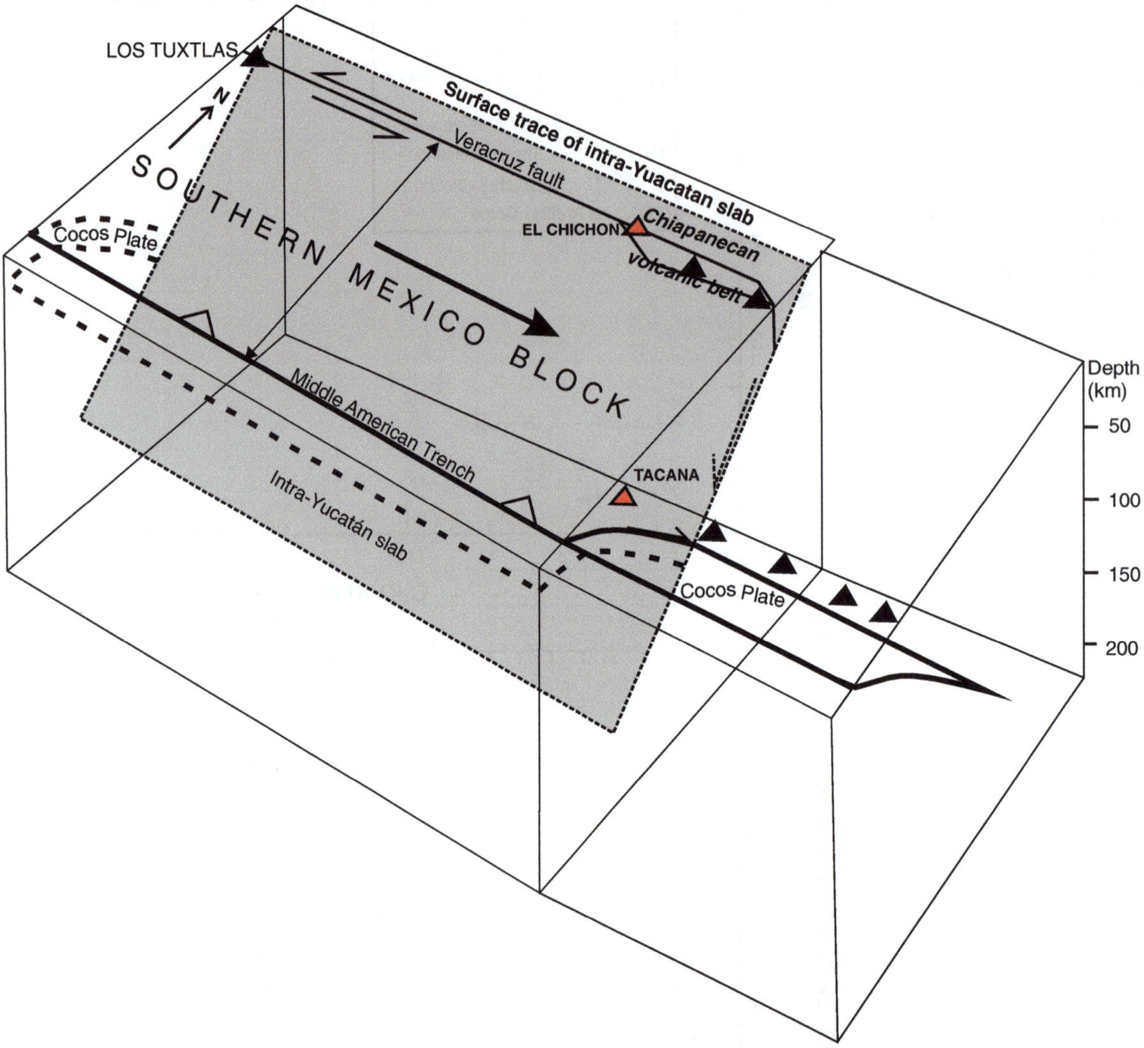

Fig. 2.9 3D tectonic model for southern Mexico involving two subducted slabs (*Cocos* and *Yucatan*). Sinistral movements of the Veracruz fault induces a pull-apart system at the Chiapanecan Volcanic Belt (after Arce et al. 2014)

later, a more robust geochemical data for Unit B demonstrated that Sr/Y ratios are around 50 (Macías et al. 2003), that correspond to mantle melts.

An alternative explanation for the genesis of El Chichón magmas rich in potassium, sulfur, and phosphorous, could be considering such melts as the product of a "rifting" mechanism combined with melting or dehydration of the Yucatan slab (Arce et al. 2014). The Yucatan slab is an oceanic plate dipping to the SW, subducted during the Miocene (Kim et al. 2011) (Fig. 2.9), which could have provided abundant hydrothermal sulfides to the mantle wedge (Arce et al. 2014). The high contents in Ti (>1 wt%) in the magmas of Central American volcanoes have been attributed to melting due to decompression of an asthenospheric mantle (Bolge et al. 2009). This concept could be applied also for El Chichón magmatism, where extensional tectonism has been documented (García-Palomo et al. 2004; Chap. 1). A Cocos-related origin for magmatism at El Chichón is under discussion, considering a truncation of the Cocos Plate at a depth of 100 km by the subducted Yucatán Plate suggested recently by Kim et al. (2011).

The possible rift origin for El Chichón magmas would be in agreement with the lateral movements of the San Juan, Arroyo de Cal, and Caimba faults

(García-Palomo et al. 2004; Chap. 1) that combined with other faults (i.e. Chapultenango Fault) would result in a pull-apart system (Arce et al. 2014) related to the major Polochic-Motagua system (Malfait and Dinkelman 1972). An extensional setting would be characterized by strike-slip faults along a pull-apart system, where volcanic structures would be concentrated, as suggested by Aydin and Nur (1982) and tested in several places around the world (i.e. western United States, Israel, Turkey, and Guatemala). The fault systems at El Chichón suggest a similar setting (i.e. pull-apart), in which magmas would be generated and stored underneath the volcano, as proposed by García-Palomo et al. (2004) and Bursik (2009).

El Chichón trachyandesites appear to be fed by parental mafic magmas that are similar in composition to mafic enclaves found in El Chichon's volcanic rocks and similar to the Chapultenango trachybasalt. Differentiation to high K_2O/P_2O_5 ratios (Fig. 2.3i) and the relatively enriched Sr and Nd isotopic ratios compared to most volcanic rocks of the CAVA (Fig. 2.5), suggest that crustal contamination is a significant process in magmatic evolution (Feigenson and Carr 1986; Wendlandt et al. 1995; Walker et al. 1995, 2007; Cameron et al. 2003). This hypothesis is consistent with isotopic data in phenocrysts of El Chichón volcanic rocks, which suggest that crustal assimilation produce the complex zonation of the plagioclase crystals (Tepley et al. 2000; Andrews et al. 2008; Jones et al. 2008). Luhr (2008) suggested that the high SO_2 flux at El Chichón could be explained by a deep (7–9 km) degassing of a H_2O-and S-rich and highly oxidized magma (i.e. Roberge et al. 2009; Christopher et al. 2010). The existence of a deep magma reservoir below El Chichón (Fig. 2.10) would be also suggested by seismic data previous to the 1982 eruption, which indicate a seismic gap at depths between 7 and 13 km below the volcano. Such area could represent a deeper magma chamber (Jimenez et al. 1999; Chap. 5). The high frequency of eruptions during the last 8,000 years (Espíndola et al. 2000) were explained with repeated injections of mafic magma (45–51 wt% SiO_2) from a deep reservoir (13 km) able to reinvigorate sluggish and cooling trachyandesitic magmas (55–61 wt% SiO_2) at shallower depths (7 km). Pre-eruptive temperatures for the 1982 eruption of El Chichón have been estimated in a range of 750–850 °C with oxygen fugacities above of the Ni-NiO buffer and pressures of 2.6–5 kb (Luhr 1990), whereas a temperature of 880 °C, with oxygen fugacities of −11 and a pressure of 3 kb have been calculated for the eruption occurred ~550 year BP (Unit B, see Chap. 6) (Macías et al. 2003) (Fig. 2.10).

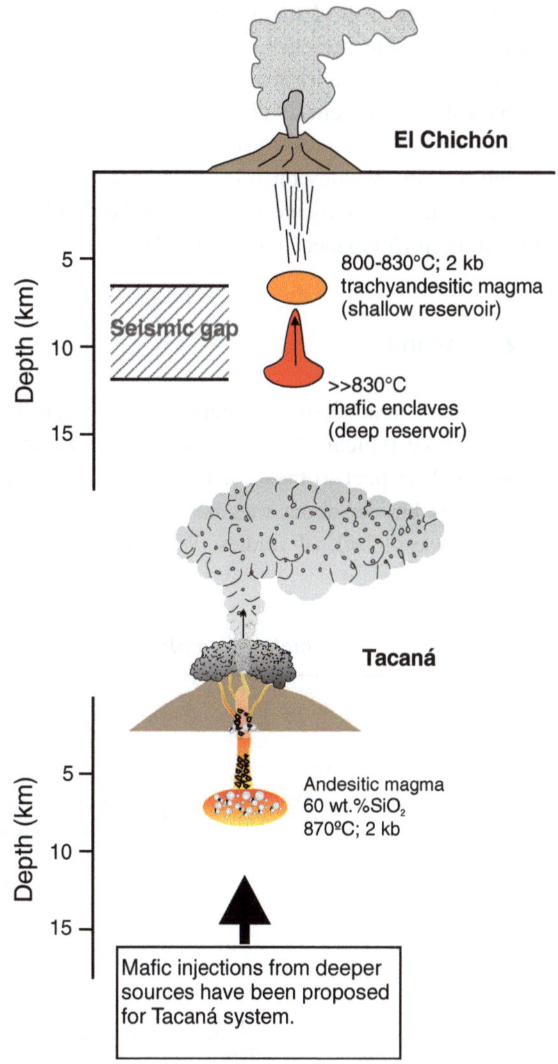

Fig. 2.10 Sketch of El Chichón and Tacaná magmatic systems based on petrologic and seismic data. Temperature and pressure estimates were taken from Luhr (1990); Macías et al. (2003), Arce et al. (2008, 2012)

As a whole, the Modern Chiapanecan Volcanic Arc lies along the northeastern border of the Southern Mexico Block proposed by Andreani et al. (2008). In this light, the genesis of the CVA could be related to the ESE movement of the Southern Mexico Block relative to North America (Andreani et al. 2008). Nevertheless, to date, no clear explanations exist, either for the genesis of trachybasaltic magmas at El Chichón, or for the presence or absence of the Tehuantepec ridge just underneath El Chichón, and its possible role on the genesis of sulphur-potassium-rich magmas. It is urgent to perform more isotopic analyses, and specific studies on trace elements, to better understand the magma genesis at El Chichón and the entire Modern Chiapanecan Volcanic Arc.

2.3.2 Tacaná

Geochemical characteristics of magmas erupted at the TVC are more typical of a subduction-related environment with respect to those of El Chichón, even if much less is known about magma genesis and evolution because of lack of studies. With the available data, magma genesis seems more clearly associated with subduction of the Cocos Plate considering that Tacaná volcanic rocks have distinct Nb, Ta, and Ti anomalies (Arce et al. 2014). The Nb and Ta anomalies seem to be related to fluids from the subducted plate (Wood et al. 1979), because released fluids from the slab contribute to the enrichment of "lithophile" elements in the magmas produced in convergent zones. Nb and Ta are not involved in this enrichment because they are not mobilized by fluids, due to the presence of insoluble phases in the subducting plates, therefore Nb and Ta would be retained in the slab (Saunders et al. 1980). Using the Ba/La ratio as a gauge of slab (fluid) contribution (i.e. Carr et al. 1990), the Tacaná magmas exhibit a slightly higher slab (fluid) signature than those of El Chichón (Fig. 2.11). This would be consistent with the shallow slab depth

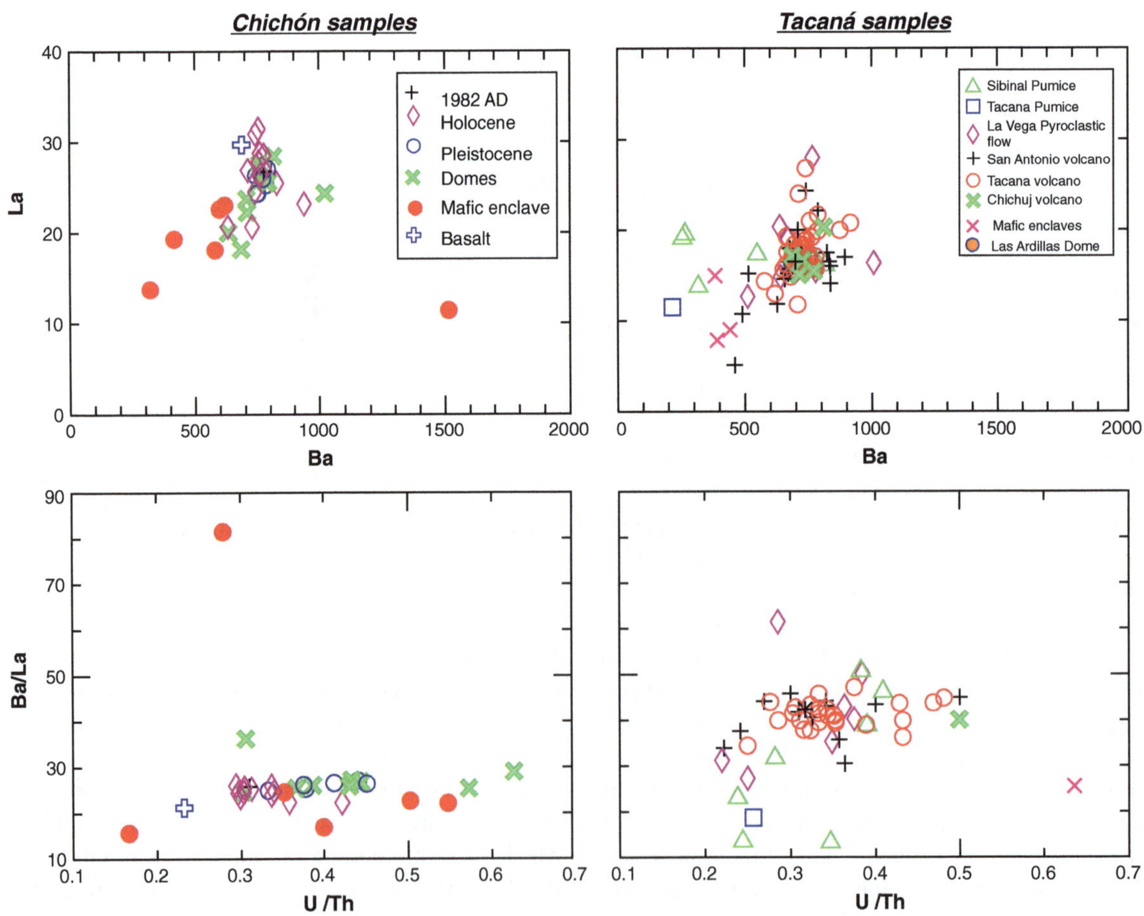

Fig. 2.11 El Chichón and Tacaná samples plotted on Ba versus La and Ba/La versus U/Th diagrams. Notice that the behavior of Tacaná rocks suggests an important slab contribution in generating magmas (Carr et al. 1990). Data from Arce et al. (2014)

beneath Tacaná, as recently proposed by Arce et al. (2014), considering that other studies suggest that slab signatures decrease with increasing slab depth (i.e. Walker et al. 2003). A greater influence of the slab would be also indicated by the higher $^{87}Sr/^{86}Sr$ ratios of the Tacaná volcanic rocks (Fig. 2.5).

Incoming mafic magmas are somewhat heterogeneous considering that a high-Ti component must also be entering the magmatic system (Fig. 2.7j). The common occurrence of amphibole would represent the evidence that Tacaná magmas, like those at El Chichón, were water-rich. As occurs at El Chichón, the relatively enriched Sr and Nd isotope ratios (Fig. 2.5) and evolution to high K_2O/P_2O_5 (Fig. 2.7i) would indicate that crustal contamination plays an important role during magmatic differentiation (Arce et al. 2014). Macías et al. (2000) already emphasized the role of magma mixing for Tacaná products. They proposed that mafic magma (54 wt% SiO_2) entered to an existing magmatic reservoir and mixed with the resident andesitic melt (60–63 wt% SiO_2). Mixing produced the observed mafic enclaves in the Mixcun pyroclastic flow deposit (see Chap. 6). Such mixing event would have been responsible for destabilizing the magmatic system and provoking the eruption ~1950 years before present (Macías et al. 2000).

The intrusion of hot magma and mixing into a cooler reservoir (870 °C) likely represents the trigger for many of Tacaná's explosive eruptions (Arce et al. 2012) by causing the over pressurization of the magmatic system (i.e. Eichelberger 1995; Sparks et al. 1997). Pre-eruptive temperature estimates for the three pumice-rich pyroclastic deposits (LVPF, SP fallout, and TP fall) range between 870–920 °C for the La Vega pyroclastic flow, and ca. 890 °C for the SP and TP fallout deposits (Fig. 2.10; Arce et al. 2008) (see Chap. 6 for stratigraphic details). An on-going petrological experimental study on the San Antonio dacitic dome suggests an equilibration pressure of ~2 kb, which would correspond roughly to depths of 7 km for the magma chamber at the TVC (Fig. 2.10) (Mora et al. 2013).

In summary, both El Chichón and Tacaná, considered subduction-related volcanoes, have emitted distinct types of magmas, one alkaline with anomalous enrichments in potassium, and sulfur (El Chichón), the other more typical calc-alkaline (Tacaná). Both volcanoes have experienced repetitive injections of mafic magmas (indicated by the presence of mafic enclaves) that have mixed in crustal magma chambers and triggered explosive eruptions. Nevertheless, both volcanoes need more detailed petrologic and geochemical studies in order to understand the genesis of high potassium and sulphur magmas at El Chichón, and the genesis of high-Ti magmas at Tacaná volcano.

References

Abers GA, Plank T, Hacker R (2003) The wet Nicaraguan slab. Geophys Res Lett 30:1098. doi:10.1029/2002GL015649

Andreani L, Rangin C, Martínez-Reyes J, Le Roy C, Aranda-García M, Le Pichon X, Peterson-Rodriguez R (2008) Neogeneleft-lateral shearing along the Veracruz fault: the eastern boundary of the southern Mexico Block. Bull Soc Géol Fr 179:195–208

Andrews BJ, Gardner JE, Housh T (2008) Repeated recharge, assimilation, and hybridization in magmas erupted from El Chichón as recorded by plagioclase and amphibole phenocrysts. J Volc Geotherm Res 175:415–426

Arce JL, Macías JL, Gardner J (2008) The ~14 Ka Plinian-type eruption of Tacaná Volcanic complex, Mexico-Guatemala. Eos Trans AGU 89(53), Fall Meeting Suppl, Abs V11C-2060

Arce JL, Macías JL, Gardner JE, Rangel E (2012) Reconstruction of the Sibinal Pumice, an andesitic Plinian eruption at Tacaná Volcanic complex, México-Guatemala. J Volc Geotherm Res 217–218:39–55

Arce JL, Walker J, Keppie JD (2014) Petrology of two contrasting Mexican volcanoes, the Chiapanecan (El Chichón) and Central American (Tacaná) volcanic belts: the result of rift- versus subduction-related volcanism. Int Geol Rev 56:501–524

Aydin A, Nur A (1982) Evolution of pull-apart basin and their scale independence. Tectonics 1:91–105

Best MG (2003) Igneous and metamorphic petrology. Blackwell Publishing, Berlin Germany, p 729

Bolge LL, Carr MJ, Milidakis KI, Lindsay FN, Feigenson M (2009) Correlating geochemistry, tectonics, and volcanic volume along the Central American volcanic front. Geochem Geophys Geosyst 10:1–15

Browne B, Gardner JE (2006) The influence of magma ascent path on the texture, mineralogy, and formation of horblende reaction rims. Earth Planet Sci Lett 246:161–176

Browne BL, Eichelberger JC, Patino LC, Vogel TA, Uto K, Hoshizumi H (2006) Magma mingling as indicated by texture and Sr/Ba ratios of plagioclase phenocrysts from Unzen volcano, SW Japan. J Volc Geotherm Res 154:103–116

Bursik M (2009) A general model for tectonic control of magmatism: examples from Long Valley Caldera (USA) and El Chichón (Mexico). Geof Int 48:171–183

Cameron BI, Walker JA, Carr MJ, Patino LC, Matias O, Feigenson MD (2003) Flux versus decompression melting at stratovolcanoes in southeastern Guatemala. J Volc Geotherm Res 119:21–50

Carr MJ, Feigenson MD, Bennett E (1990) Incompatible element and isotopic evidence for tectonic control of source mixing and melt extraction along the Central American arc. Contrib Mineral Petrol 105:369–380

Carr MJ, Feigenson MD, Patino LC, Walker JA (2003) Volcanism and geochemistry in Central America: progress and problems. In: Eiler (ed) Inside the subduction factory. Geophys Mon Ser 138:153–179

Christopher T, Edmonds M, Humphreys MCS, Herd R (2010) Volcanic gas emissions from Soufrière Hills Volcano, Montserrat 1995–2009, with implications for mafic magma supply and degassing. Geophys Res Lett 37. doi:10.1029/2009GL041325

Damon P, Montesinos E (1978) Late Cenozoic volcanism and metallogenesis over an active Benioff Zone in Chiapas, Mexico. Ar Geol Soc Dig 11:155–168

De Ignacio C, Castiñeiras P, Márquez A, Oyarzun R, Lillo J, López I (2003) El Chichón volcano (Chiapas Volcanic Belt, Mexico) transition calc-alkaline to adakitic-like magmatism: petrologic and tectonic implications. Int Geol Rev 45:1020–1028

DeLong SE, Hodges FN, Arculus RJ (1975) Ultramafic and mafic inclusions, Kanaga Island, Alaska, and the occurrence of alkaline rocks in island arcs. J Geol 83:721–736

Eichelberger JC (1995) Silicic volcanism: ascent of viscous magmas from crustal reservoirs. Annu Rev Earth Plan Sci 23:41–63

Eichelberger JC, Gooley R, Nitsan U, Rice A (1976) A mixing model for andesitic volcanism (abstract). EOS Trans AGU 57:1024

Espíndola JM, Macías JL, Tilling RI, Sheridan M (2000) Eruptive history of el Chichón volcano (Chiapas, Mexico) and its impact on human activity. Bull Volc 62:90–104

Feigenson MD, Carr M (1986) Positively correlated Nd and Sr isotope ratios of lavas from the Central American volcanic front. Geology 14:79–82

Feigenson MD, Carr M (1993) The source of Central American lavas: Inferences from geochemical inverse modeling. Contrib Mineral Petrol 113:226–235

García-Palomo A, Macías JL, Espíndola JM (2004) Strike-slip faults and K-alkaline volcanism at El Chichón volcano southeastern México. J Volc Geotherm Res 136:247–268

García-Palomo A, Macías JL, Arce JL, Mora JC, Hughes S, Saucedo R, Espíndola, JM, Escobar R, Layer P (2006) Geological evolution of the Tacaná Volcanic complex, México-Guatemala. In: Rose WI (ed) Volcanic Hazards in Central America, Spec Pap Geol Soc Am 412:39–58

Gazel E, Carr M, Hoernle K, Feigenson M, Szymanski D, Hauff F, van den Bogaard P (2009) Galapagos-OIB signature in southern Central America: mantle refertilization by arc-hot spot interaction. Geochem Geophys Geosyst 10:1–32. doi:10.1029/2008GC002246

Guzman-Speziale M, Pennington WD, Matumoto T (1989) The triple junction of the North America, Cocos, and Caribbean Plates: seismicity and tectonics. Tectonics 8:981–999

Jiménez Z, Espíndola VH, Espíndola JM (1999) Evolution of the seismic activity from the 1982 eruption of El Chichón volcano, Chiapas, Mexico. Bull Volc 61:411–422

Jones DA, Layer PW, Newberry RJ (2008) A 3100-year history of argon isotopic and compositional variation at El Chichón volcano. J Volc Geotherm Res 175:427–443

Kim YH, Clayton RW, Keppie F (2011) Evidence of a collision between the Ycatán Block and Mexico in the Miocene. Geophys J Int 187:989–1000

Layer PW, García-Palomo A, Jones D, Macías JL, Arce JL, Mora JC (2009) El Chichón volcanic complex, Chiapas, México: stages of evolution based on field mapping and 40Ar/39Ar geochronology. Geof Int 48:33–54

Le Bas MJ, Le Maitre RW, Streckeisen A, Zanettin R (1986) A chemical classification of volcanic rocks based on the total alkali-silica diagram. J Petrol 27:745–750

Luhr JF (1990) Experimental phase relations of water- and sulfur- saturated arc magmas and the 1982 eruptions of El Chichón volcano. J Petrol 31:1071–1114

Luhr JF (2008) Primary igneous anhydrite: progress since its recognition in the 1982 El Chichón trachyandesite. J Volc Geotherm Res 175:394–407

Luhr JF, Logan MA (2002) Sulfur isotope systematics of the 1982 El Chichón trachyandesite: an ion microprobe study. Geochim Cosmochim Acta 66:3303–3316

Luhr JF, Carmichael ISE, Varekamp J (1984) The 1982 eruptions of El Chichón Volcano, Chiapas, Mexico: mineralogy and petrology of the anhydrite-bearing pumices. J Volc Geotherm Res 23:69–108

Macías JL, Espíndola JM, García-Palomo A, Scott KM, Hughes S, Mora J (2000) Late Holocene Peléan style eruption at Tacaná Volcano, Mexico-Guatemala: Past, present, and future hazards. Geol Soc Am Bull 112:1234–1249

Macías JL, Arce JL, Mora JC, Espíndola JM, Saucedo R, Manetti P (2003) A 550 years old Plinian eruption at El Chichón volcano, Chiapas, Mexico: explosive volcanism linked to reheating of the magma reservoir. J Geophys Res 108(B12):2569

Macías JL, Arce JL, Garduño-Monroy VH, Rouwet D, Taran Y (2010a) Estudio de prospeccion geotermica para evaluar el potencial del volcán Chichónal, Chiapas. Unpublished Report- Contract no 9400047770 IGF-UNAM-CFE

Macías JL, Arce JL, García-Palomo A, Mora JC, Layer PW, Espíndola JM (2010b) Late-Pleistocene flank collapse triggered by dome growth at Tacaná volcano, México-Guatemala, and relationship to the regional stress regime. Bull Volc 72:33–53

Malfait BT, Dinkelman MG (1972) Circum-Caibbean tectonic and igneous activity and the evolution of the Caribbean Plate. Geol Soc Am Bull 34:263–291

Manea VC, Manea M (2006) Origin of the modern Chiapanecan Volcanic arc in southern México inferred from thermal models. In: Rose WI (ed) Volcanic hazards in Central America, Spec Pap Geol Soc Am 412: 27–38

Manea M, Manea VC (2008) On the origin of El Chichón volcano and subduction of Tehuantepec Ridge: a geodynamical perspective. J Volc Geotherm Res 175:459–471

Manea M, Manea VC, Ferrar L, Kostoglodov V, Bandy WL (2005) Structure and origin of the Tehuantepec Ridge. Earth Planet Sci Lett 238:64–77

McGee JJ, Tlling RI, Duffield WA (1987) Petrologic characteristics of the 1982 and pre-1982 eruptive products of El Chichon volcano, Chiapas, Mexico. Geof Int 26:85–108

Mercado R, Rose W (1992) Reconocimiento geológico y evaluación preliminar de peligrosidad del Volcán Tacaná, Guatemala/México. Geof Int 31:205–237

Mora JC, Macías JL, García-Palomo A, Espíndola JM, Manetti P, Vaselli O (2004) Petrology and geochemistry of the Tacaná Volcanic Complex, Mexico-Guatemala: evidence for the last 40,000 year of activity. Geof Int 43:331–359

Mora JC, Jaimes-Viera MC, Garduño-Monroy VH, Layer PW, Pompa-Mera V, Godinez ML (2007) Geology and geochemistry characteristics of the Chiapanecan volcanic arc (central area), Chiapas, Mexico. J Volc Geotherm Res 162:43–72

Mora JC, Layer PW, Jaimes-Viera MC (2010) New 40Ar/39Ar ages from the central part of the Chiapanecan Volcanic Arc, Chiapas, México. Geofis Int 51:39–49

Mora JC, Gardner JE, Macías JL, Meriggi L, Santo AP (2013) Magmatic control son eruption dynamics of the 1950 yr B. P. eruption of San Antonio Volcano, Tacaná Volcanic Complex, Mexico-Guatemala. J Volc Geotherm Res 262:134–152

Patino LC, Carr MJ, Feigenson MD (2000) Local and regional variations in Central American arc lavas controlled by variations in subducted sediment input. Contrib Mineral Petrol 138:265–283

Rebollar CJ, Espíndola VH, Uribe A, Mendoza A, Pérez-Vertti A (1999) Distribution of stress and geometry of the Wadati-Benioff zone under Chiapas, Mexico. Geof Int 38:95–106

Roberge J, Delgado-Granados H, Wallace PJ (2009) Mafic magma recharge supplies high CO_2 and SO_2 gas fluxes from Popocatépetl volcano, Mexico. Geology 37:107–110

Rose WI, Bornhorst TJ, Halsor SP, Capaul WA, Plumley PS, De la Cruz-Reyna S, Mota R (1984) Volcán El Chichón Mexico; pre-1982 S-rich eruptive activity. J Volc Geotherm Res 23:147–167

Saunders AD, Tarney J, Weaver SD (1980) Transversal geochemical variations across the Antarctic Peninsula: implication for the genesis of calc-alkaline magmas. Earth Planet Sci Lett 46:344–360

Schmincke HU (2004) Volcanism. Springer, Heidelberg, p 324

Sparks RSJ, Sigurdsson H, Wilson L (1997) Magma mixing: a mechanism for triggering acid explosive eruptions. Nature 267:315–318

Stimac JA, Pearce TH (1992) Textural evidence of mafic-felsic magma interaction in dacite lavas, Clear Lake, California. Am Mineral 77:795–809

Sun S, McDonough W (1989) Chemical and isotopic systematics of oceanic basalts: implications for mantle compositions and processes. In: Saunders A, Norry M (eds) Magmatism in ocean basins, Spec Pap Geol Soc London 42:313–345

Syracuse EM, Abers GA (2006) Global compilation of variations in slab depth beneath arc volcanoes and implications. Geochem Geophys Geosyst 7:Q05017. doi:10.1029/2005GC001045

Tatsumi Y, Kosigo T (1997) Trace element transport during dehydration processes in the subducted oceanic crust: 2. origin of chemical and physical characteristics in arc magmatism. Earth Planet Sci Lett 148:207–221

Tepley FJ, Davidson JP, Tilling RI, Arth J (2000) Magma mixing, recharge and eruption histories recorded in plagioclase phenocrysts from El Chichón Volcano, Mexico. J Petrol 41:1397–1411

von Huene R, Aubouin J, Azema J, Blackington G, Carter JA, Coulbourn WT, Cowan DS, Curiale JA, Dengo CA, Eaas RW, Harrison W, Hesse R, Hussong DM, Ladd JW, Muzylov N, Shiki T, Thompson PR, Westberg J (1980) Leg 67: the deep sea drilling Project Mid-America Trench transect off Guatemala. Geol Soc Am Bull 91:421–432

Walker JA (1981) Petrogenesis of lavas from cinder cone fields behind the volcanic front of Central America. J Geol 89:721–739

Walker JA, Carr MJ, Patino LC, Johnson CM, Feigenson MD, Ward RL (1995) Abrupt change in magma generation processes across the Central American arc in southeastern Guatemala: flux-dominated melting near the base of the wedge to decompression melting near the top of the wedge. Contrib Mineral Petrol 120:378–390

Walker JA, Roggensack K, Patino LC, Cameron BI, Matias O (2003) The water and trace element contents of melt inclusions across an active subduction zone. Contrib Mineral Petrol 146:62–77

Walker JA, Mickelson JE, Thomas RB, Patino LC, Cameron BI, Carr MJ, Feigenson MD, Edwards RL (2007) U-series disequilibria in Guatemalan lavas, crustal contamination, and implications for magma genesis along the Central American subduction zone. J Geophys Res 112:B06205. doi:10.1029/2006JB004589

Wendlandt RF, Altherr R, Neumann ER, Baldridge WS (1995) Petrology, geochemistry, isotopes in Continental Rifts: evolution, structure, tectonics. In: Olsen KH (ed) Developments in geotectonics 25, Elsevier, pp 47–60

Winter (2001) An introduction to igneous and metamorphic petrology. Prentice Hall, Bergen, p 697

Wood DA, Joron JL, Treuil M, Norry M, Tarney J (1979) Elemental and Sr isotope variations in basic lavas from Iceland and the surrounding ocean floor. the nature of mantle source heterogeneities. Contrib Mineral Petrol 70:319–339

El Chichón Volcano: Eruptive History

3

Teresa Scolamacchia and Lucia Capra

Abstract

El Chichón volcano became known worldwide after it erupted catastrophically in 1982 and killed an estimated 2,000 people. A cumulative mass of almost 8 million tons of sulfur dioxide was injected into the stratosphere. The first reconstruction(s) of the stratigraphic successions carried out at the volcano shortly after the eruption suggested a long record of eruptive activity in the geologic past. Many studies completed in the past decade, aided by $^{40}Ar/^{39}Ar$ and K/Ar dating, have confirmed that the eruptive activity indeed was long lasting, occurring from eruptive centers located several km apart, before migrating at its present position. Within this context, El Chichón might better be described as a Volcanic Complex rather than a single volcano. During the Holocene at least 12 explosive eruptions originated from the same vent that was reactivated in 1982 producing pyroclastic density currents (PDCs) along many of the same pathways as those in 1982. This chapter summarizes the spatial and temporal evolution of this volcanic complex, integrating some previously unpublished data with the existing information. The time sequence of the 1982 eruptive events has been re-examined considering records from far-away stations and eyewitnesses' accounts that were not taken into account in previous studies, providing a more coherent time-stratigraphic framework for the 1982 eruptive sequence. Considering El Chichon's history of past eruptions, the existence of a wide, shallow, groundwater table and the re-establishment of an active hydrothermal system after the 1982 eruption, the probability of a new phreatic/phreatomagmatic eruption may still exist.

3.1 Previous Works

El Chichón was recognized as an active volcano in the early 20th century by Müllerried (1932, 1933), who visited the area following a period of seismic unrest at a small hill known as Cerro la Unión (Fig. 3.1). He reported the presence of stalactites of native sulfur near fumaroles discharging H_2S, and hot springs between a 1,260 m high dome and a major crater structure, which later would be recognized as the Somma crater.

T. Scolamacchia (✉)
Department of Earth and Environmental Sciences, LMU, Theresienstrasse 41, Munich, 80333, Germany
e-mail: tscolamacchia@gmail.com

L. Capra
Campus UNAM Juriquilla, Centro de Geociencias, Boulevard Juriquilla, no. 3001, Juriquilla, Queretaro 76230, Mexico

T. Scolamacchia and J.L. Macías (eds.), *Active Volcanoes of Chiapas (Mexico): El Chichón and Tacaná*,
Active Volcanoes of the World, DOI 10.1007/978-3-642-25890-9_3, © Springer-Verlag Berlin Heidelberg 2015

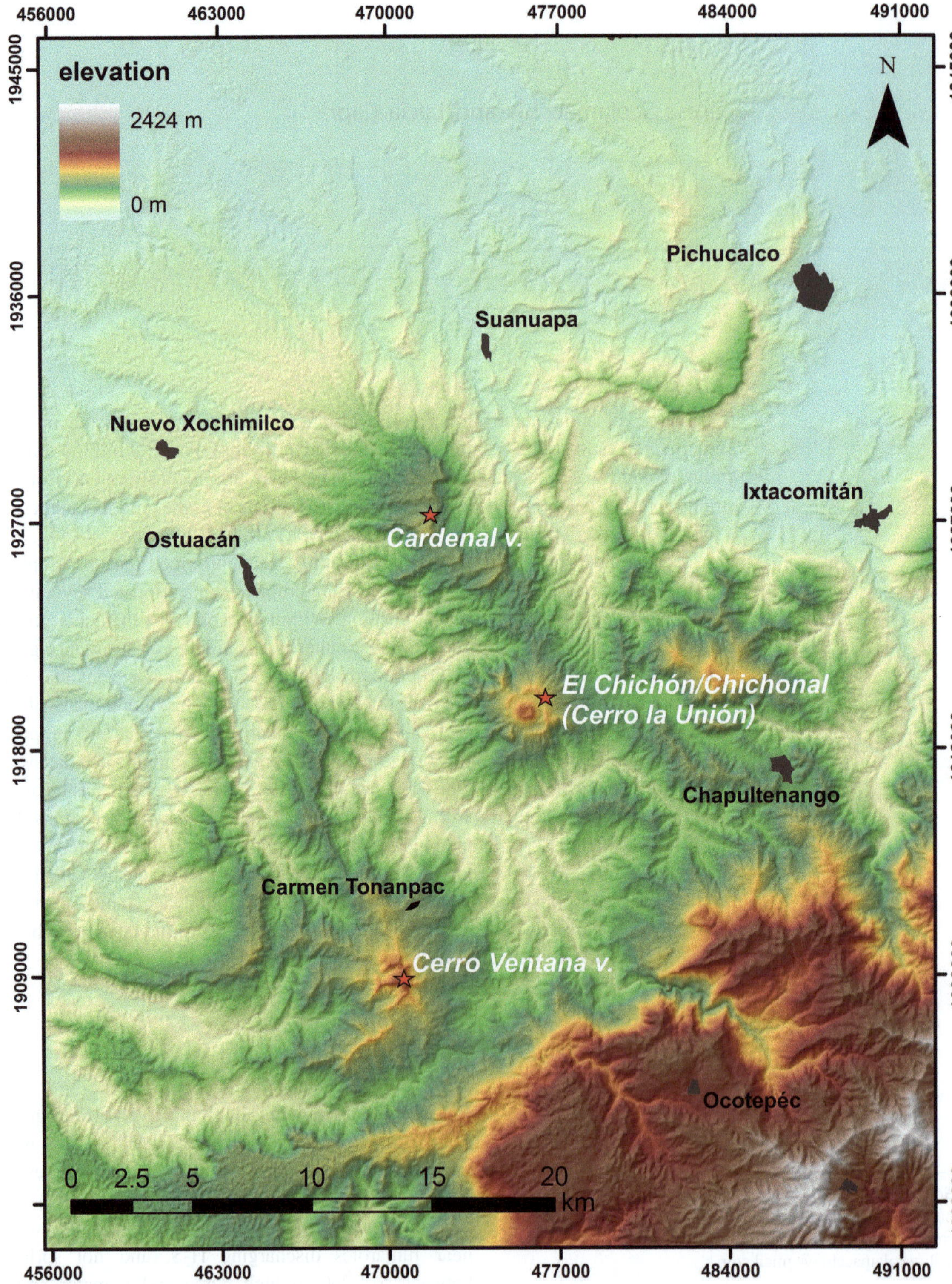

Fig. 3.1 Digital elevation model of El Chichón volcano (also known as Chichonal or La Unión) and surrounding areas illustrating the locations of other now-dormant volcanic structures (e.g. Catedral volcano and Cerro Ventana volcano) and the regional morphology controlled by the main folds and faults in the area

After this early study, El Chichón was included in the list of active volcanoes of the world of the International Association of Volcanology and Chemistry of the Earth's Interior (IAVCEI) (Mooser et al. 1958).

Further studies were carried out only 50 years later by Damon and Montesinos (1978), who were searching for ore deposits, and were the first to obtain K-Ar ages of 209 ± 19 ka for rocks that crops out in the eastern sector of the volcanic edifice (Somma crater). Following their work, Mexico's National Power Company—Comisión Federal de Electricidad (CFE)—began the evaluation of the area for its geothermal energy exploitation. During these studies, which produced the first geological map of the volcano (Canul and Rocha 1981), Templos et al. (1981) reported the presence of thermal waters on the southern flanks of the volcano. Canul and Rocha (1981) described fumarolic activity and altered ground around the central dome, and along a fault cutting the dome, warning about a potential volcanic risk in the area, after being witnesses of earthquakes and rumbling sounds when doing fieldwork. Unfortunately, both reports remained unpublished and were not taken into account, thereby contributing to the disastrous outcome of the 1982 eruption (see Chap. 8).

The 1982 catastrophic eruption motivated the research in the area to determine the eruptive history of the volcano. The dense tropical vegetation cover around the volcano had been stripped by the eruption, allowing excellent exposures of the pre-1982 deposits.

A first reconstruction of the eruptive history of the volcano was provided by Canul et al. (1983), however, these authors did not present any age constraints for the eruptive deposits. Several studies followed aimed at reconstructing the stratigraphic succession of the 1982 eruption (e.g. Sigurdsson et al. 1984, 1987; Carey and Sigurdsson 1986) and its past eruptive history. Duffield et al. (1984) defined the volcano as a "tuff cone surrounded by a ring of domes". The use of the term "tuff cone", a volcanic landform typically associated to the emplacement of pyroclastic sequences related to hydromagmatic activity involving abundant quantities of water (Wohletz and Sheridan 1983), suggests that the abundance of deposits from such kind of activity was already recognized in this early study. Rose et al. (1984) mapped the newly formed crater, and recognized several fault trends within the crater floor, defining El Chichón as a "complex of domes emplaced around a ring fracture". Tilling et al. (1984) provided the first reconstruction of the eruptive activity during the Holocene. Further works in the 1990s allowed to refine the 1982 stratigraphic succession, and recognized the presence of older structures (Macías 1994; Macías et al. 1997, 1998).

The intense erosion in the region, due to the very high (>4,000 mm) average annual rainfall (Atlas del Agua 1976), created deep gullies following 1982, exposing older deposits around the volcano. This situation favored detailed stratigraphic studies of the Holocene deposits (Espíndola et al. 2000). This work inspired other studies in the following 10 years, aimed either at refining the 1982 and previous stratigraphic successions, or to understand better how the structural setting of the region influenced the volcanic activity (Macías et al. 2003, 2004, 2008; García-Palomo et al. 2004; Scolamacchia and Macías 2005; Scolamacchia et al. 2005; Sulpizio et al. 2008; Scolamacchia and Schouwenars 2009).

A more complete scenario of the temporal evolution of El Chichón's eruptive history, including pre-Holocene activity, was obtained recently (Layer et al. 2009; Macías et al. 2010) combining photo geological interpretation, stratigraphic records, and K–Ar and C^{14} ages from previous works, with new ^{40}Ar-^{39}Ar ages and structural studies. All these studies evidenced the presence of several volcanic structures, mostly domes and ancient craters, which were emplaced along major faults oriented E-W and NNW–SSE (see Chap. 1). For these reasons, El Chichón has been recently defined a volcanic complex (Layer et al. 2009; Macías et al. 2010).

Very recent interpretations of ground-based geomagnetic and aeromagnetic surveys (Juetzler et al. 2011) made possible the characterization of underground magnetic bodies and hydrothermal vents and to relate them in comparison with results from previous studies.

The following sections summarize the spatial and temporal evolution of the El Chichón Volcanic Complex, combining the data reported in previous work, and proposing a different interpretation of some events for which discrepancy exists in the literature.

Inset Box: Nomenclature Used

The eruptive history of El Chichón has been characterized by the occurrence of explosive eruptive events. It is therefore important to clarify the terms that are used in this chapter to refer to the products of such activity.

Pyroclastic density currents (PDCs, Fisher 1990) are multiphase flows of gases and volcanic fragments that flow under the influence of gravity, which may originate in different ways (Branney and Kokelaar 2002). Considering that different types of PDCs produce similar depositional facies (Branney and Kokelaar 2002), the simple observation of deposits, give convoluted data from which to infer eruptive mechanisms. For this reason, sedimentological analyses, and the observation of the morphology of juvenile particles are often necessary. Early works classified PDCs deposits based on their lithology and sedimentary structures, into pumice and ash flows (ignimbrites), block and ash flows, and pyroclastic surges (e.g. Fisher and Schmincke 1984; Cas and Wright 1987). The currents produced by explosive hydromagmatic activity were named "base surges" (Richards 1959; Moore et al. 1966; Moore 1967; Fisher and Waters 1970; Waters and Fisher 1971) due to their similarity with the ones produced in nuclear explosions, observed moving radially at the base of a collapsing low-altitude column (Brinkley et al. 1950; Young 1965). Base surges (fully dilute pyroclastic density currents of Branney and Kokelaar 2002) are envisaged to have a lower solid concentration (≤0.1–1 vol%, Valentine and Fisher 2000), with respect to other currents (i.e. pyroclastic flows, ≥10 vol%, Freund and Bursik 1998).

The probability of occurrence of an explosive interaction between magma and external water, and the intensity of such interaction, is determined by mixing conditions. Experiments with silicate melts (Zimanowski et al. 1997a, b; Büttner and Zimanowski 1998), and others studies (e.g. Colgate and Sigurgeisson 1973; Kokelaar 1986), indicated that the probability is higher when water gets entrapped into magma. Under such confined conditions, the heat transfer between magma and water is limited by the formation of a vapor film, which isolates the two media allowing the water to persist in a metastable state. The collapse of the film (i.e. due a seismic wave) causes the direct contact between the two media enhancing the heat transfer rate (up to 10^6 K/s, Zimanowski 1998; Zimanowski and Wohletz 2000), and causing a very fine fragmentation of the melt. For this reason the particles produced by this high energetic process typically have very small grain sizes (32 μm < d < 130 μm), and represent a small fraction (e.g. 5–10 wt%) of the entire deposit, which consists mostly of fragmented country rocks (Sheridan and Wohletz 1983a; Zimanowski and Wohletz 2000). The particles produced by this interaction show a distinctive morphology (i.e. *blocky shapes* of Heiken and Wohletz 1985), attesting the occurrence of a brittle fragmentation.

The amount of water entering in contact with magma (e.g. Heiken 1971; Wohletz and Sheridan 1979; Sheridan and Wohletz 1981, 1983a; Wohletz 1983; Kokelaar 1986; Zimanowski et al. 1997a), and the critical interface between the two media (Zimanowski et al. 1997a, b; Büttner and Zimanowski 1998; Austin-Ericksson et al. 2008), represent critical factors for the occurrence of an explosive interaction. For values close to an optimum water/magma ratio (i.e. 1:10 in volume Zimanowski and Wohletz 2000), all the water involved in the process is converted into superheated steam, producing mixtures with temperatures ≥100 °C (i.e. "dry" pyroclastic surges) while, for water values exceeding the optimum ratio, the water is not completely vaporized, and remains as liquid droplets inside the clouds, producing "wet" pyroclastic surges (Moore 1967; Waters and Fisher 1971; Wohletz and Sheridan 1979; Sheridan and Wohletz 1981; Wohletz and Sheridan 1983). Experiments using different water/magma mass ratios to reproduce the conditions existing during "dry" and "wet" hydromagmatic eruptions, have shown the efficiency of the interaction is the same (Büttner et al. 1999).

The structural and textural characters of "wet" surge deposits, such as the irregular contacts between beds produced by the load of damp deposits on others still plastic beneath, the presence of vesicles and aggregates, the plastering against overhanging surfaces, the abundant content of fine ash, the poor sorting, and the rapid hardening of the deposits (Waters and Fisher 1971; Lorenz 1974; Wohletz and Sheridan 1983) would reflect the influence of condensing water vapor on the depositional mechanisms of such PDCs. Such characters are absent in "dry" surge deposits, which are generally coarser in grain-size, with minor amounts of medium to fine ash, better sorted, and display variable structures evidencing a transporting medium rich in super-heated steam (Walker 1971; Sheridan and Wohletz 1983a; Wohletz 1983).

3.2 Eruptive History

3.2.1 Pre-holocene Activity

The oldest evidence of volcanism in the area is represented by a dissected crater structure located ~4 km NW of the present crater on the trace of a major fault that runs NNW–SSE (Chichón-Catedral fault, see Chap. 1). Recent studies (Macías et al. 2010) indicated that this crater would represent a portion of a former volcanic edifice (Catedral volcano) collapsed toward the SE (Fig. 3.1). Undifferentiated pyroclastic products related to the activity of this ancestral edifice have been recognized to the NW. $^{40}Ar/^{39}Ar$ data on lithic blocks embedded in such deposits, yielded ages of 1.64 Ma (Chap. 1), indicating that this structure could represent the oldest volcanic edifice in the area but further studies are needed in order to constrain this old period of activity.

The volcanism shifted then 14 km to the SW, with the emplacement of a trachybasaltic dike (46–49 wt% SiO_2) that crops out in the outskirts of Chapultenango village, 10 km E from the actual crater. The emplacement of this dike, dated at 1.1 Ma (K–Ar), occurred along the eastern tip of the E–W strike slip San Juan fault, which runs across the volcanic complex (García-Palomo et al. 2004).

The onset of the construction of El Chichón's volcanic edifice at its present location, is inferred to have occurred in the Middle Pleistocene, based on a $^{40}Ar/^{39}Ar$ age of 372 ± 5 ka obtained on a lithic block collected inside 1982 pyroclastic products (Layer et al. 2009). Nevertheless, no other field evidences have been found on the early stages of the edification of the Somma (*Pre-Somma* of Layer et al. 2009) until the present (Fig. 3.2a).

3.2.1.1 Edification of Somma (276–55 Ka)

The Somma edifice with a maximum elevation of 1,150 m a.s.l., is a 1.5 × 2 km wide crater. It consists of annular steep dome extrusions of porphyritic trachyandesites (**Unit O** of Espíndola et al. 2000) surrounded by aprons of indurated deposits of trachyandesitic blocks immersed in a matrix of lapilli and ash, exposed as far as 2.5 km W (**Unit N** of Espíndola et al. 2000; **Unit E** of Tilling et al. 1984) interbedded with volcaniclastic deposits (Layer et al. 2009). These deposits are widely distributed around the volcano and significantly determine its morphology (Fig. 3.2a). The Somma edifice occupies an area of 40 km^2, with an estimated volume of 18 km^3 (Layer et al. 2009). Trachyandesitic dome rocks (57.8 wt% SiO_2) collected on the eastern rim of the Somma crater yielded K–Ar ages between 276 ± 6 ka (Duffield et al. 1984) and 209 ± 19 ka (Damon and Montesinos 1978), which would indicate that dome extrusion occurred in a lapse of time of approximately 77 ka (Layer et al. 2009). This activity was accompanied by dome destruction with the generation of several pyroclastic density currents (PDCs) rich in non vesicular blocks (i.e. block and ash flows), as indicated by $^{40}Ar/^{39}Ar$ isochron ages of 55 ± 6 ka on blocks of porphyritic trachyandesites from these deposits collected on the W flank (Macias et al. 2010, Fig. 3.2a), and were subsequently remobilized as lahars (Layer et al. 2009).

A gray, porphyritic lava flow (58.99 wt% SiO_2) found 2 km to the NE flank of the Somma crater (**Unit M** of Espíndola et al. 2000), was probably related to a minor episode of effusion following domes emplacement in this sector, considering their similar composition. The lapse of time during which this lava flow was emplaced is not clear since the $^{40}Ar/^{39}Ar$ ages of 15 ± 7 ka obtained on lava flow samples show large standard deviation due to excess of Ar (Layer et al. 2009), and similar ages of 12 ± 6 ka, were also obtained for samples of dome remnants on the E-SE Somma rim (Layer et al. 2009). Apparently, a mayor eruption destroyed the central part of the Somma dome complex during Late Pleistocene and formed a 1.5 km wide crater (Fig. 3.2a). The type and magnitude of the eruption are still unknown (Layer et al. 2009).

3.2.1.2 Lateral Dome Extrusion (217–44 Ka)

Dome extrusion continued during the Middle and Upper Pleistocene on peripheral areas of the Somma

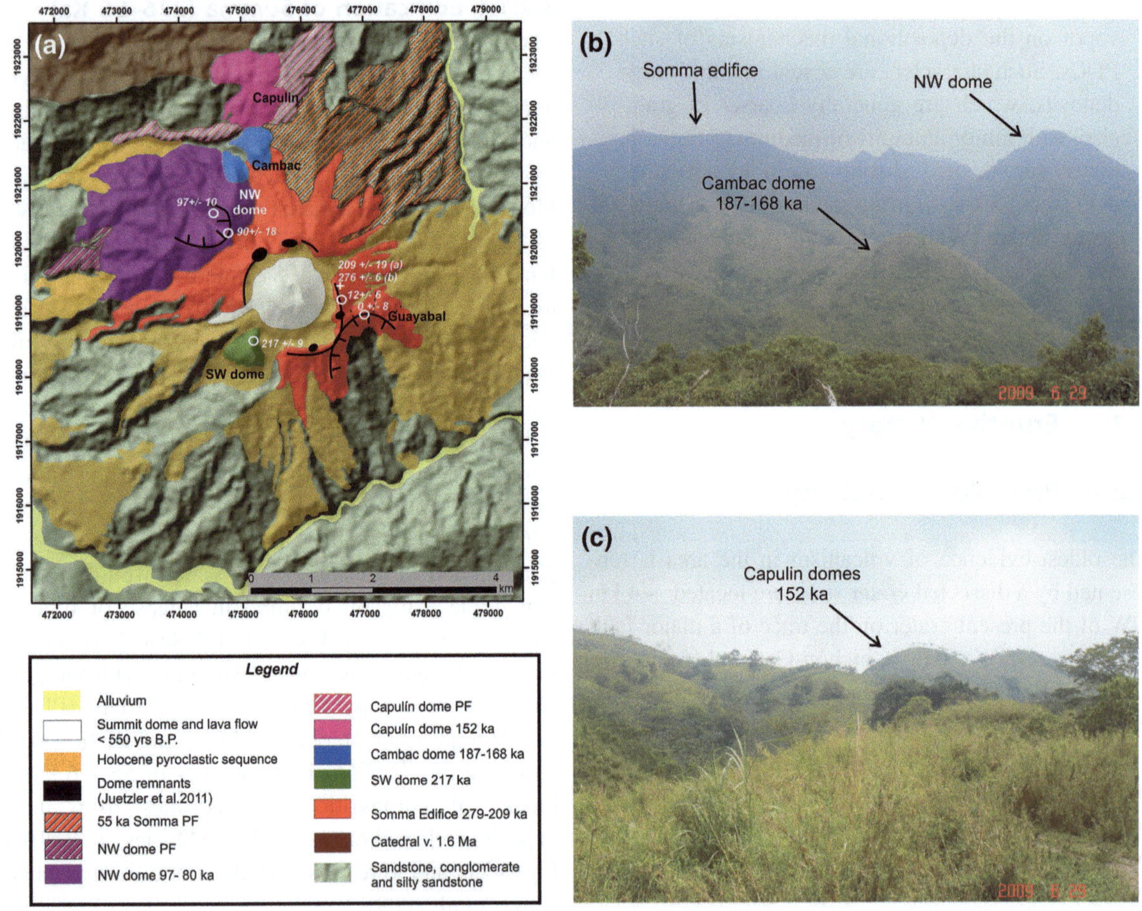

Fig. 3.2 **a** Digital elevation model showing the distribution of the folded/faulted sedimentary basement, and the structures and deposits of the different stages of the development of El Chichón Volcanic Complex. The lava dome extruded after the 550 years B.P. eruption represents the most recent constructive landform. Note the tree main notches that dissect the Somma crater to the E, N and SW. Samples dated by Layer et al. (2009) are indicated by *white circles*, while **a** and **b** indicated samples dated by Damon and Montesinos (1978) and Duffield et al. (1984), respectively; **b** panoramic view of the domes Cambac and NW, on the northern flanks of the volcano. **c** Photograph of the Capulin domes taken from the N (*Photo* by JL Arce)

edifice. To the SW, a weakly incised dome, with a maximum elevation of 900 m a.s.l. and 300 m high walls occupies an area of 0.32 km^2 with an approximate volume of 0.1 km^3 (Fig. 3.2a) (Layer et al. 2009). Canul and Rocha (1981) described for the first time this structure as "lateral dome". A sample of a massive, poorly altered trachyandesite (58.2 wt% SiO_2), collected from the dome summit yielded an $^{40}Ar/^{39}Ar$ age of 217 ± 9 ka.

Duffield et al. (1984) proposed that the SW dome was associated with a 1.2 km wide crater. Nevertheless, no evidence of deposits associated to this activity exists. According to Layer et al. (2009), the extrusion of the SW dome determined a partial disruption of part of the southwestern rim of the Somma crater, forming a 1.2 km wide collapse scar. This interpretation agrees with recent data presented by Jutzeler et al. (2011). The disruption of the southwestern Somma rim would indicate that dome extrusion occurred sometime after or was contemporaneous with the formation of the Somma crater (Layer et al. 2009). Alternatively, the scarp around the SW dome could be a feature formed by the differential erosion around the dome in contact with the rocks of the sedimentary basement. Such rocks crop out in the coalescing escarpments of the Tuspac river. This interpretation would be supported by the similar morphology of the Somma crater to the NE, discontinuous as the one to the SW.

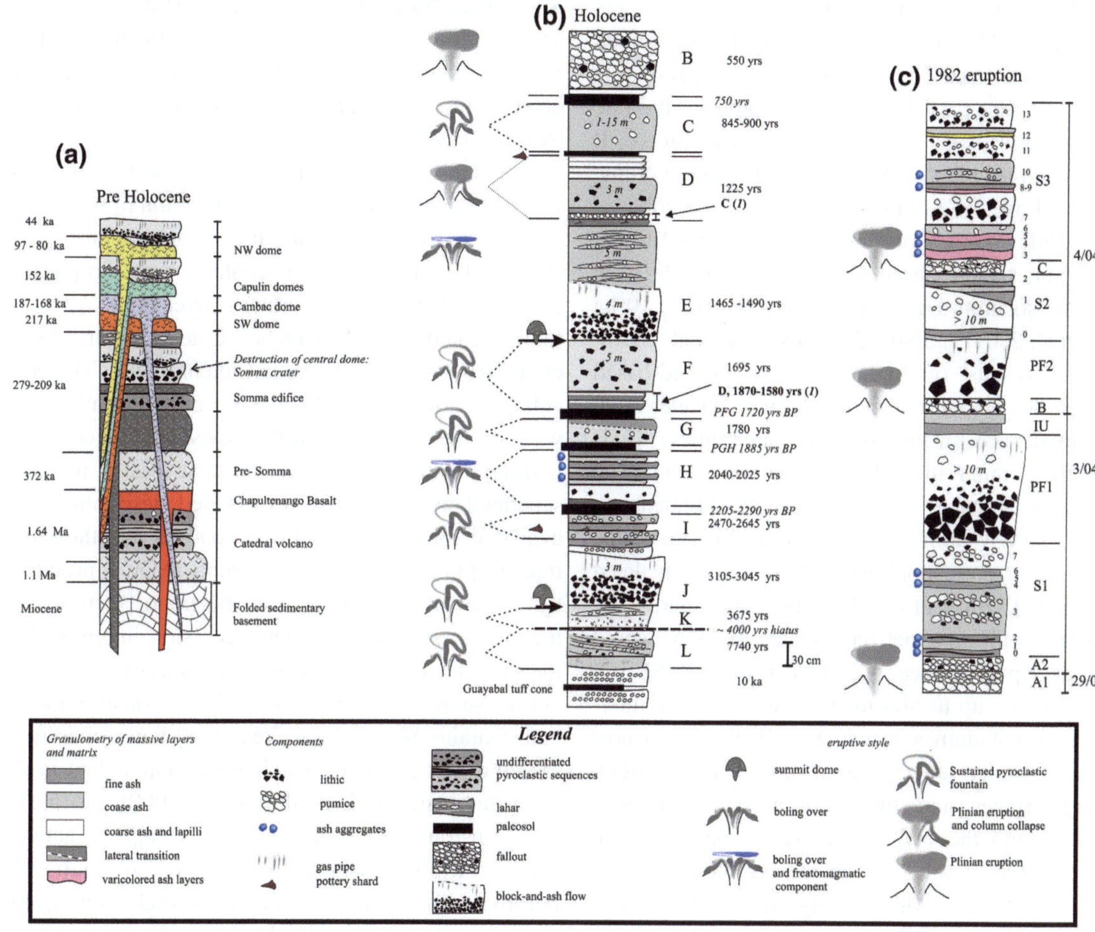

Fig. 3.3 Composite stratigraphic sections illustrating the eruptive history of El Chichón Volcanic Complex. **a** Pre-holocene stratigraphy. **b** Holocene stratigraphy showing the inferred eruptive episodes based on studies by Tilling et al. (1984) and Espíndola et al. (2000). The interpretation presented in this work about the eruptive style is also shown. **c** Stratigraphic composite section of the 1982 eruption, the timing of different phases is based on this work, see text for more detail. (1) Unit name according to Tilling et al. (1984) Table 3.1 and Fig. 3.2

Other two low-altitude (720 m a.s.l.) domes of porphyritic trachyandesites (56 wt% SiO_2) were extruded on the NW flanks of the Somma between 187 ± 13 and 168 ± 37 ka (Fig. 3.2a, b) according to $^{40}Ar/^{39}Ar$ isochron ages (Cambac domes, Macías et al. 2010). The trachyandesites from these domes contain biotite crystals beside a common association of plagioclase, amphiboles, and clinopyroxenes phenocrysts immersed in a glassy matrix (see Chap. 2).

Dome extrusion continued ~3 km NNE of the present crater, around 152 ± 21 ka, emplacing other 6 domes alineated in an E–W direction (Capulin domes, Macías et al. 2010). These domes consist of porphyritic trachyandesites (57.8 wt% SiO_2), with a maximum height of 740 m a.s.l. (Figs. 3.2a–c, and 3.3a).

Dome extrusion was apparently followed by their explosive destruction, as attested by the presence of block and ash flow deposits, consisting of trachyandesitic blocks (56 wt% SiO_2), with a common association of plagioclase, hornblende, clinopyroxenes phenocrysts in a glassy matrix, immersed in coarse ash, which reached distances of 3 km to NE and to the SW. Blocks from these deposits yielded $^{40}Ar/^{39}Ar$ ages between 102 ± 7 ka (to the NE) and 48 ± 3 ka (to the SW). The activity of dome destruction probably continued until 44 ± 10 ka as indicated by $^{40}Ar/^{39}Ar$ isochron ages on accidental lithics found in the 1982 deposits.

As the explosive destruction of Capulin domes was taking place, another dome was extruded to the NW flank of the Somma. The 1,048 m a.s.l. NW dome (Macías 1994; Macías et al. 1997), is a highly eroded structure characterized by a deep drainage, and consists of gray to green partially altered porphyritic basaltic trachyandesitic lavas (55.84–56.28 wt% SiO_2). It occupies an area of ≈5 km^2, with an estimated volume of 3 km^3 (Layer et al. 2009). A 700 m wide collapse structure left a horseshoe shaped crater opened to the NW. Dome samples used to date this episode gave $^{40}Ar/^{39}Ar$ isochron ages ranging between 97 ± 10 ka (Layer et al. 2009) and 80 ± 23 ka (Macias et al. 2010).

Recent results of high-resolution ground-based geomagnetic and aeromagnetic surveys (Jutzeler et al. 2011) evidenced strong magnetic anomalies (TMI) around the Somma rim and inside the 1982 crater. The largest TMI anomalies are located in correspondence to the NW and SW domes, and on the E–SE rim of the crater Somma, corresponding to dome remnants dated by Layer et al. (2009) at 12 ± 6 ka (Figs. 3.2a, and 3.3a).

These highly magnetized bodies, which would extend at depths between <200 m (for the E–SE rim and SW dome) up to 500 m (NW dome), were interpreted as cryptodomes in which only the uppermost part was extruded (Jutzeler et al. 2011). In this light, the large ridges and the topographic depressions formed around the NW dome would be related to sector collapse associated with the growth of cryptodomes (Fig. 3.2a). Alternatively, the collapse scar observed near the NW dome, could also have been caused by the movement of a major fault oriented NNW to SSE (Chichonal-Catedral fault) whose trace intersects the NW dome, as suggested recently by Macías et al. (2010).

3.2.1.3 Guayabal Tuff Cone (~10 Ka)

The next locus of activity migrated on the flanks of the Somma crater, 3 km to the SE of the Somma rim, determining the formation of the Guayabal "tuff cone" (Macías 1994).

The Guayabal cone is a 700 m wide horseshoe shaped crater, with a maximum elevation of 950 m a.s.l., opened to the SE into the Agua Caliente gully (Fig. 3.4) and in direct contact with porphyritic andesites of the Somma crater (Layer et al. 2009). A fault oriented NW-SE (Chichón-Catedral fault; Macias et al. 2010) runs inside Agua Caliente gully. The cone collapsed toward the SE leaving only a semicircular wall to the NE. This wall exposes a sequence of undifferentiated pyroclastic units, at least 3 m thick, separated by paleosols. No detailed descriptions are available for these pyroclastic deposits and they may refer to pyroclastic density currents (PDCs) produced by either hydromagmatic or magmatic activity. A massive to stratified deposit succession interpreted as the product of a hydromagmatic activity (i.e. pyroclastic surge), caps the entire depositional sequence and contains boulders of Cretaceous limestones from the volcano basement, which are also abundant in the andesites beneath (Espíndola et al. 2000). No useful age constrains exist to date the eruptive activity occurred from this center. In fact, $^{40}Ar/^{39}Ar$ dating of a porphyritic andesite from the dome rock forming the substrate of this cone yielded an imprecise age of 100 ± 600 ka, due to excess atmospheric argon. Moreover, a porphyritic andesite fragment embedded in the pyroclastic surge deposit that caps the sequence yielded a $^{40}Ar/^{39}Ar$ age of 0 ± 8 Ka.

Fig. 3.4 View from the SE of the remnants of the Guayabal cone. The *dashed white line* separates the base of the cone from the porphyritic andesites extruded on the Somma flanks. Note the 700 m wide horseshoe-shaped depression of Agua Caliente gully

Based on the overall appearance of the deposits, Layer et al. (2009) suggested that the type of activity that build up the Guayabal cone was mostly hydromagmatic in character. Nevertheless, the presence of paleosols separating pyroclastic sequences suggests that several explosive eruptions occurred from this eruptive center, even if more detailed studies would be useful to determine the type of eruptive events occurred.

3.2.2 Holocene Activity

The locus of volcanic activity during the Holocene returned again to the area of the Somma crater (Tilling et al. 1984; Espíndola et al. 2000; Layer et al. 2009).

Stratigraphic relationships and radiocarbon dating indicated that at least 11 explosive eruptions and domes extrusions occurred during the past 8,000 years (Fig. 3.3b; Espíndola et al. 2000). Radiocarbon dates reported here were calibrated with the program CALIB 5.0 (Stuiver and Reimer 1986); the sigma-1 range is reported in years calibrated before present (yr cal BP) and based on the calibration curve IntCal04 (Reimer et al. 2004). *Anno Domini* calibrated (AD cal) ages are also provided when needed. According to Espíndola et al. (2000), the best outcrops for Holocene deposits are located on the eastern and southern sides of the volcano at distances between 6 and 12 km from the present crater (Fig. 3.6).

The first record of activity in the Holocene was found in a ravine 4 km to the NW of the crater. It consists of a brown deposit (30 m thick) of coarse pumice and scarce andesitic lapilli immersed in a matrix of coarse ash, characterized by a diffuse stratification of the pumice lapilli, which is strongly erosive on a debris flow deposit of rounded blocks of limestone from the local basement immersed in a clayish matrix. This deposit transforms 4 m downstream into a massive deposit of coarse ash and lapilli. Embedded charcoal fragments yielded ^{14}C ages of 7740 ± 50 year BP (8454–8558 cal year BP) (**Unit L**, Espíndola et al. 2000). The textural and structural characters of the deposits, and the absence of an underlying fallout layer, suggest that this unit was probably emplaced by a sustained current produced by the prolonged collapse of a low-altitude column (i.e. "boiling over" activity of Taylor 1958). Apparently, no other activity occurred for approximately 4,000 years, or the deposits, if any, from minor activity, were eroded and not preserved. In fact, the following ^{14}C data of 3675 ± 80 year BP (3895–4094 cal year BP), refers to charcoal fragments embedded in an altered, brown deposit (12–220 cm thick) that crops out at distances of up 10 km in the eastern and southern sectors, between Chapultenango and C. Tonapac, above basement rocks and underlies a widespread yellow pumice fallout layer (unit B, see below). In the outskirts of G. Victoria village, 8 km SE from the crater, the deposit consists of a basal massive thick layer of coarse ash with subrounded andesitic lithic lapilli and scatter pumice lapilli, which gradually become more pumice-rich toward the top, with subrounded yellow pumice layers and lenses (**Unit K** of Espíndola et al. 2000, Fig. 3.3b). The deposits characteristics and their great areal extent, suggest that Unit K was emplaced by a sustained lateral current derived from the collapse of a tall eruptive column, during which pumice fallout fell into the current, and become rounded by collision/abrasion at the contact with other clasts, without being recorded as a pumice fall layer.

A dome apparently grew inside the central crater afterward and, after a period of ~600 years was explosively destroyed; probably hydromagmatic activity also occurred (Espíndola et al. 2000). Evidences for this eruptive activity are recorded by deposits that crop out at distances of 2.5 km to the E, inside El Platanar valley, and were attributed to the same period considering their similar radiocarbon age ranging between 3105 ± 70 year BP (3241–3397 cal year BP) and 3045 ± 105 year BP (3139–3371 cal year BP) (**Unit J**, Espíndola et al. 2000). They consist of a pink to gray massive, clast-supported deposit of andesitic lithic blocks immersed in a scarce matrix of lithic and pumice lapilli, with relict vertical fumarolic pipes at one side of the valley, and of a greenish to–gray deposit (3 m thick), of alternating beds of fine and coarse ash and massive beds of ash and lapilli containing tree casts, on the opposite side. The abundance of andesitic blocks in the deposit was interpreted as indicative of dome destruction. The vertical succession of alternating massive and stratified beds was instead interpreted as the occurrence of dilute PDCs due to hydromagmatic activity (Espíndola et al. 2000).

Apparently the crater remained open for the next 600–700 years, after which another explosive

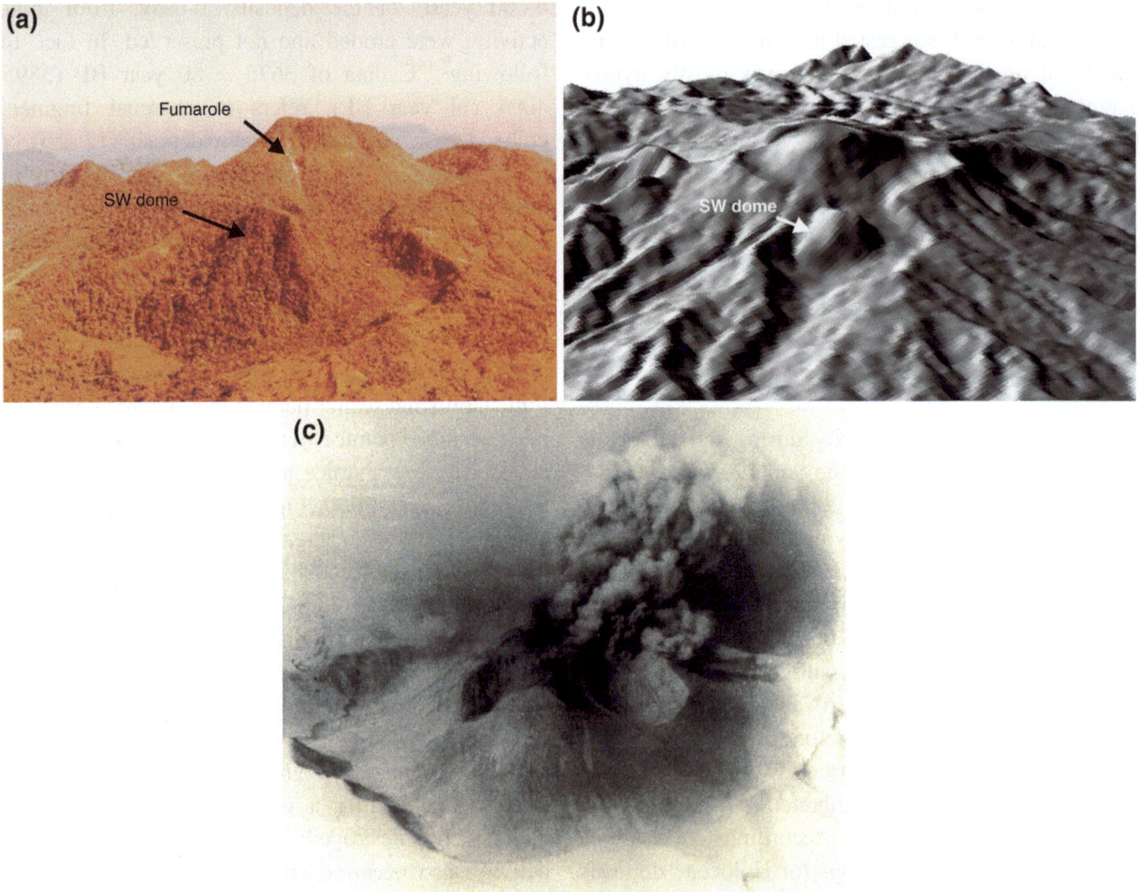

Fig. 3.5 **a** View toward the SW of the central dome inside the Somma crater, before the 1982 eruption. Note a white fumarole (*arrow*) along a vertical fracture on the central dome. The SW dome is visible in the foreground. Photo by René Canul in 1981. **b** Digital Elevation Model of the pre-eruptive topography, from the SW. **c** *Aerial view* toward the W of the central dome partially destroyed after the events occurred the night of 28 March (2315 local time). The photo was taken sometimes before April the 3rd, 1982. *Courtesy* of R. Tilling, *Geof. Int.* (2009)

Fig. 3.6 **a** *Aerial view* from the S-SE of the volcano in May 1982, showing the newly formed crater ~1 km in diameter. *Photo* by GYMSA SA de CV. **b** DEM from the S-SE of the post-1982 eruptive topography

eruption, involving the collapse of low-altitude column that fed lateral density currents, occurred. Field evidence for this period of activity consist of a massive, dark-brown, matrix-supported deposit, with scarce lithic fragments and rounded altered coarse pumice lapilli, which contain abundant pottery fragments (**Unit I**, Espíndola et al. 2000) that crops out at distance of 3 km from the present crater, inside El Platanar valley. Charcoal sampled at different locations inside the deposit, yielded ^{14}C ages between 2470 ± 50 year BP (2464–2553 cal year BP) and 2645 ± 55 year BP (2732–2797 cal year BP). A paleosol overlying this deposit yielded ages between 2205 ± 60 (2218–2311 cal year BP) and 2290 ± 250 year BP (2041–2546 cal year BP), attesting the quiescence of the volcano after this eruptive phase.

Around 2,000 year BP, eruptive activity resumed, as indicated by deposits found at distances of 2.5 and at 4 km from the crater inside El Platanar valley (**Unit H**, Espíndola et al. 2000). At lower distances (2.5 km) from the crater, a massive deposit of coarse to medium ash (10–40 cm thick) at the base is separated through an erosive surface from an overlying pink massive deposit of coarse ash (30 cm thick), containing abundant andesitic lapilli, with vertical fumarolic pipes. Charcoal inside the deposit yielded ^{14}C ages of 2040 ± 125/120 year BP (1866–2153 cal year BP). The vertical succession ends with an alternation between massive beds of coarse ash and lapilli, cross-stratified beds of coarse ash and lapilli and massive beds of fine ash displaying load structures and containing ash aggregates. At greater distances (4 km), this unit consists of alternating stratified and massive beds of coarse ash and lapilli, capped by an upper bed of fine ash containing ash aggregates. Charcoal fragments found inside this succession, yielded ages of 2025 ± 85 year BP (1889–2066 cal year BP), and were attributed to the lateral transition of the ones described at shorter distances along the same ravine by Espíndola et al. (2000). Nevertheless, the deposit distribution and their structural and textural characters do not allow a single interpretation of the events. In fact, the fine ash gray massive bed at the base of the sequence could have been deposited either by currents generated during an initial hydromagmatic activity (as suggested by Espíndola et al. 2000), but may also represent the product of an early low-concentrated ash cloud that traveled ahead of a denser, more competent portion of a pyroclastic density current, which was preserved in some places, and eroded elsewhere by the following current. The presence of elutriation pipes inside this deposit indicates the occurrence of clasts segregation after deposition caused by gas escaping following the rapid settling of clasts. This stratigraphic succession could therefore record either the product of an initial hydromagmatic activity followed by a low pyroclastic fountaining which generated dense PDC (s), when the access of water was temporarily unavailable, or the spreading of a current generated by a pyroclastic fountaining from a low eruptive column produced by magmatic activity. Apparently, the explosive interaction between magma and water followed, producing more dilute PDCs, as recorded by the uppermost portion of the stratigraphic succession closer to the crater consisting of cross-bedded deposits of coarse and fine ash that shows load structures and ash aggregates (Espíndola et al. 2000). These last currents were poorly confined by the topography, and reached up to 4 km to the E. The succession of alternating stratified and massive beds, would record the instability of the flow downstream during its final spreading, which was followed by a final settling of ash, as attested by the capping ash aggregate-bearing layer. The water source responsible for this activity derived either from a crater lake (as suggested by Macías et al. 2008) or from a phreatic layer below the crater.

Around 1,800 year BP other episodes of explosive activity occurred, as indicated by radiocarbon dating on two paleosols, which bracket a light brown ash layer at distances of 1.5 km E from the crater that yielded ages between 1885 ± 75 year BP (1723–1896 cal year BP) and 1720 ± 70 year BP (1548–1707 cal year BP). This deposit was attributed by Espíndola et al. (2000) to the emplacement of a current rich in condensing steam (i.e. wet surge), and was correlated with a brown massive deposit, rich in andesitic lithic and pumice lapilli that crops out further downstream, 4.3 km to the E, with C^{14} ages of 1780 ± 95 year BP (1602–1820 cal year BP) (**Unit G** of Espíndola et al. 2000). The stratigraphic succession alone does not allow a unique interpretation of the events. In fact, the ash-rich layer bracketed by paleosols, could alternatively represent the fallout associated to the massive deposit found downstream, produced when the PDC generated by a magmatic activity was waning, which was preserved only at some places. Another massive dark-gray deposit, 5 m thick, rich in dark gray andesitic and hydrothermalized lithic clasts (**Unit F**), exposed 2.5 km SW of the crater inside Agua Tibia gully, and dated at 1695 ± 65 year BP (1534–1636 cal year BP), would record another episode of explosive activity occurred 100 years later, under open vent conditions according to Espíndola et al. (2000). For this same lapse of time Tilling et al. (1984) reported two PDCs and associated fall layers, followed by debris flows (**Unit D**), between 1870 ± 70 (1721–1878 cal year BP) and 1580 ± 70 year BP (1395–1538 cal year BP). The occurrence of a paleosol between unit F and G (PFG of Espíndola et al. 2000) dated at 1720 ± 70 year BP would indicate that these deposits were generated during different eruptive events (i.e. around 1,800 and 1,600 year BP), separated by a significant lapse of time. The interpretation of these stratigraphic units (F and G) is not straightforward, and even if the occurrence of hydromagmatic activity was invoked (Espíndola et al. 2000; Macias et al. 2008), considering the descriptions provided by Tilling et al. (1984), the deposits would be better interpreted as the product of the collapse from low-altitude column(s). Nevertheless, for both stratigraphic successions, more detailed studies are needed to assess better the nature of events that occurred.

Another dome probably grew inside the crater and was explosively destroyed around 1,500 year BP. This activity was inferred considering the presence of a 4 m-thick gray, massive deposit of dark gray trachyandesitic and red hydrothermalized lithic blocks, in a matrix of coarse lapilli, with abundant fumarolic pipes (**Unit E** of Espíndola et al. 2000), which crops out 2 km E from the crater inside El Platanar valley. This deposit was correlated with a 5 m thick succession of cross-stratified, pumice-rich beds and massive beds, that crops out at 4.5 km E in the outskirts of the former Volcán Ch. village, dated between 1465 ± 95 (1291–1418 cal year BP) and 1490 ± 45 year BP (1327–1409 cal year BP). A similar age of 1520 ± 75 year BP (1343–1423 cal year BP), was obtained for a brown deposit of fine ash rich in pumice and carbonized tree branches found 2.5 km E, which overlies a green massive deposit of fine ash interpreted as the product of hydromagmatic activity. Based on the similar ^{14}C ages, these deposits were attributed to the same period of activity.

After 250 years, the activity resumed from an open conduit, as indicated by the stratigraphic record inside El Platanar valley and radiocarbon dating. Espíndola et al. (2000) described at distances of 2.5 km E from the crater a sequence of deposits consisting of a massive bed (20 cm thick) of brown fine ash at the base, overlain by a massive, 3 m-thick gray deposit of coarse ash rich in andesitic blocks, followed by 4 bedsets of light-gray massive coarse ash and pumice lapilli overlain by fine ash fall. Charcoal embedded in this deposit was dated at 1225 ± 105 year BP (1059–1270 cal year BP/680–891 AD) (**Unit D**, Espíndola et al. 2000; **Unit C** of Tilling et al. 1984). This vertical succession was correlated downstream, at 4 km, with a massive brown deposit of coarse ash. Tilling et al. (1984) described a Plinian pumice fall at the base of this sequence. Pottery shards found in the paleosol that overlies this deposit (Tilling et al. 1984), and small pieces of obsidian blades embedded within this unit (Espindola et al. 2000), were attributed to the Late Classic or Early Post classic period (A.D. 800–1200 probably as 1400 A.D). The fallout at the base described by Tilling et al. (1984), which was not observed in later studies (Espíndola et al. 2000), points toward the generation of a Plinian column that later collapsed to produce several PDCs, as attested by the occurrence of bedsets of massive coarse beds overlain by fine ash fallout. Apparently the crater remained open and, around 900 year BP, other pumice-rich current(s) were generated by column collapse(s) from sustained currents. Eventually pumice clasts fell inside the current without being registered as a fallout deposit. Field evidences for this kind of activity consist of massive, light-gray clast-supported deposits of lapilli and blocks of white pumice with minor coarse ash observed up to 3 km to the E, inside El Platanar valley, and up to 2 km to the N of the crater, in deep ravines, ranging in thickness between 1 and 15 m (**Unit C,** Espíndola et al. 2000). Charcoal sampled within this unit yielded ages between 845 ± 75 (686–798 cal year BP/1153–1264 AD) and 900 ± 90 year BP (739–835 cal year BP/1115–1211 AD). Ages of 795 ± 50 year BP (674–741 cal year BP) were obtained for the paleosol overlying this unit.

Explosive activity resumed around 550 ± 60 yBP (519–561 cal year BP/1389–1431 AD), as indicated by radiometric ages obtained from charcoal fragments inside a gray, massive deposit of fine ash, rich in pumice and crystals at the base of a widespread yellow pumice fall deposit (**Unit B**, Espíndola et al. 2000).

This fallout deposit covers an approximate area of 350 km^2 according to a 10 cm isopach, and overlies

Tertiary claystones to the SE and NW or, locally, a dark brown paleosol, containing abundant pottery shards, which developed on top of unit D (Espíndola et al. 2000).

Unit B is a clast-supported deposit consisting almost entirely (95 vol%) of yellow pumice blocks and lapilli (whitish on fresh surfaces), with rare gray and banded pumice lapilli, loose crystals, and gray claystones from the local basement. Dark gray mafic enclaves (44 wt% SiO_2) and amphiboles cumulates, occur either as fragments inside pumice clasts or as loose debris inside the deposits (Macías et al. 2003). All pumice samples are trachyandesitic in composition, with no compositional variation between yellow (55.1–55.4 wt% SiO_2) and gray (55.7 wt% SiO_2) pumice. The differences in color were attributed to larger proportions of hornblende in the gray pumices and a lower vesicularity (51–58 vol%) with respect to yellow pumices (63–74 vol%, Macías et al. 2003). The best outcrops are located between 5 and 10 km to the E and the S from the crater in the outskirts of Volcán Chichonal, Chapultenango and Carmen Tonapac villages (Fig. 3.7), while beyond 13 km most of the deposits have been removed by erosion. A maximum thickness of 110 cm was measured 3 km NE from the crater. Isopachs of this pumice fallout show two main dispersal axes. Close to the crater (<4 km) the dispersal is to the E, beyond 4 km, the dispersal shifts to N30°E (Macías et al. 2003). According to Macías et al. (2003), an area of at least 1,475 km^2 was covered by 1 cm of fallout during this eruptive event, which emitted an estimated volume of 2.8 km^3, corresponding to 1.1 km^3 DRE. Assuming an average density of 2.5 g/cm^3 for the lithic fragments inside the deposit, Macías et al. (2003) calculated a column height of 31 km. Based on this column height and assuming an eruption temperature of 800 °C, they estimated a mass eruption rate of 10^8 kg/s.

3.2.3 1982 Eruption

After the plinian event occurred around 550 yBP, a 1260 m a.s.l. a trachyandesitic dome was extruded within the Somma crater. A ring depression (moat) separated its basal talus breccia from the Somma crater (Fig. 3.5a, b). Apparently a lava flow flowed laterally from the dome on the western flanks of the Somma, reaching a distance of 1.5 km (Macías et al. 2010; Fig. 3.2a). This activity has been not yet well constrained in time, even though both the mineralogical association of the lava, and its stratigraphic position, suggest that a minor episode of effusive activity occurred sometime during/after dome emplacement.

Since the volcano had recorded no historical eruptions, no monitoring system existed in the area prior to 1982, except for a seismic telemetered network, installed in July 1979, at distances between 27 and 62 km from the vent, to monitor the impounding of a nearby dam (Chichoasén) by the CFE. Only the data from this local network were available to determine the onset of the eruption in later analyses (Haskov et al. 1983) before the installation of portable smoked-paper seismographs closer to the volcano by researchers of Universidad Nacional Autónoma de México (UNAM), who arrived in the area shortly after the beginning of the eruption (De la Cruz Reyna and Martín Del Pozzo 2009; Chap. 5).

A post-eruption review of available seismic records indicated that seismic precursors began in January 1980 and probably even in late 1979 (Jimenez et al. 1999). In September 1980, Canul and Rocha (1981) felt earthquakes during their fieldwork in the area. Nevertheless, their warning about a possible eruption reported to CFE remained unpublished, and therefore no preventive measurements were taken (Tilling 2009). In the months and weeks preceding the eruptions, local residents experienced several earthquakes (up to 30 events in 24 h on March 6, 1982), but the authorities did not act on their reports.

The reconstruction of eruptive events was obtained combining seismic data with available information from geostationary satellites. Eyewitness accounts provided additional observations to understand the succession of events inferred by the interpretation of the stratigraphic succession.

According to seismic records (Haskov et al. 1983), the eruption began on March 29 at 0515 UT (March 28, 2315 local time; UT = Local time-6 h) and lasted 32 min (Table 3.1). Local witnesses report this event later in time (i.e. at 0532 UT, Table 3.1). Geostationary satellites detected an eruptive column rising around 0600 UT, which according to infrared images, was 1 km above the 16.5 km-high tropopause 4 h later (Matson 1984; SEAN 1989). A dense plume drifted to the ENE, following the upper tropospheric circulation, while a less dense plume was dispersed to WSW following stratospheric winds (SEAN 1989;

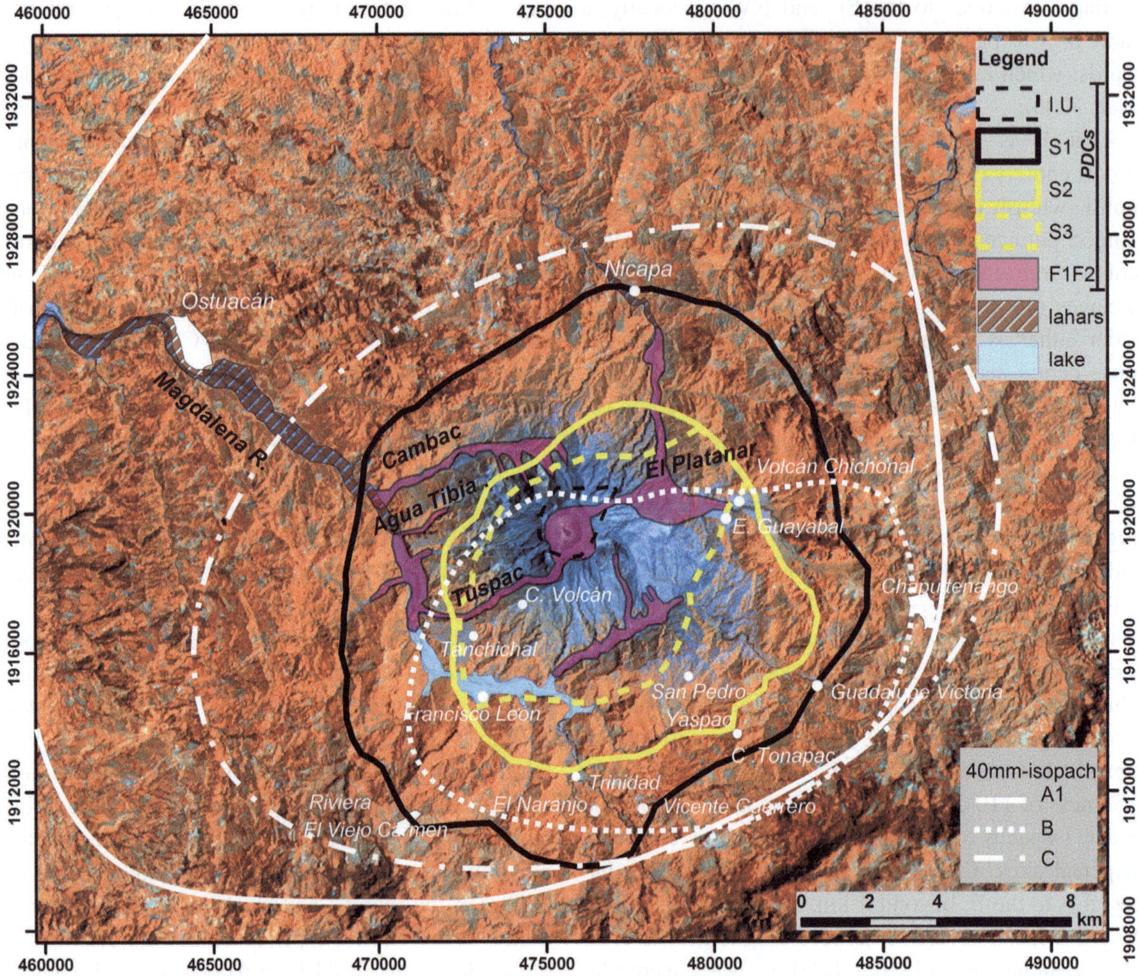

Fig. 3.7 A real distribution of the deposits generated during the 1982 eruption on a LANDSAT image. Isopachs of fallout deposits are drawn according to Carey and Sigurdsson (1986). The distribution of pyroclastic density current deposits (PDCs) is modified from Scolamacchia and Macias (2005). The extent of the temporary hot lake formed (by natural pyroclastic dams) at the confluence between the Tuspac and the Magdalena rivers is also shown. Dam failure on 26 May generated floods and lahars, the distribution of which is drawn according to Macias et al. (2003)

Table 3.1). A sulfur dioxide cloud was detected above the volcano by the Total Ozone Mass Spectrometer (TOMS) mounted on geostationary satellite (Krueger 1983; Krueger et al. 2008). A complex sequence of infrasonic signals with the occurrence of a strong gravity wave was recorded in Texas (Mauk 1983) by microbarographs and long period seismographs (Table 3.1). This event destroyed 1/4 of the dome (Fig. 3.5c), and deposited a normal-graded fallout deposit of trachyandesitic pumice lapilli, loose crystals of plagioclase, hornblende, and clinopyroxenes, with scarce lithic fragments (Fig. 3.7, fallout **A1** of Sigurdsson et al. 1984). The low content in lithic fragments, and the high degree of clast fragmentation were considered indicative of the occurrence of a phreatoplinian event (Sigurdsson et al. 1984). The polymodal grain-size distribution of this layer was attributed to the occurrence of aggregation processes during fallout (Varekamp et al. 1984). The maximum column height was estimated to be 27.3 km (Carey and Sigurdsson 1986).

Several eruptive events occurred between March 30th, and April 2, some of which reached the middle and upper troposphere according to satellite records (Matson 1984; SEAN 1989), and were seismically detected (Haskov et al. 1983), but they did not produce significant deposits according to eyewitness accounts (SEAN 1989).

The eruptive activity resumed early in the morning of April 3 between 0837 and 0840 UT as indicated by seismic records (Haskov et al. 1983). Eyewitness accounts from Chapultenango village, 10 km SE from the crater, reported lightening inside a vertical column at 0830 UT (Albarrán 1983), followed after 15 min by pumice fallout (Table 3.1).

In Texas, atmospheric pressure disturbances were registered by microbarographs in the form of complex infrasonic signals, at 0850 UT (Mauk 1983). A strong gravity wave indicated that the eruption column penetrated the tropopause (SEAN 1989). At 0900 UT (0300 local time) a rising plume was observed by geostationary satellites, and dispersed toward the E–NE (Matson 1984). According to estimates by Krueger et al. (2008) as much as 310 ktons of SO_2 were released into the upper troposphere during this eruptive event. A haze of SO_2 was visible during the day on the NE flank of the volcano (SEAN 1989). These reports attest to the occurrence of a plinian event which, according to Sigurdsson et al. (1984) deposited a minor fallout of fine ash (**A2**; Fig. 3.7 and Table 3.1), with a limited distribution and small volume. This layer was not recognized as a distinctive separate fallout deposit in the stratigraphic record in other studies (Varekamp et al. 1984; Scolamacchia and Macías 2005).

Eventually, other eruptive events occurred early in the morning of April 3, as evidenced by a complex succession of infrasonic signals registered for 27 min at Mc Kinney, Texas, starting at 0912 UT (Mauk 1983; SEAN 1989). They were also seismically detected between 0925 UT and 1040 UT (Haskov et al. 1983). The energy release for these explosive events was estimated at the equivalent of 99.5 ktons TNT and was characterized by a pulse of high frequency, high mode number infrasonic signals "similar to those produced by atmospheric nuclear weapons tests" (Mauk 1983). According to stratigraphic records, this succession of eruptive events began with high energetic explosion(s) triggered by the interaction between the rising magma with an active hydrothermal system and groundwater (Scolamacchia and Macías 2005). This event likely caused the destruction of a great portion of the remnant dome. It generated a PDC rich in hydrothermally altered lithic fragments, liquid water, and hydrothermal fluids, strongly erosive on the underlying fallout deposits and the soil beneath, which reached distances up to 6.3 km to the S, and 5 km to the N. Field evidences for this event are represented by a poorly sorted massive deposit of accidental hydrothermalized lithic blocks immersed in a matrix of coarse ash and lapilli consisting of yellowish rounded pumice, poorly vesiculated glass, loose crystals of plagioclase, hornblende, augite and minor oxides (S1–0 of Scolamacchia and Macías 2005). Pieces of overturned soil were observed embedded in this deposit (Sigurdsson et al. 1984). The presence of soil stripped from the underlying surface has been described also at the base of deposits produced by high-velocity currents (i.e. layer A0 of Fisher 1990) following the sudden decompression of a dome or cryptodome ("directed blast", e.g. Gorshkov 1959, 1963; Hoblitt et al. 1981; Moore and Sisson 1981; Fisher 1990; Belousov et al. 2007). Even if the distribution of S1–0 elongated in a N-S direction, most likely reflecting topographic control, and the possibility of a lateral explosion are unlikely, the textural and structural characteristics observed suggest that the initial event was highly energetic.

S1-0 can be correlated with the "brown massive surge layer" described by Sigurdsson et al. (1984) in the Nicapa valley, ~4 km NE of the crater. This deposit also contains aggregates of different shapes (i.e. irregular clusters, and spherical pellets with a solid core of pumice or crystals, following the nomenclature of Brown et al. 2012), consisting of poorly sorted mixtures of mm-sized pumice, and 100 μm-glass and crystals, cemented by a Fe-S-rich, orange-red film that represent up to 60 wt% of size fractions between 1 and 5 mm. Shortly afterwards, closely-spaced hydromagmatic events due to the interaction of variable proportions of magma and groundwater generated diluted PDCs (i.e. pyroclastic surges) that spread almost radially from the crater (S1–1 to S1–7 of Scolamacchia and Macías, 2005), reaching distances of 9.5 to the NE, 10.5 km to the E, 4.5 km W, and 8.5 km S. They were highly destructive up to 7 km E–SE, and 8 km S (Sigurdsson et al. 1987; Scolamacchia and Macias 2005). The deposits produced by these events are in general buried by deposits of later eruptive activity, except at distances greater than 5 km, and can be observed only in trenches dug under the present surface (Fig. 3.8a). The structural and textural characteristics of the deposits that open the sequence (S1–1, S1–2 of Scolamacchia and

Table 3.1 Chronology of the major events of the 1982 eruption, according to seismic and satellites records, eyewitness accounts, microbarographic acoustic signals recorded from far away stations, and stratigraphic analyses

Date	Seismic records	Eyewitness accounts	Satellite records	Other records (registered worldwide)	Inferred deposits
March 29	0515[a]	**0532** (1132 March the 28th, local time)	Eruptive plume 1 km above the 16.5 km-high tropopause, 4 h after the beginning of the event[g]. **1816** SO_2 cloud observed on the NW side of the volcano, spreading E and W, by TOMS[h]. Total mass of SO_2 in the clouds estimated at 0.72 Tg[h]	**0532** complex sequence of infrasonic signals registered in Texas (event 088[e]) and Antarctica[d]. Energy released 0.26 MT[e]. Occurrence of a strong gravity wave indicated that the er. plume penetrated the tropopause	Fallout A1
April 3	0840[a]	**0830** from Chapultenango (10 km SE) lightening observed inside a vertically rising eruptive cloud, followed 15 min after by fallout of pumice lapilli (up to 3 cm)[c]. Fallout continued for 45 min with a decrease in particle dimension[c]	Eruptive column emerging from the volcano at **0900**, and blew to the NE and SW[d, g] Total mass of SO_2 estimated to be 310 k tons[h]	**0850** Pressure disturbances in Texas (event 093A[e]) Gravity waves indicated that the eruption plume penetrated the tropopause Energy released 1.06 MT[e]	Fallout A2 (minor)[f]
April 3	0925–1040 1003 major event[a]	The inspection at Volcán Ch. village (4.7 km E) ≈8 h after from the previous event (around 1650, April 3) indicate downed trees and houses in E–SE direction; presence of a 80 cm thick deposit; T at the surface 60 °C, increasing toward the center[c]; 7.5 cm of new ash reported at Nicapa (7.5 km NE)[d]. At **0030** April 4 only small gas plume observed	No data available at 1200[g]	**0912** (event 093B[e]) complex acoustic infrasonic signals similar to nuclear tests. Total duration 27 min[e]. Energy released estimated at 99.5 kT[*e]. No evidence of gravity waves	S1, PF1, IU
April 4	0139 major event[a]	**0130**[c] from **Chapultenango** (10 km SE) "mushroom shaped cloud", with lightening vertically rising. Fallout of pumice blocks (40 cm) and lithics (10 cm) began after 10 min. Two "nuée ardentes" observed, one moving toward the W[c] **0135** from **Ostuacán** (11.5 km NW) heavy electrical activity inside a vertical cloud. Seismographs saturated for 20 min. Moderate fallout from >1 h.[a, b] Advancing flow toward W[b] A pumice flow deposit, observed 5 km from the summit to the NE, terminating 2 km from	**0400** NOAA geostationary satellites detected eruptive plume at heights <16.9 km[g]. Plume spread NE, following tropospheric winds and SW following stratospheric winds[d, g]. Ash drifted toward Guatemala and Belize for the next 5 h[d]	**0200** (event 094[e]) infrasonic signals registered in Texas, exciting initial gravity waves, and a complex series of acoustic wave for ∼48 min, attributed to the occurrence of distinct explosions every 2–3 min[d] Energy released estimated at 1.14 MT[*e]	Fallout B, PF2, S2

(continued)

Table 3.1 (continued)

Date	Seismic records	Eyewitness accounts	Satellite records	Other records (registered worldwide)	Inferred deposits
		Nicapa. At the distal end was about 3 m thick. Pumice blocks up to 1 m. On April 8, T measured by thermocouples at 40 cm depth averaged 360 °C, high as 402 °C			
		The pumice flow appeared to have been emplaced by 2 events in rapid succession[d]			
April 4	1110 (start)	**1122** plume reported from ground observers[d].	**1130** first appearance of the plume on geostationary satellites[d]	**1122** (094B event[e]) Infrasonic signal registered in Texas and Antarctica. Duration of the signal > 150 min	Fallout C, S3
	1215 (end)[a, b]	**1132** from Ostuacán onset of the eruption due to continuous rumbling. Felt earthquakes. Pumice fallout started at **1143** and continued until **1200**[a]	**1144** eruptive plume recorded by NOAA polar orbiting satellites[g]	Complex signal (similar to 093B) composed of several modes of infrasonic excitation, indicative of two major explosions separated early in the sequence by 8 min, followed by a series of distinct events separated in time by 3–5 min[i]. Energy release two orders of magnitude >093B. Bichromatic LP Rayleight waves and coupled air waves, registered worldwide by gravimeters and very long period seismic stations, after 1100 UT[e]. Main energy release occurs in less than 1 h[e]	
	Tremor-like activity recorded in Mexico City (≈700 km N)	In Pichucalco (20 km NE) incandescent tephra rising from the volcano. The ash cloud darkened the sky during the morning	Infrared image 3.5 h later, show the top of the plume at 16.8 km		
		Ash flow downed the trees in the Nicapa valley, and left a relatively thin layer of ash, with a T of 94 °C at 10 cm depth, measured 3 days later[d]	Total Ozone Mass Spectrometer (TOMS), detected a great sulfur dioxide cloud on the volcano and trailing off to the E. Estimated mass of SO_2 at least 3.5 Tg[h]		
		Extreme heat radiating from the deposits made impossible to reach the village of F. Leon (5 km SW from the crater). Between Ostuacán and F. Leon a river boiling and downed trees could be seen upslope[d]			
April 5		Activity lasting 3 h. No incandescent tephra ejected[d]	**1730** plume rising observed by satellites		S4, S5[j]

All times are UT (Universal time = local time + 6 h) * 1kT ~4.22 × 10^{19} ergs

[a] Haskov et al. (1983)

[b] De la Cruz and Martin Del Pozzo 2009

[c] Albarrán (1983)

[d] SEAN (1989)

[e] Mauk (1983)

[f] Sigurdsson et al. (1984)

[g] Matson (1984)

[h] Krueger et al. (2008)

[i] Widmer and Zürn (1992)

[j] Macías (1994)

Macías 2005), such as irregular contacts between beds produced by the load of damp deposits on other ones still plastic beneath, laterally discontinuous lamination, the presence of vesicles, the plastering against steep walls indicate that the events occurred at the beginning involved significant quantities of water in contact with magma, producing currents rich in condensing steam (i.e. wet surges), as observed elsewhere (e.g. Lorenz 1974; Sohn and Cough 1992; Cole et al. 2001; Sohn et al. 2012; See Inset Box). Also in S1–1 and S1–2, 10–40 wt% of components in grain-sizes between 8 and 1 mm consist of aggregates of different shapes (irregular clusters, spherical and elongated pellets with or without a core of crystals, pumice, or organic fragments). Individual components of the aggregates are crystals and glass fragments hundreds of microns in size, and mm-sized pumice, cemented by a red-orange film with an overall poor sorting. Gradational variations in grain size, determine a weak internal layering (Scolamacchia et al. 2005). The composition of the cementing film consisting of abundant S and Fe with minor P was attributed to the interplay between magmatic and hydrothermal fluids in different amounts (Scolamacchia et al. 2005).

The availability of water entering in contact with magma apparently was reduced thereafter, (e.g. Sheridan and Wohletz 1981, 1983a, b; Kokelaar 1986; Zimanowski et al. 1997a; Zimanowski 1998; Büttner et al. 1999; Zimanowski and Wohletz 2000) as attested by the textural and structural characters of the deposits, generating a PDC (i.e. dry surge) that was mainly dispersed up to 5 km to the eastern and northern sectors of the volcano and up to 7.5 km to the S. This PDC was able to carbonize wood, attaining temperatures of 360–400 °C (Sulpizio et al. 2008), and to partially or completely erode deposits from previous activity, including topsoil and sedimentary rocks of the Tertiary basement (Sigurdsson et al. 1984). It deposited a coarse-grained reddish bed of coarse pumice lapilli and minor hydrothermalized lithics from the central dome, immersed in abundant ash (S1–3, Fig. 3.8b–c). Small-magnitude events due to an increase in the amount of water entering in contact with magma followed, producing dilute PDCs that deposited a succession of multiple laminae, slightly different in color, grain-size and degree of vesiculation with a patchy distribution up to 3 km to the E-SE (S1–4–5–6). Aggregates (spherical pellets and irregular clusters) either cemented or not by a red-orange film, occur as components (10–40 wt%) in S1–5 and S1–6 in size fractions between 1 and 8 mm, beside white pumice, poorly vesiculated glass, and crystals (Scolamacchia et al. 2005). The following eruptive event involved less amount of water producing a dilute PDC (dry surge) with a high erosive power at distances between 3.5 and 4.7 km from the crater to the E-SE that reached the village of El Naranjo 8 km S (Fig. 3.7). Field evidences for this event consist of a light gray deposit of pumice lapilli and blocks in a matrix of coarse ash, with minor accidental lithics from the former dome (S1–7, Fig. 3.8d), with a maximum thickness of 90 cm in topographic lows at 3.5 km SE from the vent. At Esquípula Guayabal, 3.5 km SE from the crater, this current was able to pick up and displace at least locally, concrete blocks 47 × 15 cm (Scolamacchia and Shouwenaars 2009). Other objects (e.g. steel reinforced bars of houses foundations), embedded in this deposit, were observed bent in an E-SE direction. This current was able to erode completely 6 cm of fallout A1, at El Naranjo, 8.5 km S. The preservation of a fine ash layer on top of this deposit due to the fallout from the waning portions of the current (Fig. 3.8d) would indicate its cessation (e.g. Walker 1984; Branney and Kokelaar 2002) and a pause in the eruptive activity.

According to Sigurdsson et al. (1984) and subsequent studies to date, the succession of eruptive events described above started on April 4 around 0135 UT.

However, combining available eyewitness accounts with seismic, infrasonic records, and stratigraphic data, several evidences support the hypothesis that these powerful hydromagmatic events actually occurred a day earlier, during the early morning hours of 3 April (Scolamacchia 2012; Table 3.1):

(A) People returning to Volcán Chichonal (4.8 km E), from Chapultenango (10 km SE), several hours after the plinian event that occurred at ~0830 UT on 3 April, observed that trees had been downed in a ESE direction along the small road that connect the two villages; artifacts and houses in Volcán Ch. had been downed in the same direction (Table 3.1). The directional downing of trees and houses reported by eyewitness accounts, implies the passage of an energetic lateral load, which can be likely associated to the movement of the diluted PDCs described above.

(B) According to eyewitness reports, the village of Volcán Chichonal was already buried by a 80 cm-thick deposit made of "several light and dark

Fig. 3.8 **a** View of the former village of Volcán Chichonal, located 4.8 km E from the crater. Two trenches dug below the present ground surface are seen in the foreground. In the background are visible the eastern walls of the Somma crater. **b** Partial view of the deposits in one of the trenches (*Left* in **a**) showing the stratigraphic succession from *A1* to the lower subunits of *S1* (partial). **c** Detail of the lower portion of the trench in **b** showing fallout *A1*, overlain by the laminated roof of the collapsed church and by subunits *S1–1* and *S1–3*. **d** View of the complete succession of deposits in one of the trenches dug at Volcán Chichonal village (*left* in **a**). From the *base* to the *top* are visible subunits *S1–3* (*red*), *S1–7* (*gray*), fallout B, unit S2, fallout C, and several varicolored horizons of *S3*. Shovel for scale measures 67 cm. **e** Detail of the uppermost succession visible in the trench on the *right side* in **a**, showing the varicolored laminated deposits produced by the initial hydromagmatic events that alter the emplacement of fallout **c**, characterized by a high proportion between magma and water. Note the subtle discontinuous lamination of some of the layers and the subsequent erosion (2 cm) by a subsequent (*S3–7*) current ("dry" pyroclastic surge). **f** View of the succession of unit S3 deposited at Esquípula Guayabal, 3.5 km SE of the crater. The varicolored deposits at the *bottom*, reflecting the deposition from "wet" pyroclastic density currents, are partially eroded by the "dry" current *S3–7*. This situation is similar to what observed at Volcán Chichonal (in **e**). Note the succession of stacked inverse-graded pumice layers suggesting unstable, waxing conditions of the current at this location

horizons" some of which showed vesicular texture (Table 3.1); the temperature measured at the deposit surface was 60 °C (Albarrán 1983). These accounts suggest that other deposits, beside fallouts layers A1/2, had already been emplaced during early morning of 3 April. The descriptions of the textural and structural characteristics point toward the occurrence of diluted PDCs produced by magma-water interaction, as also suggested by the temperature measured at the deposit surface only few hours after their emplacement, which would indicate that the parent currents had temperatures <100 °C. The total thickness of fallout A1 and unit S1 measured recently in a trench dug at the former Volcán Chichonal village, is 49 cm (Fig. 3.8a–d). The difference in thickness with what was measured in 1982 can be likely attributed to the post-depositional compaction following the subsequent burial by later deposits.

(C) At 0912 UT on 3 April, microbarographs in Texas registered infrasonic signals that resemble those

Fig. 3.8 continued

produced by nuclear explosions (Mauk 1983; Table 3.1). Similar infrasonic signals were recorded also during most of the hydromagmatic events that occurred during the 1991 eruption of Pinatubo (Hoblitt et al. 1996; Tahira et al. 1996). The similarities existing between base surges generated during nuclear explosions (Young 1965) and those generated during hydromagmatic explosions, have been discussed in several studies (e.g. Richards 1959; Moore et al. 1966; Moore 1967), suggesting that the signals registered in Texas can be related to the hydromagmatic events described above.

Eventually, the removal of a great portion of the dome by the initial flashing of the hydrothermal system and the following phreatomagmatic activity promoted magma vesiculation by releasing the overlying pressure, and caused the generation of a low-altitude column, which collapsed with the formation of dense, lithic-rich, PDC(s) (PF1 of Macias et al. 1997; "lithic debris flow" of Sigurdsson et al. 1984). Such current(s) were partially confined by the Somma walls, where they left a 10 m-thick massive deposit rich in hydrothermal lithics at the base (Macías et al. 1997), but they were able to spread along major drainages to the N (Cambac), the SW (Tuspac), and the E-SE (El Platanar). At distances of 1–1.5 km E, a decrease in the Somma slopes from 11° to 3° resulted in a reduction in flow speed and the deposition of the coarser load of the current, as attested by a massive breccia deposit rich in juvenile and lithic blocks up to 1.2 m in diameter (Macias et al. 1998). The flows were blocked at ~4 km distance by the Susnubac river (Fig. 3.7) perpendicular to flow direction (Sigurdsson et al. 1984). This event probably occurred around 1030 UT on April 3 (0430 local time), as suggested by a major peak observed in the seismic signals (Haskov et al. 1983). According to the stratigraphic record in proximal areas (Macías et al. 1997; Scolamacchia and Macías 2005), other minor eruptive events occurred shortly after the emplacement of PF1, due to the contact between groundwater and magma, producing PDCs that waned at distances of few hundred meters remaining confined inside the Somma crater (Lower IU). These events were followed by a more concentrated PDC, which emplaced a small-volume lithic-rich deposit (UI-3), which was followed again by a close succession of other hydromagmatic events (Upper IU) whose products, mainly contained by the Somma walls, reached maximum distances of 1.5 km to the NE. The types of aggregates found in the PDC deposits of Upper IU, and the degree of alteration of juvenile fragments, has been interpreted as the evidence that hydrothermal fluids were involved also during this phase of the eruption (Scolamacchia et al. 2005). However, no eyewitness accounts exist about these events, as they occurred early in the morning and were restricted to proximal areas, which were neither accessible nor explored.

Another eruptive phase began at about 0130 UT on April 4, according to eyewitness accounts from Chapultenango that described a vertical rising plume (Albarrán 1983), which also was observed from Ostuacán at 0135 UT (De la Cruz-Reyna and Martín del Pozzo 2009, Fig. 3.9) and registered by seismographs at 0139 UT (Haskov et al. 1983). Heavy fallout of pumice and lithic lapilli followed, lasting 1 h according to both reports (Haskov et al. 1983; Albarrán 1983; Table 3.1). Shortly after, currents moving on the ground toward the west were observed from Ostuacán (De La Cruz Reyna and Martín del Pozzo 2009) and from Chapultenango "moving laterally at high-velocity…without rising material" (Albarrán 1983). At 0200 UT, few initial gravity waves, and a complex succession of acoustic signals were registered in Texas for ~48 min and attributed to the occurrence of distinct explosions every 2–3 min (SEAN 1989; Table 3.1). At 0400 UT, geostationary satellites

Fig. 3.9 View of the crater of El Chichón from Ostuacán village (10.5 km NW) the evening of April 3, 15–30 min after 1935 local time (0135 UT). A *vertical* plume is glimpsed rising vertically from the crater, attesting the occurrence of what will be later identified as the Plinian event B. *Photograph* by Servando de La Cruz Reyna. Courtesy of the author

reported a plume elongated in a NE-SW direction (Matson 1984; SEAN 1989). Eventually, this eruptive phase began with the generation of a plinian column that penetrated the tropopause and emplaced a widespread gray to reddish deposit of slightly normal-graded white pumice, rich in hydrothermally-altered lithic lapilli (fallout **B**, Varekamp et al. 1984; Sigurdsson et al. 1984). Grain-size data and isopach maps, indicated that fallout layer B has the coarser proximal components among all fallout deposits of the eruption, and a moderate distal fine ash component. Based on the maximum lithic distribution, Carey and Sigurdsson (1986) concluded that this plinian event had the largest mass eruption rate (6×10^7 kg/s) among the fallouts produced in 1982, and calculated a maximum column height of 31.6 km. However, their calculated altitude is not confirmed by satellite observations (SEAN 1989), and the volume mapped on the ground would indicate a much smaller column height (21 km, according to Matson 1984). More recent estimates (Bonasia et al. 2012) using the Buoyant Plume Theory (Bursik 2001) implemented in the FALL3D numerical code (Costa et al. 2006; Folch et al. 2009) obtained a column height of 28 km for this plinian event.

The collapse of this eruptive column generated dense PDCs that were channeled through major valleys to the W, NE and N, following the volcano's drainages, as indicated by eyewitness accounts from Ostuacán and Chapultenango (Table 3.1). The deposits of this activity consist of 2–3 flow units of massive, dark-gray, deposits of pumice lapilli and ash, more rich in lithic blocks inside the moat (Macías et al. 1997), where they partially ponded. The Magdalena river to the W, and the Susnubac river to the S, were able to block these flows (Sigurdsson et al. 1984), but no obstacles existed toward the NE, and the flows advanced up to 5 km along the El Platanar river (SEAN 1989; Table 3.1). Contemporaneous phreatomagmatic activity accompanied the collapse of the column, as indicated by the successions of one to four bedsets made up of a fining upward sequence of beds rich in white pumice lapilli and blocks with different structures, which become progressively finer-grained from the base toward the top that are intercalated to (S2-0), or directly overlie (S2-1, S2-2), pyroclastic flow PF2 (Scolamacchia and Macias 2005). They crop out discontinuously from the crater at distances greater than 1.3 km, covering an approximate area of 57.1 km^2 (Fig. 3.7). According to deposit distribution, these currents spread mostly to the eastern sector, following El Platanar river, and one affluent of the Susnubac river, and to the SW, along the Tuspac river, being strongly erosive on underlying deposits at distances between 3.5 and 5 km (Scolamacchia and Macías 2005; Scolamacchia and Schouwenaars 2009). To the E, they consist of a maximum of four bedsets, each one capped by a fine massive ash layer, attesting the emplacement of 4 different currents some of which were able to displace locally small tree trunks up to distances of 5.2 km E (Scolamacchia and Macías 2005). Components of finer fractions are white pumice and dense dark-gray trachyandesitic glass, loose crystals of plagioclase, augite, hornblende and minor sphene in different proportions as occur in other PDC deposits of the same eruption, beside a low percentage (2–3 wt%) of accidental altered clasts from the old dome. Weakly vesiculated glass with planar surfaces intersecting at high angles, (*blocky shapes*, Sheridan and Wholetz 1983a; Heiken and Wholetz 1985) is common in fractions finer than 125 μm (Scolamacchia and Macías 2005) indicating the occurrence of a brittle fragmentation of the melt at the contact with the water.

A final eruptive sequence started on April 4 at 1110 UT, according to seismic records (Haskov et al. 1983). Eyewitness accounts reported the start of this event between 1122 UT (SEAN 1989) and 1132 UT (Haskov et al. 1983). A rising plume was first recorded around 1130 UT by a NOAA geostationary satellite (Table 3.1). Heavy fallout of pumice lapilli started at 1143 UT, and continued for about an hour (Haskov et al. 1983). Infrared images indicated that an eruptive column penetrated the tropopause 3.5 h later (SEAN 1989). As a result of this activity, a widespread fall deposit of normal-graded white trachyandesitic pumice lapilli and crystals, with minor lithic content, was deposited in an ENE direction across 3,000 km^2 (fallout **C** of Sigurdsson et al. 1984). Based on a model of different transport direction in the tropospheric and stratospheric circulation, Carey and Sigurdsson (1986) calculated a column height of 28.8 km, attributing this event to a phreatoplinian activity. The vertical rising of this Plinian eruptive plume was registered worldwide by gravimeters and very long period seismic stations as bichromatic long-period Rayleigh waves and coupled air- waves (Widmer and Zürn 1992; Zürn and Widmer 1996). Based on infrasonic records, Mauk (1983) calculated that this event liberated an energy equivalent of 2.6 MT TNT in a complex sequence; gravity waves, induced by the thermal perturbations associated to the vertical rising of the plume, (Zürn and Widmer 1996), were followed by several high-frequency infrasonic pulses between 1309 UT and 1435 UT, which were attributed to the occurrence of distinct explosive events separated in time by a 3–5 min. Stratigraphic evidences indicated that after the deposition of fallout C, hydromagmatic activity was dominant, producing a series of diluted PDCs (unit S3 of Sigurdsson et al. 1984). Downed trees due to a "flow that left a thin layer of ash" were reported at distances of about 5 km to the NE after this event (SEAN 1989; Table 3.1). According to Sigurdsson et al. (1984, 1987) the deposits resulting from this activity were rapidly eroded due to heavy rains. Nevertheless, subsequent studies recognized deposits from these currents dispersed across an area of 44 km^2 (Scolamacchia and Macías 2005). This activity involved different proportions of water and magma. The resulting deposits consist respectively of varicolored, vesiculated layers of fine-ash and lapilli with discontinuous lamination, and of coarser-grained deposit of coarse ash, lapili and blocks which were strongly erosive on the deposits beneath and reached distances up to 4.8 km from the vent (Fig. 3.8e). Some of these currents were short-lived and vanished at distances of 2.5 km (e.g. S3-1-2 of Scolamacchia and Macías 2005) or were eroded by following flows. Others were of higher magnitude and traveled distances as far as 5 km E (e.g. S3-7). Irregular clusters, consisting of pumice up to 6 mm in size and crystals, and minor spherical pellets (with or without a solid core) of glass and crystals cemented by a S-Fe-rich film, represent up to 90 wt% of the components of size fractions coarser than 4 mm in S3-3. They are abundant in deposits that crop to the E-NE between 2 and 3.6 km, and reduce consistently at distances greater than 4 km. To the S, aggregates are free of a cementing film, and characteristically smaller in dimensions (≤2 mm), with individual components ranging between 10 and 300 microns maximum (Scolamacchia et al. 2005). The hydromagmatic activity continued, producing alternating "dry" (T > 100 °C) PDCs, which were erosive up to 5 km to the ESE (e.g. S3-7-11-13 of Scolamacchia and Macías 2005) or "wet" (T < 100 °C) PDCs whose deposits, otherwise eroded, were preserved mostly in topographic lows at distances <2 km to the E (e.g. S3-8-9-10 and S3-12). These discrete, numerous, closely-spaced explosions generated surge-type pyroclastic density currents (e.g. Moore et al. 1966; Moore 1967; Kokelaar 1986; White 1991) that can be likely correlated with the infrasonic signals registered in Texas on April 4 after 1309 UT (Mauk 1983; Table 3.1), even if the time of eruptive events is only approximate considering the time lag existing between the eruptive events and the registered infrasonic signals. Similar infrasonic signals were registered during most of the hydromagmatic events preceding the climactic caldera collapse during the 1991 eruption of Pinatubo volcano (Tahira et al. 1996). Such signals were generally accompanied by an abrupt onset of high-amplitude tremor but, in some cases, no seismic signals were recorded (Hoblitt et al. 1996).

Minor phreatomagmatic events probably occurred during the following days, but their distribution remained confined inside the newly formed crater (Units S4 and S5 of Macías 1994, Table 3.1), and they were reworked/eroded during the following raining season.

The massive emplacement of pyroclastic deposits at the confluence between the Magdalena and the Tuspac rivers caused a rise of 30 m of the base of the thalweg of the Magdalena river (Macías et al. 2004). A 55-m thick dam formed 3.5 km downstream of the village of F. León, and extended for approximately 1 km downstream. A hot lake with a volume of $\sim 26 \times 10^6$ m^3 formed behind the dam in late April (Fig. 3.7), and increased up to $\sim 40 \times 10^6$ m^3 in early May, when total rainfall was estimated to be ~70 mm (Quintas 2000). It extended 4 km upstream to the confluence between the Magdalena and Susnubac rivers (Medina-Martínez 1982, Fig. 3.7). This dam failed by overtopping on May the 26th at 0130 UT. A hot flow, with temperatures of ~90 °C at Xochimilco and ~82 °C at Ostuacán, mixed with the waters of the Grijalva river, 28 km downstream, and traveled up to 35 km downstream reaching the Peñitas hydroelectric dam, which was in construction, killing 1 person and injuring other 3. Lahar deposits from the event of May 26th are exposed along the Platanar river as four flat terraces and cover a minimum area of 1.1×10^6 m^2 with an average thickness of 4 m (Macías et al. 2008). Only the Moba river, to the East, was unaffected by this dam failure.

In summary, ≈2.2 km^3 of trachyandesitic (56–57.5 wt% SiO_2) tephra, corresponding to 1.1 km^3 DRE at 2.6 g/cm^3 were erupted during the 1982 eruption (Carey and Sigurdsson 1986). No significant compositional or mineralogical differences were observed among the three fallout layers A1, B and C, characterized by a mineral association consisting of plagioclase (An_{32-80}) + hornblende + augite + apatite + biotite + anhydrite ± Ti-magnetite ± sphene and pyrrhotite (Luhr et al. 1984). Anhydrite crystals (up to 2 wt%) were recognized for the first time in fresh pumice from the eruption, occurring both as discrete phenocrysts and inclusions in other silicate crystals, indicating that they precipitated from a S-rich (2.6 wt% SO_3) melt (Luhr et al. 1984; Varekamp et al. 1984; Luhr and Logan 2002; Luhr 2008). This sulfur content, considered too large to have been dissolved in the melt at pre-eruptive magmatic temperatures of 800–850 °C (see Chap. 2), suggest that a significant fraction of the erupted S was present as a separate gas phase prior to eruption (Luhr et al. 1984). Part of this sulfur was inferred to be released as an oxidized vapor phase in the eruptive clouds (Varekamp et al. 1984), and absorbed into the ash after oxidation, determining premature fallout of particles of different grain-sizes and producing a polymodal distribution in ash fallout layers (Varekamp et al. 1984).

Nevertheless, the occurrence of different types of aggregates cemented by a film rich in sulfur, similar (i.e. internal grain-size distribution, sorting, and morphology) to sulfur aggregates described on the slopes of small sulfur cones, formed on the crater floor at Poás volcano when the lake dried out (Oppenheimer 1992), suggests that also at El Chichón liquid sulfur acted as a binder between particles (Scolamacchia and Dingwell 2014).

Liquid sulfur was eventually explosively ejected during most of the eruptive events, being more abundant in deposits produced at the onset of activity that caused the flashing of the hydrothermal system on April 3 at 0912 UT (S1-0), and in those produced shortly after the onset of the phreatoplinian activity C (S3-3). The abundance of cemented irregular clusters in deposits produced during the last phases of the eruption (i.e. unit S3) suggests that the supply of liquid sulfur was long lasting. Therefore, it is reasonable to think that native/elemental sulfur was probably slowly accumulated in the hydrothermal system at El Chichón, by precipitation from condensing magmatic components (i.e. SO_2 and H_2S), emitted from the degassing trachyandesitic magma, similar to what reported at other andesitic volcanoes with active hydrothermal systems, where vapor-dominated areas surrounded by brine solutions, characterize the hydrothermal systems (e.g. Bennet and Raccicchini 1978; Oppenheimer and Stevenson 1989; Oppenheimer 1992; Christenson and Woods 1993; Christenson 2000; Christenson et al. 2010).

Supporting this hypothesis, Mülleried (1933) described the presence of stalactites of native sulfur near fumaroles discharging H_2S almost 50 years before the 1982 eruption; similar features were also observed in ejecta from Poás (Bennet and Raccicchini 1978; Francis et al. 1980) and Ruapehu (Nairn et al. 1979), and interpreted as an evidence of the presence of liquid sulfur.

Between 5 and 9 million tons of SO_2 were emitted into the atmosphere during the 1982 eruption (Krueger et al. 1995), corresponding to a cumulative mass of 7.5 million tons, according to more recent estimates (Krueger et al. 2008). This amount was only exceeded by the eruption of Pinatubo in 1991. The massive injection of sulfur in the stratosphere caused a 5–6 °C warming in the tropical lower stratosphere, already in October 1982 (Parker and Brownscombe 1983).

3.3 Characteristic Activity and Hazard Assessment

The stratigraphic record for the El Chichón Volcanic Complex gives an idea of the most frequent kind of eruptions, as well as the approximate maximum distribution of related deposits.

Beside the frequent dome extrusions that occurred during the construction of the Somma edifice (see Sect. 3.2), accompanied in some cases by their explosive destruction, the stratigraphic record available for the Holocene, indicate that the eruptive activity was mostly explosive in character. Only two episodes of effusive activity are documented during the Holocene, and were apparently related to a subordinate activity associated to dome emplacement. The maximum run-out of lava flows was limited to the first 1–2 km from the present crater on the Somma flanks to the NE (unit M of Espíndola et al. 2000) and the W, associated to the pre-1982 dome (Fig. 3.2a).

According to past records, pyroclastic density currents generated by the collapse of eruptive columns of different altitudes seem to represent one of the most frequent eruptive phenomena. The deposits from such activity are generally confined to main ravines draining the volcano, flowing maximum distances of 4 km to the NW (at ~7,700 year BP), 3 km to the NE (~2,040 year BP, and ~900 year BP), and 4.3 km E (~2,400 and ~1,800 year BP). An exception is represented by the deposits dated at ~3,675 year BP, which were able to travel as far as 10 km to the E and S, according to Espíndola et al. (2000). The areal extent of this deposit suggests that this event was the greatest in magnitude of those that have occurred at El Chichón, but more detailed studies would be necessary to confirm this hypothesis.

Block and ash flows were produced by the explosive disruptions of central domes that grew inside the Somma crater, around 3,045 year BP, 1,500 year BP and during the last eruption. Stratigraphic records indicate that this kind of activity produced flows that were mainly confined in major ravines at distances between 2 (unit J), and 2.5 (unit E) km to the E. Their areal distribution resembles that of PF1 and PF2 produced during the 1982 eruption, which were partially confined by the Somma walls, and followed the path of major tributaries around the crater, reaching up to 4 km S and W, where they were blocked by the Susnubac and Magdalena rivers. Apparently, some of them were able to travel to the NW and NE along El Platanar Valley up to 4.5 km (**Unit E**).

The only possible evidence for eruptive activity involving the collapse of a plinian column is suggested by the presence of a fallout layer at the base of a pumice-rich deposit (Tilling et al. 1984). This event generated several pyroclastic density currents, which were recognized up to 2.5 km to the East (~1,250 year BP, **Unit D**). The apparent lower extent of these deposits can be attributed to post-depositional erosion or to a lack of outcrops, due to subsequent burial from younger deposits.

According to the information available, the interaction between rising magma with external water (likely groundwater, see below) occurred at least five times in the last 10,000 years. The Guayabal tuff cone located on the flanks of Somma represents the first record of this kind of activity. The extension of deposits from this eruptive center cannot be determined as the edifice is only partly preserved. Possible evidence for hydromagmatic activity from the central crater exists for deposits dated around 3,100 (unit J), 1,900 year BP (unit G), 1,500 year BP (unit E), even if more detailed studies would be necessary to confirm the stratigraphic data, and numerous hydromagmatic events are well documented during the last eruption in 1982. During the 1,500 year BP eruption, and in 1982, this kind of activity accompanied dome destruction. The 1982 eruptive events were characterized by different water/magma ratios. This alternation of eruptive events was likely caused by the interaction of rising magma with a water table located at shallow depths. This hypothesis is consistent with the presence of springs located at different altitudes on the Somma flanks, observed before the 1982 eruption (Templos 1981; Casadevall et al. 1984), and the formation of a lake inside the crater within a few weeks from the 1982 eruption, interpreted as due to the inflow of groundwater (Casadevall et al. 1984). Recent studies on the origin of waters from the springs located on the W and SE flanks of the volcano (see Chap. 4), would confirm the presence of a widespread (ca. 3.5 km^2), shallower aquifer at depths of ~280 m beneath the crater, at the contact between the volcanic rocks and the sedimentary basement, beside a more deeper one, located at depths of ~2 km inside the Cretaceous basement.

In light of new data, the earlier hypothesis of the existence of a multiple conduit system, to explain the occurrence of phreatomagmatic activity contemporaneous to the collapse of the plinian column B (Scolamacchia and Macías 2005), now can be discarded. In fact, recent analyses of magnetic anomalies inside the crater formed in 1982 (Juetzler et al. 2011) ruled out the presence of a multiple conduit system beneath El Chichón.

The presence of fumaroles discharging H_2S, and hot springs described on the volcano flanks (Mülleried 1933; Templos 1981; Canul and Rocha 1981) suggested the presence of a hot water hydrothermal system overlain by a small vapor-dominated cap before the 1982 eruption (Casadevall et al. 1984; Rye et al. 1984). The flashing of this hydrothermal system eventually triggered one of the most energetic events, responsible for the destruction of a great portion of the dome on 3 April.

Hydrothermal fluids were involved in different amounts during several eruptive events together with liquid sulfur as attested by the presence of ash aggregates cemented by a film rich in S Fe, and minor P with variable amounts of Na, Mg, and Ca.

Eventually the binding action of liquid sulfur was particularly efficient in removing from the eruptive clouds not only particles in the size range of ash but also in the size of lapilli (>2 mm) as attested by the maximum dimensions of the fragments accreted (Scolamacchia 2014; Scolamacchia and Dingwell 2014). These dimensions exceed those (10–100 μm) commonly removed by water bridges or electrostatic attraction (e.g. Sheridan and Wohletz 1983b; Schumacher and Schmincke 1995; Brown et al. 2010, 2012).

Even if the hydrothermal system of the volcano was partially destroyed during the 1982 eruption, a new one was established shortly after as attested by the active hydrothermal areas present inside the crater (Casadevall et al, 1984; see Chap. 4). According to Juetztler et al. (2011), the main heat source for the present lake-spring hydrothermal system would be provided by the remnants of the dome destroyed in 1982, corresponding to high magnetic anomalies observed inside the 1982 crater in correspondence of active hydrothermal areas inside the present crater. Other thermal manifestations would be controlled by the upper portions of the hydrothermal system formed after the last eruption, as indicated by recent geochemical studies (Chap. 4).

The generation of lahars must have been frequent during El Chichón eruptive history, considering the high precipitation rate in the area (4,000 mm/year, Atlas del Agua 1976), and the abundance of unconsolidated volcaniclastic material. Lahar deposits were recognized intercalated to deposits of blocks and ash during the early stages of construction of the Somma edifice (see Sect. 3.2), and a precise record of events exists for the 1982 eruption, which deposits are exposed along the Platanar river (see Sect. 3.2). Numerical simulations, using post-1982 topography, indicated that future lahars (1×10^6 m^3 in volume) could inundate villages along the Platanar and Magdalena rivers, and an extraordinary flooding event such as the 1982 lake break-out (3×10^6 m^3) could potentially reach the town of Ostuacan (12.5 km NW) along the Magdalena river, also affecting the Grijalva river system (Macias et al. 2008).

Pyroclastic fall deposits, from plinian eruptive column(s), were recognized only in 3 of the 12 explosive eruptions during the Holocene, corresponding to the 1,250 year BP (Unit D Tilling et al. 1984), the 550 year BP (unit B) and the 1982. The areal distribution of unit B was greater than the 1982 fall products, and dispersed ashes across an area of 240 km^2, according to the 20 cm isopach (Macías et al. 2003). Taking into account the prevailing wind directions during different seasons, directed to the E during spring, summer and autumn, more dispersed in all direction during winter, and the isopach of the most dispersed fallout layer of the 1982 eruption (A1), an area of 45,000 km^2 to the NE would be covered by a distal fallout 1 mm, if a similar event would occur in the future (Macias et al. 2008). Based on the numerical simulation of the 1982 plinian events and using daily wind records spanning the last 20 years (1991–2010), Bonasia et al. (2012) indicated that for a moderate eruption (e.g. fallout A1, VEI > 4), an area of 4,000 km^2 would have a probability greater than 2 % of being covered by a ash load exceeding 100 kg/m^2 (minimum value considered critical for structural damages). This probability would increase to more than 5 % for a VEI > 5 eruption (e.g. fallouts A1 + B +C), across an area of 12,000 km^2 including the city of Villahermosa, capital of Tabasco. This last scenario, would affect more than 1 million people and the air traffic in the area.

The most likely eruptive scenarios within a given time span was evaluated by Mendoza-Rosas and De la Cruz-Reyna (2010) using a probabilistic analysis considering Holocene and historical eruption time-series based on their magnitudes (VEI 2 to 5). Their study evaluated the probabilities of occurrence of at least one eruption exceeding a given VEI in a determined time interval (20, 50, 100 and 500 years). In these terms eruptions with a VEI > 4, such as the 29 March at 0532 UT and 4 April at 1122 UT during the 1982 eruption (Macías et al. 2008), would have a probability of 10 % in 100 years, and one of 39 % in 500 years, to occur. This second scenario would be confirmed by the 550 year BP plinian eruption. The most probable event with a VEI > of 2 or 3, calculated for a 500 years interval, would have probabilities of 74 and 57 % respectively to occur. A similar scenario was envisaged by Espindola et al. (2000), who used a statistical method to have a rough estimate of the eruption rate based on the ten repose periods of the past 3,700 years. This analysis yields a 22 % probability for an eruption similar to the 1982 within the next 100 years to occur (Espindola et al. 2000).

For hazard assessments, it is necessary to take into account that the main topographic irregularities related to the regional folds and fault system (García-Palomo et al. 2004), or ancient structures such as the Somma crater (Figs. 3.1 and 3.2a), because they can influence the courses, hence distribution, of different types of gravity-driven flows, in a radius between 1.5 and 2 km, and between 3.5 and 5 km from the crater.

The three notches that dissect the Somma crater to the E, N and SW, corresponding to the river source of El Platanar to the E, S. Pablo Cambac to the N and Tuspac to the S, represented main "pathways" for the initial dispersion of most of the pyroclastic density currents. The walls of this ancient crater were able to control the distribution of some PDCs (e.g. unit IU) to the SE where they are 1,100 m a.s.l. high, but not to the NE, where they reach only 900 m a.s.l. The best exposures of Holocene deposits, and the most complete sections of the last eruption, occur in the eastern sector, reflecting the smooth topography determined by the presence of the San Juan fault (García-Palomo et al. 2004), which favored a greater dispersion of most of the gravity driven flows to the E-SE (Sigurdsson et al. 1984; Espíndola et al. 2000; Scolamacchia and Macías 2005). Variations of a few degrees in the topographic gradient at ~2 km from the crater, contributed to destabilize most of the flows, as suggested by the structural features in the exposed PDCs deposits (i.e. units PF1 and S2, Macías et al. 1998; Scolamacchia and Macías 2005), indicating that the currents dropped their coarser load at this distance. Nevertheless, diluted PDCs continued to travel with high momentum for almost 3 km into the 2–2.5 km wide El Platanar plain, where the topography gently declined from 620 to 500 m a.s.l., destroying the village of Volcán Chichonal between 3.6 and 4.7 km E. It is noteworthy that a new settlement was established there a few years ago, almost at the same distance (from the volcano summit) of site buried during the 1982 eruption.

Considering the areal distribution of the deposits of the 1982 eruption, it is reasonable to believe that diluted PDCs, with a low solids concentration, (i.e. pyroclastic surges, 0.1–1 vol%, Wilson and Houghton 2000), represent the most destructive phenomena, because they were poorly controlled by the topography and were able to reach greater distances with respect to other types of flows. The extension, textural and structural characters of the deposits and the eyewitness accounts of the 1982 eruption, indicate that some of these currents reached distances of 8.5 km to the S, 9.5 km NE and 10.5 km E, being able to down trees at distances of $\sim$7 km to the E. All villages at a radial distance of 5 km from the crater were completed buried by the pyroclastic products of the eruption, but apparently these kinds of currents (i.e. base surges produced by magma-water interaction), caused the greater damages. Steel reinforced bars of house foundations embedded in these deposits, were observed bent in the same directions of the currents attesting their high dynamic pressure (Valentine 1998) at distances between 3.5 and 4 km (Scolamacchia and Schouwenaars 2009).

In the village of Esquípula Guayabal 3.5 km SE from the crater (Fig. 3.7), the impacts caused by ash particles hundreds of microns in size were found on a steel basketball pole remaining at its original position. Based on the deformation observed in the steel structure, a range of particle's velocities between 710 and 980 m/s were obtained (Scolamacchia and Schouwenaars 2009). Such velocities, much higher with respect to the speeds considered typical for pyroclastic density currents (tens to few hundreds m/s e.g. Druitt 1998; Wilson and Houghton 2000; Morrisey and Mastin 2000; Branney and Kokelaar 2002) are in the range of those observed for the initial air shocks produced during nuclear explosions ($\sim$1,000 m/s, Wohletz 1998; Valentine 1998), and those predicted theoretically and numerically for shock waves accompanying explosive eruptions (Wohletz et al. 1984; Wohletz and Valentine 1990). The impacts were attributed to an acceleration induced by shock waves, due to an efficient momentum coupling between a gas phase inside the clouds and the particles (max 280 μm in size) such that a sudden expansion of the gas caused by shock wave(s) was able to drag the particles up to high speeds (Scolamacchia and Schouwenaars 2009). Shock waves have been recorded and observed several times during explosive eruptions (e.g. Gorshkov 1959; Nairn 1976; Ishihara 1985), and their occurrence is to be expected in eruptions characterized by an unsteady discharge of material (Wohletz and Valentine 1990). Such conditions would occur during short-lived explosions (e.g. Vulcanian activity), the rapid decompression of a dome or cryptodome (e.g. Kieffer 1981a), the initial stages of caldera forming eruptions (Wohletz et al. 1984; Valentine and Wohletz 1989), or hydromagmatic eruptions, where shock waves generation has been experimentally observed and attributed to the heat energy transfer from magma to water during the fragmentation process (Zimanowski 1998).

The probability of atmospheric shock waves generation due to an unsteady discharge of material at the vent is realistic for the eruptive events occurred in 1982 at El Chichón (both during magmatic and phreatomagmatic phases of the eruption).

Moreover, when the reservoir pressures exceed atmospheric pressure by a factor of 5 (i.e. $P_{res/amb} > 5{:}1$, which are likely in explosive volcanic eruptions, a secondary system of shocks may form within the discharging supersonic eruptive plume, almost immediately after the passage of the flow head ("underexpanded jet" of Kieffer 1981a, b; Kieffer and Sturtevant 1984).

Such complex system of secondary shocks, consisting of crossed oblique shocks, and a perpendicular Mach disk shock, may propagate at distance of several vent diameters, reaching a steady location before collapsing back when the pressure of the reservoir decreases to sonic values. Recent scaled experiments and numerical simulations (Orescanin et al. 2010) indicated that in unsteady eruptive events (e.g. Vulcanian and blast-type), a Mach disk shock could generate at the vent, and propagate downstream until reaching equilibrium distances that, for $P_{res/amb}$ 150:1, can be in the order of 240 m from the vent, for vulcanian events resembling those occurred at Soufrière Hills volcano in 1997, and up to 7 km for a blast-type event such the one occurred at Mount St Helens in 1980. For a blast-type event with ratios $P_{res/amb}$ between 100:1 and 250:1, such distances would vary between 5.7 and 9 km, and the Mach disk would remain at its equilibrium location, for periods between 84 and 104 s before collapsing back toward the vent (Orescanin et al. 2010).

In this light, the distance from the crater at which the basketball pole impacted by ash was found (3.5 km) it is not unrealistic and the establishment of Mach disk shock structure, could have likely occurred also during the 1982 eruption of El Chichón.

In any of these scenarios the occurrence of such phenomena would have important implications for hazard assessments, but is not taken into account (Scolamacchia and Schouwenaars 2009).

3.4 Appraisal of the Eruptive History of El Chichón Volcano

By combining previous studies with unpublished K–Ar and $^{40}Ar/^{39}Ar$ dating on newly mapped stratigraphic units it was now possible to delineate a more complete scenario of the volcanic activity in time and space at El Chichón.

This recent information contributes to fill in the temporal gaps in the reconstruction of the eruptive history of the volcano, indicating that volcanic activity in the area was long-lasting, and occurred from several eruptive centers, controlled by a regional system of conjugated dextral strike-slip N-S, and a sinistral strike slip E-W faults.

Volcanism at El Chichon began around 1.64 My on the trace of a major NNW-SSE fault (the recently mapped Chichón–Catedral fault), building a major edifice, Catedral volcano, which later collapsed to the SE. The volcanism migrated 14 km to the SE around 1 My at the eastern tip of the E-W strike-slip San Juan fault with the emplacement of a trachybasaltic dike. The onset of activity at its present position began apparently already around 372 ka, according to $^{40}Ar/^{39}Ar$ ages obtained on a lava fragment from 1982 pyroclastic deposit, but no field evidence of this activity has been found until now. The construction of the Somma edifice was characterized by frequent extrusions of domes since about 276 ka. A major eruption destroyed the central part of the Somma dome complex apparently during late Pleistocene, forming a 1.5 km wide crater. According to new data, dome-building activity was contemporary to the Somma edification, resulting in the construction of the SW dome around 217 ka, the Cambac dome, on the NW sector, around 187–168 ka, and the Capulin domes, on the NNE, around 152 ka. Both Cambac and Capulin domes were subsequently explosively destroyed with the formation of block and ash flows, and following lahars, which are widely dispersed on the Somma flanks. During this explosive activity, another dome was extruded to the NW flank of the Somma, about 97–80 ka.

Explosive activity, mostly hydromagmatic in character built the Guayabal cone on the SE Somma flanks, on the trace of the major Chichón-Catedral fault, around 10 ka. This poorly preserved structure, collapsed toward the SE, was apparently active also during the Holocene.

The eruptive activity during the Holocene has been almost exclusively explosive in character, occurring from the same vent that was reactivated in 1982. Even with no substantial changes in the composition of the products, which have been always trachyandesitic, periods of dome growth followed by their explosive disruption alternated with eruption under open crater conditions, during which the interaction between magma and external water occurred at least five times. The stratigraphic record of the last 4,000 years, including the 1982 eruption, suggests a recurrence interval from a minimum of 100 years to a maximum of 600 years.

A better definition in time of the different phases of the 1982 eruption, was possible thanks to the analysis of the records from far away stations and eyewitnesses accounts that were not analyzed in previous works. Such interpretation is more consistent with the existing data, suggesting that the most destructive events, with the generation of PDCs, had already occurred by the early morning of 3 April (after 0312, local time), instead of on 4 April at 0139 UT as reported in previous eruption chronologies (i.e. Sigurdsson et al. 1984 and following studies). At this time, in fact, all existing records point toward the generation of the second Plinian event B.

The occurrence of aggregates cemented by a sulfur-rich film unevenly distributed among wet surge deposits of the 1982 eruption, being more abundant in those produced at the beginning of the events occurred on April 3, and after the phreatoplinian eruption on April 4, suggest that liquid sulfur was explosively ejected during several eruptive phases. Therefore it is likely that the accumulation of sulfur layers occurred also at El Chichón, as observed at other volcanoes with active hydrothermal systems. Considering the maximum sizes of individual fragments accreted (>2 mm), and the distribution of cemented aggregates, the binding action of such substance was able to remove coarser particles, in the size of lapilli from eruptive clouds already at distances of few km from the vent.

Such evidences have important implications on the volumetric estimates of different grain sizes inside eruptive clouds, and should be considered in models of particles dispersion and sedimentation from eruptive plumes.

No dome began to grow after the 1982 eruption, and the absence of a large positive magnetic anomaly inside the newly formed crater suggests that it will not likely grow any time soon in the near future. Nevertheless, the precipitation of elemental sulfur, and other alteration minerals, from acid solutions generated by condensing volatiles exsolved from andesitic magma bodies, has been invoked to reduce rock permeability. Such process should be critically considered for its possible role in sealing the active hydrothermal system, determining its overpressurization, and leading to phreatic eruptions. In addition, due to the presence of a shallow groundwater table, also the probability of phreatomagmatic eruptions should be taken into account.

References

Albarrán J (1983) Experiencias de campo y reseña de la actividad del volcán Chichonal en Abril de 1982. In: IdG UNAM (ed) El volcán Chichonal VI Convención Geologica Nacional- Sociedad Geologica Mexicana. UNAM, Mexico DF, Mexico, pp 57–67

Austin-Erickson A, Büttner R, Dellino P, Ort MH, Zimanowski B (2008) Phreatomagmatic explosions of rhyolite magma: experimental and field evidence. J Geophys Res 113: B11201. doi:10.1029/2008JB005731

Atlas del Agua de la República Mexicana (1976) Secretaría de Recursos Hidráulicos, México, 253 pp

Belousov A, Voight B, Belousova M (2007) Directed blasts and blasts-generated pyroclastic density currents: a comparison of the Bezymianni 1956, Mount St Helens 1980, and Soufriére Hills, Monteserrat 1997 eruptions and deposits. Bull Volcanol 69(7):701–740

Bennet FD, Raccicchini SM (1978) Subaqueous sulphur lake in Volcan Poas. Nature 271:342–344

Branney MJ, Kokelaar BP (2002) Pyroclastic density currents and the sedimentation of ignimbrites. Geol Soc Lond Mem 27:143

Brinkley SRJ, Kirkwood JG, Lapson CW, Revelle R, Smith AL (1950) Shock from underwater and underground blasts. In: Los Alamos Scientific Laboratory: the effects of atomic weapons. U. S. Printing Office, Los Alamos, New Mexico, pp 83–113

Brown RJ, Branney MJ, Maher C, Dávila-Harris P (2010) Origin of accretionary lapilli within ground-hugging density currents: evidence from pyroclastic couplets on Tenerife. Geol Soc Am Bull 122(1/2):305–320

Brown RJ, Bonadonna C, Durant AJ (2012) A review of volcanic ash aggregation. Phys Chem Earth, Part A/B/C 45–46:65–78

Bonasia R, Costa A, Folch A, Macedonio G, Capra L (2012) Numerical simulation of tephra transport and deposition of the 1982 El Chichón eruption and implications for hazard assessment. J Volcanol Geotherm Res 231–232:39–49

Büttner R, Zimanowski B (1998) Physics of thermo-hydraulic explosions. Phys Rev E 57:5726–5729

Büttner R, Dellino P, Zimanowski B (1999) Identifying magma-water interaction from the surface features of ash particles. Nature 401:688–690

Bursik M (2001) Effect of wind on the rise height of volcanic plumes. Geophys Res Lett 18:3621–3624

Canul RF, Rocha VL (1981) Informe geológico de la zona geotérmica de "El Chichonal". Chiapas Comisión Federal de Electricidad, Morelia, México. Informe 32–81:38

Canul RF, Razo AM, Rocha VL (1983) Geología e historia volcanológica del volcán Chichónal, Estado de Chiapas. Revista del Instituto de Geología, U.N.A.M., México, pp 3–22

Carey S, Sigurdsson H (1986) The 1982 eruption of El Chichón volcano, Mexico (2): observation and numerical modelling of tephra-fall distribution. Bull Volcanol 48:127–141

Carey S, Sigurdsson H (1989) The intensity of Plinian eruptions. Bull Volcanol 51:28–40

Cas RAF, Wright JV (1987) Volcanic successions: modern and ancient. Chapman & Hall, London, 528 pp

Casadevall TJ, De la Cruz-Reyna S, Rose WI, Bagley S, Finnegan DL, Zoller WH (1984) Crater lake and post-eruption hydrothermal activity, El Chichón Volcano, Mexico. J Volcanol Geotherm Res 23:169–191

Christenson BW (2000) Geochemistry of fluids associated with the 1995–1996 eruption of Mt. Ruapehu, New Zealand: signatures and processes in the magmatic-hydrothermal system. J Volcanol Geotherm Res 2000:1–30

Christenson BW, Wood CP (1993) Evolution of a vent-hosted hydrothermal system beneath Ruapehu Crater Lake, New Zealand. Bull Volcanol 55:547–565

Christenson BW, Reyes AG, Young R, Moebis A, Sherburn S, Cole-Baker J, Britten K (2010) Cyclic processes and factors leading to phreatic eruption events: insights from the 25 September 2007 eruption through Ruapehu Crater Lake, New Zealand. J Volcanol Geotherm Res 191:15–32

Colgate SA, Sigurgeisson T (1973) Dynamic mixing of water and lava. Nature 244:552–555

Cole PD, Guest JE, Duncan AM, Pacheco JM (2001) Capelinhos 1957–1958, Faial Azores: deposits formed by an emergent surtseyan eruption. Bull Volcanol 63:204–220

Costa A, Macedonio G, Folch A (2006) A three-dimensional Eulerian model for transport and deposition of volcanic ashes. Earth Plan Sci Lett 241:634–647

Damon P, Montesinos E (1978) Late Cenozoic volcanism and metallogenesis over an active Benioff zone in Chiapas, Mexico. Ariz Geol Soc Dig XI:155–168

De la Cruz Reyna S, Martin Del Pozzo AL (2009) The 1982 eruption of El Chichón volcano, Mexico: eyewitness of the disaster. Geofis Int 48:21–31
Druitt TH (1998) Pyroclastic density currents. In: Gilbert JS, Sparks RSJ (eds) The physics of explosive volcanic eruptions. Geological Society of London, Special Publication, London, pp 145–182
Duffield W, Tilling R, Canul RF (1984) Geology of El Chichón volcano, Chiapas, Mexico. J Volcanol Geotherm Res 20:117–132
Espíndola JM, Macías JL, Tilling R, Sheridan MF (2000) Volcanic history of El Chichón volcano (Chiapas, Mexico) during the Holocene and its impact on human activity. Bull Volcanol 62:90–104
Fisher RV (1990) Transport and deposition of pyroclastic surge across an area of high relief: the 1980 eruption of Mount St. Helens, Washington. Geol Soc Am Bull 102:1038–1054
Fisher RV, Waters A (1970) Base surge bed forms in maar volcanoes. Am J Sci 268:157–180
Fisher RV, Schmincke, HU (1984) Pyroclastic rocks. Springer, Berlin, 472 pp
Folch A, Costa A, Macedonio G (2009) FALL3D: a computational model for transport and deposition of volcanic ash. Comput Geosci 35(6):1334–1342
Francis PW, Thorpe RS, Brown GC, Glasscock J (1980) Pyroclastic sulphur eruption at Poás volcano, Costa Rica. Nature 283:754–756
Freundt A, Bursik M (1998) Pyroclastic flow transport mechanisms. In: Freundt A, Rosi M (eds) From magma to tephra. Elsevier, Amsterdam, pp 173–231
García-Palomo A, Macías JL, Espíndola JM (2004) Strike-slip faults and K-alkaline volcanism at El Chichón volcano, southeastern Mexico. J Volcanol Geotherm Res 136:247–268
Gorshkov GS (1959) Gigantic eruption of the Volcano Bezymianny. Bull Volcanol 20:77–109
Gorshkov GS (1963) Directed volcanic blasts. Bull Volcanol 26:83–88
Heiken GH (1971) Tuff rings: examples from the Fort Rock - Christmas Lake Valley basin, south-central Oregon. J Geophys Phys Res 76:5615–5626
Haskov J, De la Cruz Reyna S, Sing SK, Medina F, Gutiérrez C (1983) Seismic activity related to the March–April 1982 eruption of El Chichón volcano, Chiapas, Mexico. Geophys Res Lett 10(4):239–296
Heiken GH, Wholetz K (1985) Volcanic ash. University of California Press, Berkeley, p 246
Hoblitt R, Miller CD, Wallance JW (1981) Origin and stratigraphy of the deposit produced by the May 18 directed blast. In: Lipmann PW, Mullineaux DR (eds) The 1980 eruption of Mount St. Helens, Washington. Washington D. C., pp 401–419
Hoblitt RP, Wolfe EW, Scott WE, Couchman MR, Pallister JS, Javier D (1996) The Preclimactic eruptions of Mount Pinatubo, June 1991. In: Newhall CG, Punongbayan, RS (eds) Fire and mud. University of Washington Press, Washington D.C., pp 457–511
Ishihara K (1985) Dynamic analysis of volcanic explosions. J Geodyn 3:327–349
Jimenez Z, Espíndola VH, Espíndola JM (1999) Evolution of the seismic activity from the 1982 eruption of El Chichón Volcano, Chiapas, Mexico. Bull Volcanol 61:411–422
Jutzeler M, Varley N, Roach M (2011) Geophysical characterization of hydrothermal system and intrusive bodies, El Chichón volcano (Mexico). J Geophys Res 116:B04104
Kieffer SW (1981a) Blast dynamics at Mount St. Helens on 18 May 1980. Nature 291:568–570
Kieffer S (1981b) Fluid dynamics of the May 18 blast at Mount St. Helens. In: Lipman PW, Mullineaux, DR (eds) The 1980 eruptions of Mount St. Helens, Washington. U.S. Geological Survey, pp 379–400
Kieffer S, Sturtevant B (1984) Laboratory studies of volcanic jets. J Geophys Res 89:8253–8268
Kokelaar BP (1986) Magma-water interaction in subaqueous and emergent basaltic volcanism. Bull Volcanol 48:275–289
Krueger A (1983) Sighting of El Chichón sulfur dioxide clouds with the Nimbus 7 Total Ozone Mapping Specrometer. Science 220(4604):1377–1379
Krueger AJ, Walter LS, Bhartia PK, Schnetzler CC, Krotkov NA, Sprod I, Bluth GJS (1995) Volcanic sulfur dioxide measurements from the total ozone mass spectrometer instruments. J Geophys Res 100:14057–14076
Krueger A, Krotov N, Carn S (2008) El Chichon: the genesis of volcanic sulfur dioxide monitoring from space. J Volcanol Geotherm Res 175:408–414
Layer PW, García-Palomo A, Jones D, Macías JL, Arce JL, Mora JC (2009) El Chichón volcanic complex, Chiapas, México: stages of evolution based on field mapping and $^{40}Ar/^{39}Ar$ geochronology. Geofís Int 48:33–54
Lorenz V (1974) Vesiculated tuffs and associated features. Sedimentology 21:273–291
Luhr JF (2008) Primary igneous anhydrite: progress since its recognition in the 1982 El Chichón trachyandesite. J Volcanol Geotherm Res 175:394–407
Luhr JF, Logan MAV (2002) Sulfur isotope systematics of the 1982 El Chichón trachyandesite: a ion micropobe study. Geochim Cosmochim Acta 66(18):3303–3316
Luhr JF, Carmichael ISE, Varekamp JC (1984) The 1982 eruption of El Chichón volcano, Chiapas, Mexico: mineralogy and petrology of the anhydrite bearing pumices. J Volcanol Geotherm Res 23:69–108
Macías JL (1994) Violent short-lived eruptions from small-size volcanoes: El Chichón, Mexico (1982) and Shtyubel, Russia (1907). Ph.D. dissertation, State University of New York, Buffalo
Macías JL, Sheridan MF, Espíndola JM (1997) Reappraisal of the 1982 eruption of El Chichón volcano, Chiapas, Mexico: new data from proximal deposits. Bull Volcanol 58:459–471
Macías JL, Espíndola JM, Bursik M, Sheridan MF (1998) Development of lithic-breccias in the 1982 pyroclastic flow deposits of El Chichón Volcano, Mexico. J Volcanol Geothermal Res 83:173–196
Macías JL, Arce JL, Mora JC, Espíndola JM, Saucedo R, Manetti P (2003) A 500-year-old Plinian eruption at El Chichón volcano, Chiapas, Mexico: explosive volcanism linked to reheating of the magma reservoir. J Geophys Res 108(B12):2569 doi:10.1029/2003JB002551
Macías JL, Capra L, Scott KM, Espíndola JM, García-Palomo A, Costa JE (2004) The 26 May 1982 breakout flows derived from a failure of a volcanic dam at El Chichón, Chiapas, Mexico. Geol Soc Am Bull 116(1–2):233–246
Macías JL, Capra L, Arce JL, Espíndola JM, García-Palomo A, Sheridan MF (2008) Hazard map of El Chichón volcano,

Mexico: Constraints posed by eruptive history and computer simulations. J Volcanol Geotherm Res 175(4):444–458

Macías JL, Arce JL, Garduño-Monroy VH, Rouwet D, Taran Y, Layer P, Jiménez A, Álvarez R (2010): Estudio de prospección geotérmica para evaluar el potencial del volcán Chichonal, Chiapas. Unpublished Report no. 9400047770 IGF-UNAM-CFE

Matson M (1984) The 1982 El Chichón volcano eruption-a satellite perspective. J Volcanol Geotherm Res 23:1–10

Mauk FJ (1983) Utilization of seismically recorded infrasonic acoustic signals to monitor volcanic explosions: the El Chichón sequence 1982- a case study. J Geophys Res 88:10385–310401

Medina-Martínez F (1982) El Volcán Chichón. GEOS 2(4):19

Mendoza-Rosas AT, De la Cruz-Reyna S (2010) Hazard estimates for El Chichón volcano, Chiapas, Mexico: a statistical approach for complex eruptive histories. Nat Haz Earth Syst Sci 10:1159–1170

Moore JC, Nakamura K, Alcaraz A (1966) The 1965 eruption of Taal volcano. Science 151:955–960

Moore JC (1967) Base surges in recent volcanic eruptions. Bull Volcanol 30:337–363

Moore JG, Sisson TW (1981) Deposits and effects of the May 18 pyroclastic surge. In: Lipmann PW, Mullineaux DR (eds) The 1980 eruptions of Mount St. Helens, Washington. Washington D.C., pp 421–438

Mooser F, Meyer-Abich, H, Mc Birney AR (1958) Catalogue of the active volcanoes of the world including Solfatara fields. Part VI, Central America. International Volcanological Association, Napoli, pp 26–30

Morrisey MM, Mastin L (2000) Vulcanian eruptions In: Sigurdsson H, Houghton B, Rymer H, Stix J, Mc Nutt S (eds) Encyclopedia of volcanoes. Academic Press, New York, pp 463–475

Mülliered FKG (1932) Der Chichón, ein bischer unbegannter tätiger Vulkan im nordlichen Chiapas, Mexiko. Zeit Vulkanol XIV:191–209

Mülliered F (1933) "El Volcán", único volcán en actividad descubierto en el estado de Chiapas. Mem Rev Acad Ciencias "Antonio Alzate" 54(11/12):411–416

Nairn IA (1976) Atmospheric shock waves and condensation clouds from Ngauruhoe explosive eruptions. Nature 259:190–192

Nairn IA, Wood CP, Hewson CAY (1979) Phreatic eruptions of Ruapehu: April 1975. NZ J Volcanol Geophys 22(2):155–173

Oppenheimer C (1992) Sulphur eruptions at Volcán Poás, Costa Rica. J Volcanol Geotherm Res 49:1–21

Oppenheimer C, Stevenson D (1989) Liquid sulphur lakes at Poás volcano. Nature 342:790–793

Orescanin MM, Austin JM, Kieffer SW (2010) Unsteady high-pressure flow experiments with application to explosive volcanic eruptions. J Geophys Res 115:B06206

Parker DE, Brownscombe JL (1983) Stratospheric warming following the El Chichón volcanic eruption. Nature 301:406–408

Quintas I (2000) Eric II. Documento de la base de datos climatológica y del programa extractor. Instituto Mexicano de Tecnología del Agua. Morelos, Mexico, 75 pp

Reimer PJ, Baillie MGL, Bard E, Bayliss A, Beck JW, Bertrand CJH, Blackwell PG, Buck CE, Burr GS, Cutler KB, Damon PE, Edwards RL, Fairbanks RG, Friedrich M, Guilderson TP, Hogg AG, Hughen KA, Kromer B, McCormac FG, Manning SW, Ramsey CB, Reimer RW, Remmele S, Southon JR, Stuiver M, Talamo S, Taylor FW, van der Plicht J, Weyhenmeyer CE (2004) IntCal04 Terrestrial radiocarbon age calibration, 26–0 ka BP. Radiocarbon 46:1029–1058

Richards AF (1959) Geology of the Islas Revillagigedo, Mexico. 1. Birth and development of Volcán Bárcena, Isla San Benedicto. Bull Volcanol 22:73–123

Rye RO, Luhr JF, Wasserman MD (1984) Sulfur and oxygen isotopic systematics of the 1982 eruptions of El Chichón volcano, Chiapas, Mexico. J Volcanol Geotherm Res 23:109–123

Rose WI, Bornhorst TJ, Halsor SP, Capaul WA, Plumey PS, De la Cruz Reyna S, Mena M, Mota R (1984) Volcán El Chichón, Mexico: pre S-rich eruptive activity. J Volcanol Geotherm Res 23:147–167

Scolamacchia T, Macías JL (2005) Distribution and stratigraphy of deposits produced by diluted pyroclastic density currents of the 1982 eruption of El Chichón volcano, Chiapas, Mexico. Rev Mex Cienc Geol 22:159–180

Scolamacchia T, Macías JL, Sheridan MF, Hughes SR (2005) Morphology of ash aggregates from wet pyroclastic surges of the 1982 eruption of El Chichón volcano, Mexico. Bull Volcanol 68:171–200

Scolamacchia T, Schouwenaars R (2009) High-speed impacts by ash particles in the 1982 eruption of El Chichón (Mexico). J Geophys Res 114:B12206

Scolamacchia T. (2012) "El Chichón 1982: reinterpretación de la cronología de los eventos eruptivos". Unión Geofísica Mexicana (UGM) General Meeting, Puerto Vallarta, México, October 28–November 2, 2012. Puerto Vallarta, Jalisco, Mexico

Scolamacchia T (2014) Another look to the mechanisms of formation of ash aggregates in pyroclastic deposits. In: Rocha R, Pais, J, Kullberg, JC, Finney S (eds) At the cutting edge of Stratigraphy. STRATI First International Congress on Stratigraphy, Lisboa, Portugal, 1–7 July 2013. Springer Geology, vol XLV. 1343 pp

Scolamacchia T, Dingwell DB (2014) Sulfur as a binding agent of aggregates in explosive eruptions. Bull Volcanol 76 (10):871. doi:10.1007/s00445-014-0871-1

Schumacher R, Schmincke HU (1995) Models for the origin of accretionary lapilli. Bull Volcanol 56:626–639

SEAN Smithsonian Institution (1989) Global volcanism 1975–1985. Prentice Hall, Englewood Cliffs, p 657

Sheridan MF, Wohletz K (1981) Hydrovolcanic explosions: the systematics of water-pyroclast equilibration. Science 212:1387–1389

Sheridan MF, Wohletz KH (1983a) Hydrovolcanism: basic considerations and review. J Volcanol Geotherm Res 17:1–29

Sheridan MF, Wohletz, KH. (1983b) Origin of accretionary lapilli from the Pompeii and Avellino deposits of Vesuvius. In: Gooley R (ed) Microbeam analysis. San Francisco Press, Phoenix, p 336

Sigurdsson H, Carey SN, Espíndola JM (1984) The 1982 eruption of El Chichón volcano, Mexico: stratigraphy of pyroclastic deposits. J Volcanol Geotherm Res 23:11–37

Sigurdsson H, Carey SN, Fisher RV (1987) The 1982 eruptions of El Chichón volcano, Mexico (3): physical properties of pyroclastic surges. Bull Volcanol 49(2):467–488

Sohn YK, Chough SK (1992) The Ilchulbong tuff cone, Cheju Island, South Korea: depositional processes and evolution of an emergent, Surtseyan-type tuff cone. Sedimentology 39:523–544

Sohn YK, Cronin SJ, Brenna M, Smith IEM, Németh K, White JDL, Murtagh RM, Jeon YM, Kwon CW (2012) Ilchulbong tuff cone, Jeju Island, Korea, revisited: a compound monogenetic volcano involving multiple magma pulses, shifting vents, and discrete eruptive phases. Geol Soc Am Bull 124 (3/4):259–274. doi:10.1130/B30447.1

Stuiver M, Reimer PJ (1986) A computer program for radiocarbon age calibration. Radiocarbon 28:1022–1030

Sulpizio R, Zanella E, Macías JL (2008) Deposition temperature of some PDC deposits from the 1982 reuption of El Chichón volcano (Chiapas, Mexico) inferred from rock magnetic data. J Volcanol Geotherm Res 175(4):494–500

Tahira M, Nomura M, Sawada Y, Kosuke K (1996) Infrasonic and acoustic-gravity waves generated by the Mount Pinatubo eruption of June, 15, 1991. In: Newhall CG, Punongbayan RS (eds) Fire and mud. University of Washington Press, Washington D.C., pp 601–612

Taylor GA (1958) The 1958 eruption of Mount Lamington, Papua. Austr Bur Min Resour Geol Geophys Bull 38:1–117

Templos LA, Munguia Bracamontes F, Barrera VM (1981) Observaciones geoquímicas en la zona geotérmica del Chichonál, Chiapas, Mexico. Unpublished report. Comisión Federal de Electricidad, Mexico. Informe 33–81:32

Tilling RI (2009) El Chichón "surprise" eruption in 1982: lessons for reducing volcano risk. Geofís Int 48(1):3–19

Tilling R, Rubin M, Sigurdsson H, Carey S, Duffield W, Rose WI (1984) Holocene eruptive activity of El Chichón volcano, Chiapas, Mexico. Science 224:747–749

Valentine GA (1998) Damage to structures by pyroclastic flows and surges, inferred from nuclear weapons effects. J Volcanol Geotherm Res 87:117–140

Valentine G, Wohletz KH (1989) Numerical models of Plinian eruption columns and pyroclastic flows. J Geophys Res 94:1867–1887

Valentine GA, Fisher RV (2000) Pyroclastic surges and blasts In: Sigurdsson H, Houghton B, McNutt SR, Hazel R, Stix J (eds) Encyclopedia of volcanoes. Academic Press, New York, pp 571–580

Varekamp JC, Luhr JF, Prestegaard KL (1984) The 1982 eruption of El Chichón volcano (Chiapas, Mexico): character of the eruptions, ash-fall deposits and gas phase. J Volcanol Geotherm Res 23:39–68

Walker GPL (1971) Grain-size characteristics of pyroclastic deposits. J Geol 79:696–714

Walker GL (1984) Characteristics of dune-bedded pyroclastic surge bedset. J Volcanol Geoth Res 20:281–296

Waters AC, Fisher RV (1971) Base surges and their deposits: Capelinhos and Taal volcanoes. J Geophys Res 76:5596–5614

White JDL (1991) Maar-diatreme phreatomagmatism at Hopi Buttes, Navajo Nation (Arizona), USA. Bull Volcanol 53:239–258

Widmer R, Zürn W (1992) Bichromatic excitation of long-period Rayleigh and air waves by the Mount Pinatubo and El Chichón volcanic eruptions. Geophys Res Lett 19(8):765–768

Wilson CJN, Houghton BF (2000) Pyroclastic transport and deposition In: Sigurdsson H, Houghton B, McNutt SR, Hazel R, Stix J (eds) Encyclopedia of Volcanoes. Academic Press, New York, pp 545–55

Wohletz KH (1983) Mechanisms of hydrovolcanic pyroclast formation: grain-size, scanning electron microscopy, and experimental studies. J Volcanol Geotherm Res 17:31–64

Wohletz KH, Sheridan MF (1979) A model for pyroclastic surge. Geol Soc Am Spec Pap 180:305–318

Wohletz KH, Sheridan MF (1983) Hydrovolcanic explosions II. Evolution of basaltic tuff rings and tuff cones. Am J Sci 283:385–413

Wohletz KH, Valentine GA (1990) Computer simulations of explosive volcanic eruptions. In: Ryan MP (ed) Magma transport and storage. Wiley, New York, pp 113–135

Wohletz KH, McGetchin TR, Sanford MT II, Jones EM (1984) Hydrodynamics aspects of caldera-forming eruptions: numerical models. J Geophys Res 89:8269–8285

Young GA (1965) The physics of base surge. In: U.S. Naval Ordenance Lab NOLTR 64-103, AD-618733, Wake Oak, Maryland, p 294

Zimanowski B (1998) Phreatomagmatic explosions. In: Freundt A, Rosi M (eds) From magma to tephra: modelling physical processes of explosive volcanic eruptions Elsevier Science Publications, Amsterdam, pp 25–54

Zimanowski B, Wohletz KH (2000) Physics of phreatomagmatism-I. Terra Nostra 6:515–523

Zimanowski B, Frölich G, Lorenz V (1991) Quantitative experiments on phreatomagmatic explosions. J Volcanol Geotherm Res 48:341–358

Zimanowski B, Büttner, R, Lorenz V (1997a) Premixing of magma and water in MFCI experiments. Bull Volcanol 58:491–495

Zimanowski B, Büttner R, Lorenz V, Häfele H G (1997b) Fragmentation of basaltic melt in the course of explosive volcanism. J Geophys Res 102(B1):803–814

Zürn W, Widmer R (1996) Worldwide observation of bichromatic long-period Rayleigh waves excited during the June, 15, 1991, eruption of Mount Pinatubo. In: Newhall CG, Punongbayan RS (eds) Fire and mud. University of Washington, Press Seattle, pp 615–624

Fluid Geochemistry of El Chichón Volcano-Hydrothermal System

4

Loïc Peiffer, Dmitri Rouwet, and Yuri Taran

Abstract

El Chichón volcano hosts an intense hydrothermal system with surface manifestations consisting of an acid lake, steam vents, steam-heated boiling pools, mud pools and boiling springs in the crater, as well as several hot springs located on the outer slopes. This chapter reviews previous studies of the El Chichón volcano-hydrothermal system and proposes a conceptual model of the aquifer structure based on more than 15 years of fluid geochemical monitoring (major and rare-earth elements, $\delta^{18}O$-δD, $^{87}Sr/^{86}Sr$). This model contains two aquifers: (1) Aquifer 1, located beneath the crater in the volcanic deposits, produces a total thermal water discharge of 220 L/s and feeds the flank 'Agua Caliente-Agua Tibia' spring group; (2) Aquifer 2, much deeper and with a lower total discharge of 7 L/s, is located in the evaporite-limestone basement and feeds the flank 'Agua Salada-Agua Salada new' spring group. The deep waters from Aquifer 2 have a much higher salinity than Aquifer 1 waters (25,000 vs. 2,200 mg/L Cl) and can be associated with oil-field brines. The crater lake chemistry and dynamics are mainly controlled by the steam condensation from Aquifer 1 waters and by the activity of the Soap Pool springs. Their chemical and isotopic composition can be associated with the volcanic Aquifer 1 water by a model of a single step liquid-vapor separation. Finally, El Chichón volcano is located in a non-classic volcanic arc and rather peculiar local and regional tectonic setting, as supported by CO_2 flux surveys and He and C isotope systematics of emitted gases.

4.1 Introduction

Hydrothermal activity at El Chichón, now 33 years after the eruption, is still very intense but differs from a common post-eruptive stage at other volcanoes. In the volcano crater there are no high-temperature fumaroles; instead, all thermal manifestations are typical of the upper part of a boiling, meteoric-hydrothermal system with a steam cap, mainly represented by steam vents, steam-heated boiling pools, mud pools and small boiling springs discharging the steam condensate. The most representative feature of such hydrothermal

L. Peiffer (✉) · Y. Taran
Instituto de Geofísica, Universidad Nacional Autónoma de México, Ciudad Universitaria, México D.F, 04510, Mexico
e-mail: loic.peiffer@gmail.com

L. Peiffer
Instituto de Energías Renovables, UNAM, Privada Xochicalco S/N, Centro, Temixco, Morelos, 62580, Mexico

D. Rouwet
Istituto Nazionale Di Geofisica E Vulcanologia, Sezione Di Bologna, Via Donato Creti 12, Bologna, 40128, Italy

T. Scolamacchia and J.L. Macías (eds.), *Active Volcanoes of Chiapas (Mexico): El Chichón and Tacaná*,
Active Volcanoes of the World, DOI 10.1007/978-3-642-25890-9_4, © Springer-Verlag Berlin Heidelberg 2015

activity is a large, shallow, acidic crater-lake. This lake is a giant steam-heated pool where acidity (pH $\sim$ 2.5) is provided by the surface oxidation of H_2S (Taran and Rouwet 2008). But not only steam vents and steam-derived manifestations exist in the crater: the most puzzling feature of the El Chichón crater hydrothermal activity is the presence of boiling neutral saline springs with an important but variable outflow rate. This is a very rare case where, after the explosive ejection of a huge amount of juvenile material, neutral Na-Cl water starts to discharge in the newly formed eruptive crater. A similar situation has been reported for the Pinatubo post-eruption hydrothermal activity after the 1991 catastrophic eruption (Stimac et al. 2004). Both volcanoes are similar in rock composition (i.e. derived from sulphur-rich magmas), eruptive style, and host large active hydrothermal systems characterized by an acid crater lake and numerous acid to near-neutral hot and warm springs on their slopes. In both cases, the eruptions did not destroy the long-living hydrothermal systems associated with the previous magmatic activity. El Chichón volcano also represents a special case of a volcano situated within a large regional oil-gas field. Its magmatic plumbing system cuts thick sedimentary strata of evaporites and oil-gas-bearing limestone, and this structure inevitably affects the chemistry of the hydrothermal system.

This chapter reviews a number of studies of El Chichón volcano-hydrothermal system since the first published work by Casadevall et al. (1984), who described the initial stage of the volcano crater after the 1982 eruption, when the newly formed crater lake was ultra-acidic and the fumarolic vents had temperatures above the local boiling point. The geochemical features of the volcano-hydrothermal system, including crater and flank thermal vents are discussed throughout this chapter: water and gas chemical and isotopic compositions (δD, δ^{18}O, δ^{13}C, $^{87}Sr/^{86}Sr$, $^{3}He/^{4}He$), solute and gas fluxes. Finally, a conceptual model of the system and its relationship with local and regional tectonics is proposed.

4.2 Hydrothermal Manifestations

4.2.1 Crater Lake and Associated Thermal Manifestations

Before the March–April 1982 eruption, intense fumarolic activity with altered ground and hot springs occurred in the area between the inner dome and the Somma Crater (Müllerried 1933; Molina-Berbeyer 1974; Canul and Rocha 1981; Templos et al. 1981; see Chap. 3). Fumarole temperatures from 93 to 98 °C and hot spring temperatures from 20 to 71 °C were registered (Molina-Berbeyer 1974).

The 28 March and 3–4 April 1982 eruptions destroyed the dome and created a 150 m-deep and 1 km-wide crater hosting four individual lakes, numerous fumaroles and steaming ground. The initially separated lakes coalesced into one lake by November 1982 (Espíndola et al. 2000). Based on a visual estimation from the east crater rim, the lake surface area was estimated to be 1.4×10^5 m^2 and the corresponding volume 5×10^6 m^3, assuming an average depth of 40 m (Casadevall et al. 1984). The large water volume of the lake was due to an input of 4–5 m of meteoric waters between April and November 1982, the rainy season in the El Chichón area. This lake has never since disappeared, despite the fact that it suffered drastic changes in volume (Rouwet et al. 2004; Taran and Rouwet 2008; Rouwet 2011).

In January 1983, the lake had an ultra-acid pH of 0.56, a temperature from 52 to 58 °C and a high total salinity (TDS = 34,000 mg/L). The fumaroles, with maximum measured temperature of 115 °C, contained more than 99 vol% of H_2O and minor amounts of CO_2, CH_4, H_2, H_2S, and SO_2 (Casadevall et al. 1984). The acidic pH of the lake was generated by the absorption of magmatic gases (HCl, HF, SO_2) into the lake filled with meteoric waters. The lake chemical composition was characterized by an unusual low SO_4/Cl ratio ($\sim$24,000 mg/kg of Cl and only $\sim$3,500 mg/kg of SO_4) compared with other ultra-acid crater lakes (Varekamp et al. 2000), and contrasting with the high sulfur content in solid products of the eruption ($\sim$2 wt% of magmatic anhydrite). By October 1983, the pH of the lake water had increased to 1.8 while temperature had dropped to 42 °C, probably due to dilution and cooling by rainwater. Maximum fumarole temperatures at that time had decreased to 99 °C.

After three years of existence, the lake's pH and temperature stabilized around 2.5 and 30 °C, respectively, indicating a decrease in dissolution of magmatic gases into the lake and therefore a transition from magmatic to hydrothermal stage (Armienta and De la Cruz-Reyna 1995; Taran et al. 1998). The lake composition during the following years was studied by

Armienta and De la Cruz-Reyna (1995), Taran et al. (1998), Tassi et al. (2003), Rouwet et al. (2004, 2008, 2009), Taran and Rouwet (2008) and Peiffer et al. (2011). Despite stable pH and temperature, the salinity of the lake as well as its volume varied considerably with time.

A bathymetric survey of the lake conducted in 1998 showed a relatively flat bottom morphology with depths ranging from 1.3 to 3.3 m and a few funnel-shaped depressions of 6 m deep (Taran and Varley 1999). These results are in contrast with the 1982 lake depth estimation of 40 m by Casadevall et al. (1984). The drastic variation in depth could be explained by the burying of a large proportion of the lake by rockfalls and landslides from the crater walls. The high annual rainfall (~4,000 mm) could have caused strong erosion and the filling of the crater (Inbar et al. 2001). However, Jutzeler et al. (2011) suggest that the level of the crater floor did not change since the eruption.

In 1995, Taran et al. (1998) discovered a group of geyser-like boiling springs located some 20–100 m from the lake in the north part of the crater (Figs. 4.1a and 4.2b). These springs, called 'Soap Pools' (SP) for their soap-like bubble and foam appearance at that time, have a peculiar behavior, which consists in periods of pure vapor exhalation (fumarole emission) alternating with periods of water discharges. The hot and neutral discharged waters feed the lake with a variable flow rate up to 30 L/s. The changes in salinity and volume of the lake were attributed not only to precipitation variations, but mainly to the activity of the Soap Pool springs (Taran and Rouwet 2008). Besides the SP springs, Taran et al. (1998), and Tassi et al. (2003) also described several low salinity boiling springs and drainless pools of acid-sulfate composition.

Inside the crater, the main northern fumarolic field (FN) is located to the northeast of the lake, (Fig. 4.1b) and becomes partially or completely flooded when the lake surface widens (Taran and Rouwet 2008). Other small fumarolic fields are located to the south and northwest of the lake (FS and FW). Bubbling gases escaping from sub-lacustrine fumaroles are visible at the lake surface in E–W and NW–SE alignments (Mazot and Taran 2009; Mazot et al. 2011). Bubbling gases are also observed in several hot springs and mud pools (MP) (Taran et al. 1998; Mazot et al. 2011).

4.2.2 Hot Springs on the Volcano Slopes

Müllerried (1933) briefly mentioned four groups of thermal springs located on the south and northeast

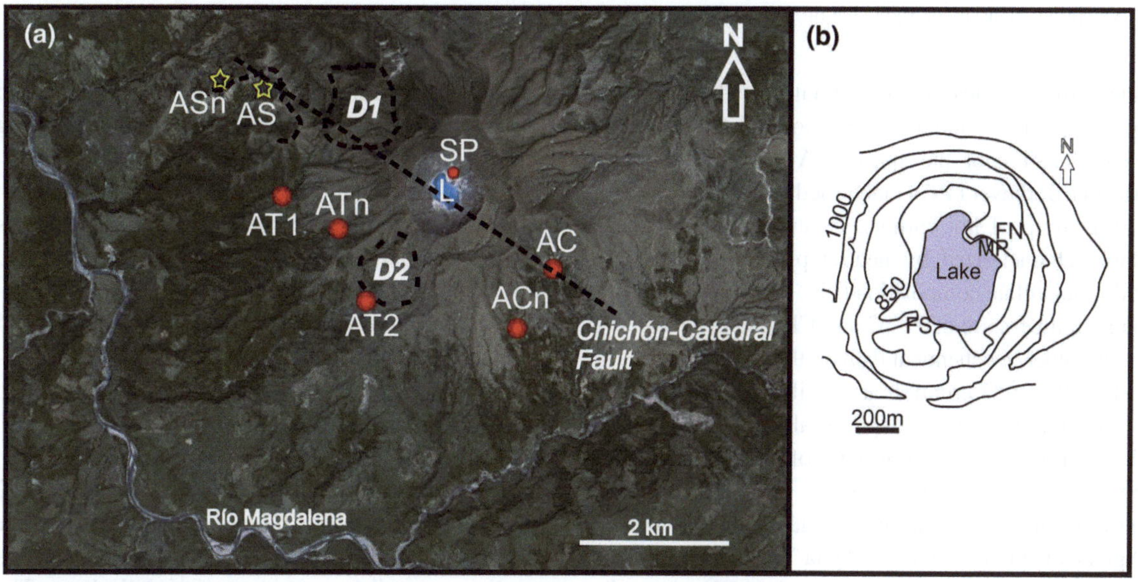

Fig. 4.1 **a** Map of thermal waters manifestations at El Chichón volcano (Google Earth image). AC, ACn, AT1, AT2 AT1n, AS, ASn and SP are thermal springs. L lake. D1 and D2, the NW and SW domes. The curved dotted line west of D1 corresponds to the collapse structure of the NW dome. The straight dotted line is the Chichón-Catedral Fault. **b** Main fumarolic fields (FS and FN) and mud-pool (MP) location within the 1982 El Chichón crater

Fig. 4.2 Some pictures of the El Chichón thermal manifestations. **a** Crater lake (January 2011). **b** Soap pool springs. **c** ACn spring. **d** AS spring. Note on the right side of the picture the funnel collecting the AS bubbling gas

flanks of the volcano, but without providing any data on temperature and chemical composition, nor the exact location of the springs. A few decades later, Molina-Berbeyer (1974) identified the 'Agua Caliente' (AC) group of hot springs, located on the southeast flank, characterized by neutral pH and bicarbonate-chloride-sulfate composition (Fig. 4.1a). These springs had a temperature of 71 °C and Cl content of 456 mg/L. No other geochemical data on the El Chichón fluids before the 1982 eruption are available.

In addition to the AC springs already known before the eruption, six other groups of springs have now been identified and characterized. They are all located on the south-east to north-west flanks of the volcano: Agua Caliente New (ACn), Agua Tibia 2 (AT2), Agua Tibia 1 new (AT1n), Agua Tibia 1 (AT1), Agua Salada (AS) and Agua Salada new (ASn) (Fig. 4.1a, Taran et al. 2008; Taran and Peiffer 2009; Peiffer et al. 2011). These springs are generally located in deep canyons with luxuriant vegetation, greatly impeding access.

The AC-ACn-AT1-AT2-AT1n-AT1 (AC-AT in abbreviation) springs discharge neutral hot waters (36–78 °C) at an altitude between 600 and 650 m asl, at the contact between Middle Miocene claystones-siltstones-sandstones of the sedimentary basement and pyroclastic deposits of the volcanic edifice. The AS-ASn springs are located at a lower altitude of ~550 m and discharge neutral to acid hot waters (53–79 °C). The AC-AT springs are closest to the crater (<1 km), and AS-ASn are the furthest (~2.2 km). The AT2 springs are located at the base of the SW dome, while AS and ASn springs emerge inside a canyon corresponding to a collapse structure of the NW dome. The AC springs are found inside a horseshoe-shaped canyon related to the Guayabal tuff cone (Macías 1994).

The 'Chichón-Catedral' fault, described by Macías et al. (2010), passes through the AC springs, the crater, AS springs and the NW dome (Fig. 4.1a, see also Chap. 1). Not coincidentally, the only flank springs manifesting gas bubbling are the AC and AS springs. This fault could be the main control of actively degassing of the volcano (Mazot et al. 2011).

Another type of springs, called 'Agua Roja' (for 'red waters', AR) after their reddish color caused by the high content of Fe-oxyhydroxide precipitates, are found all over the volcano and discharge cold (~25 °C) and slightly acidic water (pH 3–6).

Some images of the lake, SP, ACn and AS springs are shown in Fig. 4.2a–d.

4.3 Chemical and Isotopic Composition of the Crater Lake and Thermal Springs: 1995–2011 Evolution

4.3.1 Water Chemistry

Since 1995, the El Chichón crater-lake and Soap Pool crater springs and the AC-AT flank springs have periodically been sampled to monitor the volcanic activity (Taran et al. 1998; Rouwet et al. 2004, 2008; Rouwet 2006; Peiffer et al. 2011). A compilation of the chemical composition for the 1995–2011 period is presented in Table 4.1.

During these years, the lake has suffered drastic changes in surface area, as volume has ranged between 3 and 23 × 10^4 m^3 (Rouwet et al. 2008; Taran and Rouwet 2008). These variations in volume are accompanied by chemistry changes (Cl/SO_4 = 0.5–20 molar ratio, Table 4.1, Fig. 4.3) and can occur over short periods of a few weeks making the lake very dynamic (Rouwet el al. 2004, 2008): the residence time of the lake water is estimated to be as low as 2 months (Taran and Rouwet 2008). There is no systematic correlation between the yearly (average of 4,000 mm/year with the rainy season from June to November) and the variations in volume and salinity of the crater lake (Rouwet et al. 2008).

The SP water showed a steady decrease in Cl content from 11,780 mg/L in 1995 to 4,118 mg/L in 2004. Since then, the Cl content varied from 1,842 to 3,412 mg/L (Table 4.1, Fig. 4.3). Using a mass and Cl balance, Rouwet et al. (2004, 2008) confirmed that the water discharged by the SP springs represents the only source of Cl for the lake. When the SP does not discharge water (vapor exhalation), the lake water rapidly becomes low in Cl leading to a SO_4-dominated composition, typical for steam-heated pools. Therefore, the changes in lake chemistry and volume are directly related to the dynamics of these SP springs. If the lake were not recharged by the SP springs, it would probably disappear during the dry season (Taran et al. 1998; Rouwet et al. 2004, 2008).

Rouwet et al. (2008) interpret the 1995–2004 decreasing trend in Cl content of the SP springs as the result of progressive dilution by meteoric water of the shallow aquifer feeding the SP springs. Extrapolating this Cl dilution trend back to the year 1983 (Fig. 4.3), they found a hypothetical Cl concentration for the 1983 SP springs of 24,000 mg/L, (i.e., exactly the Cl concentration of the initial crater lake, according to Casadevall et al. 1984). It was suggested by the same authors that the water feeding the SP springs could be constituted by the initial 1982 lake buried by landslides that occurred during the next months following the eruption (Inbar et al. 2001), while Jutzeler et al. (2011) believe that this buried lake is just the contact of the water table with an older brecciated magmatic system resulting from the previous ~550 year BP eruption (Espíndola et al. 2000; Chap. 2). This initially acidic brine acquired its neutral pH by water-rock interactions and mixing with infiltrated meteoric waters. Furthermore, the cyclic alternation of water and vapor discharge of the Soap Pool springs is explained by the saturation state of the aquifer: when the aquifer is well-recharged, SP springs discharge water, and when the aquifer level is lower, a steam cap develops and SP springs liberates steam instead of water.

Following the 1995–2004 dilution trend, the Cl content of Soap Pool would have reached 0 mg/L by 2008. However, the Soap Pool springs show a more constant concentration in Cl (3,000 ± 1,000 mg/L) since the year 2004; similar to the Cl concentrations of the AC-AT springs (Peiffer 2011).

The waters discharged by the AC, ACn, AT1, AT1n springs are of Na–Ca–Cl–SO_4 type with near-neutral pH (Table 4.1). During the last 18 years, their Cl content remained relatively constant between 1,459 and 2,330 mg/L (Fig. 4.3) as did their pH of 6.43–7.74. The AT2 springs are similar in composition although more diluted, with a Cl content from 583 to 1,454 mg/L. The Cl content of the sample of Molina-Berbeyer

Table 4.1 Major elements and $^{87}Sr/^{86}Sr$ in El Chichón waters

Sample	Date	T(°C)	pH	SiO_2	Na	K	Ca	Mg	Al	Fe	Sr
L	Jan-83	56	0.56	257	607	232	2,110	424	745	914	15.6
L	May-95	33	2.25	111	400	115	228	38	7.2	26	na
L	Abr-98	34.4	2.15	199	1,233	191	545	98	na	5	1.6
L	Jun-04	32	2.22	263	813	126	365	39	1.1	8	2
L	Mar-05	31	2.1	na	248	41	119	18	na	16	0.7
L	Nov-06	26	2.48	108	79	14	45	8	3	7	0.3
L	Mar-07	29.2	2.57	138	488	68	216	21	6	9	1.8
L	Nov-09	na	na	85	15	4	15	2	3	1.4	0.1
L	Jun-10	34.6	2.25	79	209	48	120	22	9	16	4
L	Nov-10	26.8	2.65	55	82	15	42	6.6	na	na	na
L	Jan-11	28	2.33	106	233	42	90	10	na	na	na
SP	May-95	99	3.3	439	4,450	1,200	1,994	348	7.4	0.1	na
SP	Mar-96	99	5.18	104	5,780	777	2,014	208	na	na	na
SP	Apr-98	99	6.48	96	4,013	604	1,615	214	0	0	na
SP	Apr-01	88	6.47	na	2,407	359	940	351	0	0	na
SP	Jun-04	98	6.31	284	1,775	277	591	93	0.04	1.3	4.4
SP	Mar-07	98	7.07	105	1,401	215	365	64	na	na	na
SP	Nov-09	na	7.41	na	954	146	216	27	0.03	1.9	2.3
SP	Jun-10	81	6.87	120	1,193	229	480	33	0.008	3.4	4.3
SP	Nov-10	72	7.34	59	1180	222	292	16.6	na	na	na
SP	Jan-11	78	6.91	72	1145	206	237	12.4	na	na	na
AC	Jan-97	71	6.42	172	816	109	623	51	0.9	0.3	3.5
AC	Jun-04	67	6.41	147	749	94	530	47	na	0.04	3.1
AC	Nov-06	70.6	5.77	251	708	93	462	49	0.01	5.2	4.4
ACn	Jun-08	57	7.45	159	656	60	368	49	0.12	5.7	4.2
AT1	Jun-04	78.2	5.61	152	806	111	483	71	na	4.2	3.1
AT1n	Jun-08	50	7.74	191	704	90	514	49	<0.02	4.9	2.6
AT2	Apr-98	49	5.98	107	702	55	666	61	na	na	na
AT2	Nov-06	46.5	5.85	133	334	30	427	27	0.01	5.5	3.9
AT2	Nov-06	36	5.35	99	246	21	369	22	0.03	0.01	3.1
AT2	Mar-07	35.4	5.15	93	349	25	477	26	0.12	0.03	3.5
AS	Jun-04	78.7	5.46	134	3,486	357	848	28	na	0.93	30
AS	Mar-05	73	5.56	na	3,312	337	851	22	na	1.6	50
AS	Jun-04	58.2	2.18	225	4,830	488	1,181	40	na	3.4	46
AS	Mar-05	73	2.27	na	4,231	429	1,092	45	na	14	60
AS	Jun-09	51	2.7	na	2,124	233	488	19	42	18	27
AS	Jun-09	58.5	2.56	na	1,180	156	272	0.5	12	2.1	15
AS	Jun-09	53.2	3.03	na	1,927	215	450	4	17	4.6	23
AS	Jun-10	53	2.5	171	4,560	497	1,089	4.3	9	25	68
ASn	Jun-09	75.4	7.54	na	4,711	446	1,173	2	0.004	na	78
ASn	Nov-09	71.7	6.38	122	2,750	279	579	11	0.003	2.6	33
ASn	Nov-09	70.4	6.24	118	2,480	227	402	13	0.01	2.5	32

(continued)

Table 4.1 (continued)

Sample	Date	T(°C)	pH	SiO_2	Na	K	Ca	Mg	Al	Fe	Sr
ASn	Jun-10	78.2	6.21	100	4,795	441.8	1,427	<1	0.008	2.1	67
ASn	Jun-10	78.2	6.5	53	4,890	447	1,344	3.6	na	2.3	na
AR	Jan-97	23.5	3.7	43	16.3	12.7	277	16	1.2	0.33	3.6
AR	Jan-97	24	6.07	32	13.7	11.7	151	18	0.48	0.08	0.9
AR	Nov-09	26	4.35	113	11	12	311	2	2.8	52	1.4

Sample	Date	Ba	HCO_3	Cl	SO_4	F	B	Br	$^{87}Sr/^{86}Sr$	Ref
L	Jan-83	na	0	24,030	3,550	0.16	433	na	na	1
L	May-95	na	0	980	671	0.5	10	na	na	2
L	Abr-98	0.11	0	2,795	755	0.35	36	na	na	3
L	Jun-04	0.21	0	1,970	467	0.42	na	2.3	na	3
L	Mar-05	0.06	0	535	605	0.38	bdl	0	na	3
L	Nov-06	0.04	0	135	327	0.45	4.4	na	na	4
L	Mar-07	0.16	0	1,149	332	0.95	16	2.4	0.70407	4
L	Nov-09	0.14	0	25	132	bdl	0.4	na	0.70407	5
L	Jun-10	0.049	0	483	1304	1.2	7	na	na	5
L	Nov-10	na	0	175	462	bdl	3	na	na	6
L	Jan-11	na	0	499	470	bdl	8	na	na	6
SP	May-95	na	na	11,780	1,272	0.44	143	na	na	2
SP	Mar-96	na	16	13,100	131	0.26	152	na	na	2
SP	Apr-98	na	53	9,580	174	0.41	121	na	na	3
SP	Apr-01	na	21	6,286	373	3.04	na	15.2	na	3
SP	Jun-04	0.84	60	4,118	206	0.2	40	5.3	0.70409	3
SP	Mar-07	na	43	3,028	264	0	na	6.4	0.70407	4
SP	Nov-09	0.13	na	1,842	316	bdl	35	na	0.70408	5
SP	Jun-10	0.42	42	3,412	329	bdl	16	na	na	5
SP	Nov-10	na	30	2,668	249	bdl	39	na	na	6
SP	Jan-11	na	29	2,500	222	bdl	40	na	na	6
AC	Jan-97	0.13	191	2,330	451	0.47	23	19	0.70414	2
AC	Jun-04	0.12	0	1,643	698	0.65	14	1.8	0.70413	7
AC	Nov-06	0.09	293	1,459	733	0.57	18	3.2	na	4
ACn	Jun-08	0.08	196	1,504	456	na	17	na	na	8
AT1	Jun-04	0.09	201	1,812	746	0.63	18	2.8	na	7
AT1n	Jun-08	0.07	103	1,827	525	na	16	na	na	8
AT2	Apr-98	na	85	1,454	900	1.3	14	na	na	2
AT2	Nov-06	0.05	122	803	738	1.14	9	1.6	0.70419	4
AT2	Nov-06	0.08	76	583	618	0.38	7	1.6	na	4
AT2	Mar-07	0.10	49	911	797	0.3	9	1.5	na	4
AS	Jun-04	0.77	214	7,022	162	0.59	27	14	na	7
AS	Mar-05	0.94	195	6,858	175	bdl	bdl	21	0.70532	7
AS	Jun-04	0.55	0	10,003	590	0.61	38	27	na	7
AS	Mar-05	0.39	0	8,848	665	bdl	bdl	24	0.70531	7
AS	Jun-09	0.16	0	4,162	389	bdl	18	13	na	5

(continued)

Table 4.1 (continued)

Sample	Date	Ba	HCO_3	Cl	SO_4	F	B	Br	$^{87}Sr/^{86}Sr$	Ref
AS	Jun-09	0.28	0	2,449	376	bdl	9	na	na	5
AS	Jun-09	0.22	0	3,975	319	bdl	14	na	0.70542	5
AS	Jun-10	0.4	0	11,586	544	bdl	23	na	na	5
ASn	Jun-09	2.1	142	10,051	65	bdl	na	na	0.70534	5
ASn	Nov-09	0.59	101	5,352	93	bdl	19	na	na	5
ASn	Nov-09	0.5	na	4,680	94	bdl	21	na	na	5
ASn	Jun-10	1.8	126	11,978	140	bdl	31	na	na	5
ASn	Jun-10	na	98	11,496	137	bdl	35	na	na	5
AR	Jan-97	na	bdl	3.7	710	0.16	bdl	na	na	2
AR	Jan-97	na	12	3.7	387	0.12	0.6	na	na	2
AR	Nov-09	0.14	0	4.7	780	bdl	0.01	na	0.70407	5

Concentrations in mg/L.; *bdl* below limit detection; *na* not analyzed; Reference 1. Casadevall et al. (1984), 2. Taran et al. (1998), 3. Rouwet et al. (2008), 4. Rouwet et al. (2009), 5.Peiffer et al. (2011), 6.Peiffer (2011), 7.Taran et al. (2008), 8.Taran and Peiffer (2009). Sr isotopic data are all from Peiffer et al. (2011)

(1974) from the AC springs before the eruption strongly differs from the current Cl concentration of the springs, now ~5 times more concentrated. However, this spring could have been sampled from a more diluted vent and hence this lower Cl concentration does not necessarily reflect a change in water chemistry after the eruption (Taran et al. 1998).

The AS waters are also of Na–Ca–Cl–SO_4 type but differ from the previous spring waters by an acidic to near-neutral pH (2.18–5.56) and a much higher salinity (Cl content from 2,500 up to 12,000 mg/L, Fig. 4.3). The ASn springs have similar Cl contents but are characterized by a near-neutral pH (6.21–7.54), lower SO_4 (<140 mg/L vs. up to 665 mg/L for AS springs) and Mg (<13 mg/L vs. 45 mg/L for AS springs) (Table 4.1). Their neutral pH and high salinity make the ASn springs the most exotic springs at El Chichón. The difference in acidity between AS and ASn springs results from superficial oxidation of H_2S present in hydrothermal vapors (Peiffer et al. 2011), as also demonstrated by low pressure fumarolic exhalations in the AS canyon and much higher sulfate concentrations in AS waters comparing with ASn waters (Table 4.1).

The AR springs discharge slightly acidic water (pH 3–6) enriched in Ca and SO_4 (up to 2,000 mg/L). Following Taran et al. (1998), their composition results from the leaching of anhydrite present in the 1982 pyroclastic deposits by infiltrating meteoric waters, proven by their specific Ca/SO_4 molar ratio of ~1. AC-AT waters are probably mixed with AR waters as their enrichment in Ca and SO_4 are similar (Peiffer et al. 2011).

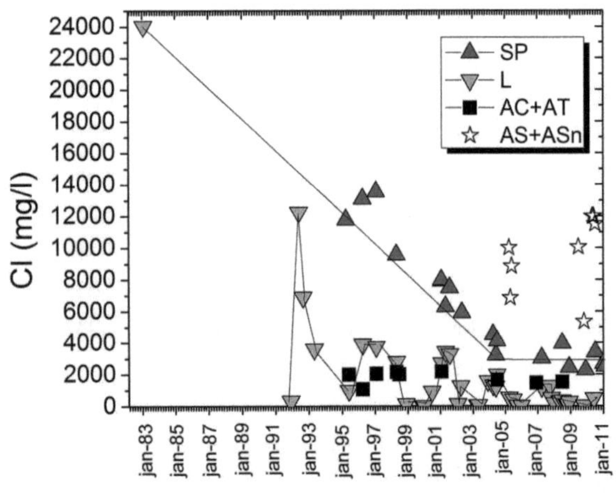

Fig. 4.3 Cl content evolution for the lake (L), SP, AC-AT and AS-ASn springs (modified from Rouwet et al. 2008). Note the decreasing trend of the SP waters Cl content until 2004 followed by relative constancy

Taran et al. (2008), Morton-Bermea et al. (2010) and Peiffer et al. (2011) published the concentrations of rare earth element (REE) in waters from El Chichón. The solubility of REE in water is mainly controlled by water pH, wall-rock nature, specific mineralogy, complexing ligands (Cl, SO_4, F) and temperature. Therefore, REE concentrations in water provide important clues to the nature of water–rock interaction (Michard 1989; Lewis et al. 1997; Wood 2006). The AC-ASn-SP springs have the lowest concentration with total REE concentration (ΣREE) less than 0.6 μg/l, while AS and AR springs as well as the lake are more concentrated. The AS waters have a ΣREE between 30 and 38 μg/l while the lake waters show a wider range of concentration between 9 and 98 μg/l. The AR waters are the most enriched in REE with a total of 207 μg/l.

4.3.2 Water and Sr Isotopes

The composition of $\delta^{18}O$ and δD in El Chichón lake and spring waters is shown in Fig. 4.4. The SP springs are characterized by nearly constant values of $\delta^{18}O$ (1.5 ± 1.5 ‰) and δD (−2 ± 10 ‰) (Rouwet et al. 2008; Taran and Rouwet 2008). The data points for the lake waters show a trend with a slope of ~4.6 corresponding to the mixing between meteoric waters and Soap Pool waters subsequently affected by evaporation. The scattering along this "evaporation trend" is higher compared to trends for most crater lakes (Varekamp and Kreulen 2000), demonstrating the importance of the mixing with the SP spring input. The $\delta^{18}O$ and δD values for AC-AT spring waters generally fall along the local meteoric water line ($\delta D = 8\delta^{18}O + 13$; Taran et al. 2008), although some samples show a positive shift in $\delta^{18}O$ of ~2 ‰. This shift probably reflects water-rock interaction inside the aquifer. The data points for the AS and ASn springs show a more significant positive oxygen shift of up to 3.5 ‰. This oxygen shift is correlated with the Cl content of AS-ASn springs, and probably reflects the mixing (dilution) between superficial meteoric waters and a deep end member rich in Cl and characterized by more positive $\delta^{18}O$ values. AR waters show a negative shift in oxygen compared to meteoric values. Taran et al. (1998) suggested that these superficial waters lose their oxygen-18 due to hydration of pumice and formation of low-temperature clay minerals (Fig. 4.5).

The El Chichón waters can be divided into three groups according to their $^{87}Sr/^{86}Sr$ ratios (Table 4.1): (1) the lake and SP from the crater as well as AR springs have $^{87}Sr/^{86}Sr$ values of 0.70407–0.70409, (2) the flank springs AC-AT show ratios of 0.70413–0.70419, while (3) the AS and ASn springs are characterized by higher values of 0.70531–0.70542 (Peiffer et al. 2011).

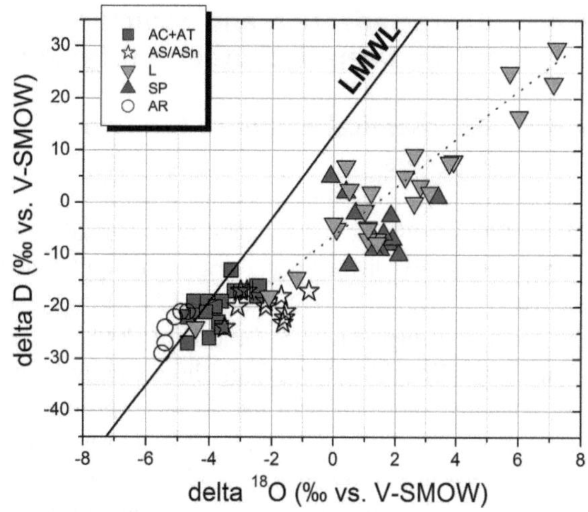

Fig. 4.4 $\delta^{18}O$-δD diagram for the lake, SP, AC-AT, AS-ASn and AR springs. LMWL local meteoric water line. The dotted line represents the trend described by the lake waters that results from the mixing between meteoric waters and SP waters and subsequent evaporation

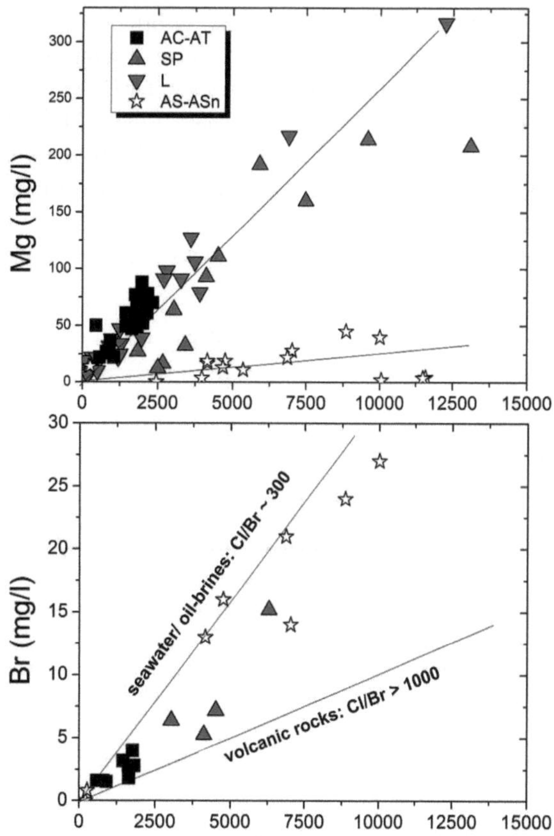

Fig. 4.5 Mixing plots for El Chichón thermal waters

4.4 Characterization and Structure of the Deep Hydrothermal System

4.4.1 Geochemical Evidence for Two Aquifers

Plots of Cl versus Mg and Br identify two trends corresponding to two distinct water origins or aquifers for the AC-AT (AC-ACn-AT1-AT1n-AT2) springs and the AS-ASn springs (Fig. 4.6; Taran et al. 2008; Peiffer et al. 2011). The AS-ASn springs show clearly lower Mg contents compared with the AC-AT springs and crater fluids. The Cl/Br of AS and ASn waters is typical of seawater values (289). This low Cl/Br could reflect the presence of evaporites from the sedimentary basement. Unfortunately, no chemical analysis of these evaporites is available to confirm this hypothesis. Another important Cl contribution could come from oil-field brines from the underlying sediments. The numerous oil wells in the area surrounding the volcano make the presence of such brines beneath the volcano highly probable. In particular, the oil brines extracted from the Catedral production well, located some 20 km north-west from the volcano, have similar Cl/Br ratios as those measured in AS/ASn waters (283–334) with Cl concentration up to 25,000 mg/L (Méndez Ortiz 2007). The Cl/Br for other springs are intermediate between seawater values and the ones observed for waters in contact with volcanic rocks (Cl/Br > 1,000;

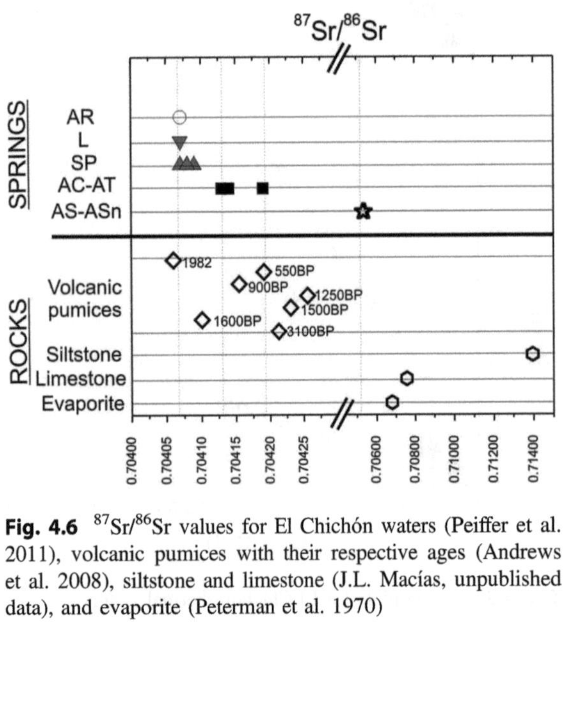

Fig. 4.6 $^{87}Sr/^{86}Sr$ values for El Chichón waters (Peiffer et al. 2011), volcanic pumices with their respective ages (Andrews et al. 2008), siltstone and limestone (J.L. Macías, unpublished data), and evaporite (Peterman et al. 1970)

Taran et al. 2008). The Cl/Br ratios of the SP springs sampled from 1995 to 2011, as well as those measured in the lake also fall on the AC-AT trend (Rouwet 2006; Peiffer et al. 2011).

Peiffer et al. (2011) characterized the nature of El Chichón springs using Sr isotopic ratios, Sr concentrations and Ca/Sr ratios. The AC-AT springs with $^{87}Sr/^{86}Sr$ values of 0.70413–0.70419 similar to the values observed in the volcanic deposits (0.704060–0.704256; Andrews et al., 2008; Fig. 4.6) are believed to originate from a volcanic aquifer (Aquifer 1). This aquifer is probably located beneath the crater at a maximum depth of 280 m, corresponding to the difference in altitude between the flank springs and the crater bottom. The neutral AC-AT springs fed by this aquifer are located at the same altitude (600–650 m asl) and at similar distance from the crater (<1 km). Such spatial distribution suggests a radial and sub-horizontal flow inside the aquifer, along the contact between the sedimentary basement and volcanic deposits (where the flank springs appear). The composition at depth of this aquifer, best approximated by the AC-AT springs composition, is of Na–Ca–Cl–SO_4 type with near-neutral pH, and Cl concentration of $\sim$2,200 mg/L. The water in this aquifer has a meteoric origin as shown by $\delta^{18}O$-δD values of the AC-AT flank springs. The SP and lake waters, as well as the AR waters have $^{87}Sr/^{86}Sr$values of 0.70407–0.70409, similar to those of younger and superficial volcanic deposits from the 1982 eruption ($^{87}Sr/^{86}Sr$: 0.70406) suggesting their equilibration at the contact with these deposits. In contrast, the AC-AT waters are equilibrated with deeper volcanic layers with ages of 900 to 550 y BP ($^{87}Sr/^{86}Sr$: 0.70416–0.70419; Fig. 4.6; Andrews et al. 2008). These data indicate that the superficial aquifer feeding the SP springs, described by Rouwet et al. (2004, 2008), corresponds to the upper part of the volcanic Aquifer 1.

The higher $^{87}Sr/^{86}Sr$ values in AS-ASn springs, from 0.70531 to 0.70542, are intermediate between the volcanic deposits and sedimentary rocks (evaporite: 0.7068, Peterman et al. 1970; limestone: 0.70757; siltstone: 0.71396; Macías unpublished data) and probably reflect a mixed sedimentary-volcanic aquifer (Aquifer 2) (Fig. 4.6). This aquifer would be located in the limestone-evaporite horizon at more than 2,000 m beneath the surface. The presence in this sedimentary horizon of volcanic rocks probably associated with the NW dome could explain the intermediate $^{87}Sr/^{86}Sr$ value of the AS-ASn waters. The discharge of Aquifer

2 waters is probably controlled by fracture permeability generated by the Chichón-Catedral fault and the collapse structure of the NW dome (Fig. 4.1). The AS-ASn spring waters fed by Aquifer 2 are of Na–Ca–Cl–SO_4 type with acid to near-neutral pH, although more saline than Aquifer 1 waters.

The Sr concentrations, and Ca/Sr ratios of the thermal waters from El Chichón also allow a distinction between Aquifer 1 and 2 (Peiffer et al. 2011). The AC-AT waters, as well as SP and lake waters from Aquifer 1 have less than 5 mg/L of Sr with Ca/Sr of ~130, typical of water in volcanic aquifers (Peiffer et al. 2011). On the contrary, the AS/ASn waters from Aquifer 2 are characterized by much higher Sr contents up to 80 mg/L, with low Ca/Sr ratios of ~17. These low Ca/Sr ratios are typical values observed for formation waters or brines associated with oil reservoirs (Peiffer et al. 2011). The oil brines from the Catedral production well present similar Ca/Sr ratios (15–22, Méndez Ortiz 2007). The origin of such low Ca/Sr ratios probably reflects a mineral control in the sedimentary horizon by a set of minerals such as anhydrite-calcite and strontianite under high temperature of 230 °C (Peiffer 2011; Peiffer et al. 2011).

Rare Earth Elements (REE) distributions in El Chichón waters reflect superficial processes and do not allow the discrimination between the two aquifers (Peiffer et al. 2011). The main factor controlling the solubility of REE in waters is acidity. Acidic pH conditions enhance mineral solubility and reduce adsorption processes, leading to a higher REE solubility (Michard 1989; Lewis et al. 1997; Wood 2006). The acidic water from the AS springs and the crater-lake show total REE concentrations between 9 and 98 μg/l while the neutral AC-ASn-SP waters have lower concentrations >0.6 μg/l. As the acidity of the AS springs and the lake water at El Chichón is a superficial process resulting from the shallow oxidation of H_2S contained in hydrothermal fluids (Taran et al. 1998), the enrichment in REE reflects the superficial leaching of volcanic rocks by the acidic waters. The REE in acidic waters mimics the composition of the volcanic rocks confirming the role of pH on REE solubility (Peiffer et al. 2011). Although characterized by irregular REE distribution, neutral waters are generally characterized by a horizontal profile compared with the volcanic rocks, corresponding to depletion in Light Rare Earth Elements (LREE: La, Ce, Pr, Nd, Sm). This depletion could reflect the alteration state of the wall-rocks. LREE are more incompatible than the Heavy Rare Earth Elements (HREE: Gd, Tb, Dy, Ho, Er, Tm, Yb, Lu), and therefore are preferentially lixiviated during early alteration stages (Takano et al. 2004). The control of complexing ligands such as Cl, SO_4, and F on the REE solubility is unlikely due to the lack of correlation between the ratio La/Yb and Cl, SO_4, F concentrations in waters from El Chichón. No correlation with temperature was observed either (Peiffer et al. 2011).

Two interesting characteristics revealed by the REE composition are their particularly high concentration in the AR spring waters, and the positive correlation between the Cl contents of the lake and its REE contents (Peiffer et al. 2011). The AR waters, although not the most acidic (pH 3–6), have the highest content in REE of all the waters analyzed, 207 μg/l, value that is comparable to contents observed in very acidic crater lakes such as Maly Semiachik (Russia), Ruapehu (New Zealand), Copahue (Argentina) and Kawah Putih (Indonesia) (Takano et al. 2004). The REE distribution in AR waters parallels the 1982-magmatic anhydrite pattern (Peiffer et al. 2011). On this basis, the high content of REE in the AR waters could be easily explained by the relatively high solubility of anhydrite at low temperature, and the high content of REE in the anhydrite (~500 mg/kg; Luhr et al. 1984). The dissolution of 1 g of $CaSO_4$ should supply the solution with ~500 μg of REE. The AR water chemistry results from the leaching of this anhydrite, dissolving higher quantity of REE than waters in contact with pure volcanic rocks.

The lake shows variable ΣREE in time from 9 to 98 μg/l. Surprisingly, the ΣREE is correlated to the Cl contents of the lake. In fact, knowing that the Soap Pool waters are the unique source of Cl for the lake and, considering their low REE content, the correlation between REE and Cl in the lake is unexpected (Taran et al. 1998). Peiffer et al. (2011) suggested that a higher salinity of the waters at low pH could induce a higher leaching potential. When the Soap Pool springs discharge water to the lake, the lake waters become more reactive with the sediments and are able to leach a higher quantity of REE. A higher salinity causes the ionic strength of the water to increase and the activity coefficients to decrease, hence mineral solubility is expected to increase.

4.4.2 Outflow Rates of Thermal Waters and Hydrogeological Parameters

The total discharge of each group of springs, and therefore the total discharge of the two aquifers, were estimated by measuring the flow rates, and analyzing the chemical composition, of the rivers draining the volcano (Taran and Peiffer 2009; Peiffer et al. 2011). Assuming a conservative behavior for Cl (i.e. no Cl is lost as a mineral or gas phase) a total discharge of 220 L/s of waters with 2,000 mg/L of Cl was calculated for Aquifer 1 and a discharge of 7 L/s of waters with 12,000 mg/L of Cl for Aquifer 2.

Using these outflow rate estimates, it is possible to calculate some hydrogeological parameters such as the residence time t (s) of the waters inside the aquifer and the hydraulic conductivity parameter K (m/s). Residence time can be calculated by the equation:

$$t = V/q \tag{4.1}$$

where V is the volume of the water inside the aquifer and q the discharge of the aquifer. This equation assumes a state of equilibrium where the discharge rate of the aquifer must be equal to the recharge rate. Annual precipitation in the area surrounding the volcano is about 4,000 mm (Atlas del Agua 1976). Considering an area of ~ 10 km^2 for the hydrographic basin of El Chichón, the precipitation rate would be of 1.3 m^3/s, well above the discharge of the aquifer (0.22 m^3/s). Although this estimate does not consider the run-off water, ~ 20 % of the total annual precipitation is enough to maintain the hydrothermal system of El Chichón in a state of equilibrium.

Assuming a pyramidal shape with a base 4 km long by 2 km wide, and a height of 280 m, the volume of pyroclastic rocks that constitute the volcanic Aquifer 1 can be crudely estimated to be 0.75 km^3. The length of the pyramid corresponds to maximum distance between the flank springs AC and AT1, and the width to maximum distance between the same springs and the Chichón-Catedral fault (Fig. 4.1), supposed to control the thermal manifestations of El Chichón (Mazot and Taran 2009; Macías et al. 2010). The height of 280 m corresponds to the difference in altitude between the flank springs and the crater bottom.

No data on the porosity is available for El Chichón pyroclastic deposits that constitute the volcanic aquifer. Therefore values of porosity between 5 and 25 %, typical for such kind of material (i.e. volcanic deposit at Poás volcano in Costa Rica, Rowe et al. 1995; Sruoga et al. 2004) are considered here. Using these porosity values, and a water volume between 3.75×10^7 and 1.88×10^8 m^3 residence times between 5 and 27 years are obtained. However, such estimates would imply that the entire volcanic horizon is fully saturated in water. Considering that only the lower half of this horizon is likely to be saturated, the residence time is reduced to intervals between 3 and 13 years. These values are close to residence times estimated for the volcanic aquifer of Poás volcano (3–17 years, Rowe et al. 1995).

The hydraulic conductivity K (m/s) can also be estimated by the relation:

$$K = \left(n_o.\Delta L^2\right)/(t.\Delta H) \tag{4.2}$$

where n_o is the porosity, ΔL is the length (m) of water flow inside the aquifer, ΔH is the difference in altitude (m) along the flow path, and t is the residence time (s) (Davis and Bentley 1982). Using the residence times derived from above and the same porosity values, a ΔL between 1,000 and 2,000 m, and a ΔH between 140 and 280 m, K is estimated to be 10^{-5} and 10^{-7} m/s. Wohletz and Heiken (1992) estimated the permeability of non-altered pyroclastic rocks to be similar to the permeability of sandstones with K values of 10^{-7} and 10^{-3} m/s. Rowe et al. (1995) calculated for the Poás volcanic aquifer similar K values of 10^{-5} and 10^{-7} m/s.

The extension of Aquifer 2 is difficult to estimate, despite the fact that the area surrounding the volcano is well-known due to petroleum exploitation. The resurgence of these waters is probably controlled by fractures along the Chichón-Catedral fault, and the collapse structure of the NW dome.

4.5 Solute Geothermometry and Re-Equilibration Processes

The chemical composition of deep aquifer waters rising to the surface can be affected by superficial processes such as dilution, mixing, and boiling, as well as by re-equilibration processes at different levels between the aquifer and the surface. Solute geothermometry helps to understand and discriminate these processes. The use of solute geothermometers is based

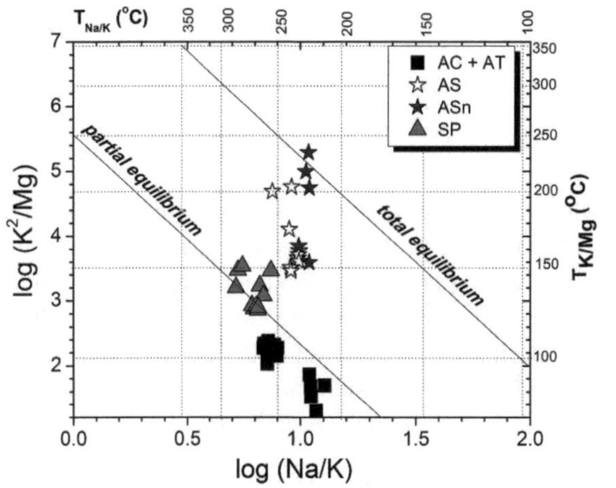

Fig. 4.7 Na–K–Mg diagram El Chichón thermal waters (modified from Giggenbach 1988)

on the assumption of the existence of a chemical equilibrium between the thermal solution and minerals of the wall rock. The temperatures calculated by Na/K and K/Mg geothermometers, derived by Giggenbach (1981, 1988) are shown in Fig. 4.7. The Na/K temperature of the AC-AT waters from Aquifer 1 is between ~220 and 260 °C while their K/Mg temperature varies between ~80 and 110 °C. The divergence observed between the geothermometers indicates that these waters are not fully equilibrated with the wall-rock. These "immature waters", as defined by Giggenbach (1988), are typical for young thermal waters from volcano-hydrothermal systems, and their composition is the result of equilibration and re-equilibration occurring in the aquifer and on the way to the surface. In general, Na and K concentrations in water are relatively conservative, while Mg and Ca concentrations are affected by temperature decrease and steam separation (Fournier 1981, 1989; Giggenbach 1988). The K/Mg temperature obtained for Aquifer 1 waters corresponds to the discharge temperature of the AC-AT springs and probably reflects the re-equilibration of the deep Aquifer 1 fluid at lower temperatures, after vapor-liquid phase separation, with Mg-rich clay minerals and subsequent mixing with superficial waters rich in Ca and SO_4 such as the AR waters (Taran et al. 1998; Taran and Peiffer 2009). The composition of the waters of Aquifer 1 could also reflect their interaction with the glass of surrounding volcanic deposits, with the possible release of significant amounts of cations in solution.

The SP waters also show discrepancy between their Na/K and K/Mg temperature, of ~260–290 °C and ~120–150 °C respectively, and are thus also considered as immature waters.

The ASn springs (Aquifer 2) sampled after the dry season (June 2009–2010) present similar Na/K and K/Mg estimated temperatures around 220–230 °C, and can be considered as fully equilibrated, or "mature waters" ($T_{Na/K} \sim T_{Mg/K}$; Giggenbach 1988). The ASn springs sampled just after the rainy season (November 2009), as well as their acidic homologue the AS springs, have lower K/Mg temperatures of 150–200 °C and fall between the field of equilibrated and partially equilibrated waters. Large amounts of rainfall during the rainy season can interact with superficial volcanic deposits, generating temporary "AR-type" (i.e. Fe–Ca–SO_4-type) springs. These spring waters can mix with rising ASn waters, causing the disequilibrium shift observed for ASn waters sampled just after the rainy season. In the case of AS springs, their acidity, interpreted to be caused by H_2S oxidation, probably leads to re-equilibration processes close to the surface.

The deep Aquifer 2 waters, best represented at the surface by the ASn springs, are the only mature waters at El Chichón. This state of maturity can only be achieved after a long residence time with steady water-rock interaction. The ASn waters, whose chemical compositions are close to those for oil-brines, could have residence times of several thousands of years, typical time range observed for oil-brines (Birkle et al. 2009). The upflow and discharge of the deep Aquifer 2 waters is most probably controlled by a fracture-permeability, and is eventually fast enough to prevent re-equilibration of these waters on their way to the surface.

Silica geothermometry (Henley et al. 1984) was also applied to El Chichón waters. The temperatures estimated using the quartz geothermometer for all water types are generally low, between ~110 and 190 °C. The amorphous silica and cristobalite geothermometers give temperatures that are respectively lower than ~75 °C, and between ~50 and 150 °C. The silica geothermometers can be affected by dilution and evaporation processes, because they are based on the absolute concentration of one element, in contrast to the Na-K geothermometer based on the ratio of two elements. Therefore, the temperature estimates obtained using silica geothermometers are less reliable than the Na-K ones.

4.6 Boiling, Cooling and Dilution Processes Inferred from δ^{18}O-δD and Chloride Concentrations

The numerous degassing features in the crater as well as the acidity of the lake provide clear evidence of a boiling hydrothermal system beneath the crater, most probably related to the Aquifer 1. The waters of this volcanic aquifer rise to the surface through a fractured zone probably corresponding to the magmatic conduit of the 1982 eruption, and boil off a vapor phase at ~100 °C generating a steam cap below the crater. The connection between the crater manifestations and the underlying Aquifer 1 is reinforced by the presence of SP springs within the crater. These springs have a similar chemical composition as the AC-AT springs (Cl-ions trends, Ca/Sr) and come from the upper part of the volcanic Aquifer 1 (^{87}Sr/^{86}Sr). However, the SP waters showed a decreasing Cl trend up to the year 2004, contrasting with the stable concentration of AC-AT springs (Fig. 4.3). Furthermore, their isotopic values remained relatively constant overtime ($\delta^{18}O = 1.5 \pm 1.5$ ‰, $\delta D = -2 \pm 10$ ‰), much different from the δ^{18}O and δD values for AC-AT springs ($\delta^{18}O = -4.7$ to -2.4; $\delta D = -27$ to -13). Since 2004, the Cl concentration of SP waters (1,842–4,118 mg/L) has been similar to the one of AC-AT (~2,000 mg/L).

The dynamics of the SP springs thus differ from that of AC-AT springs. To explain their nearly constant δ^{18}O values as well as the Cl dilution trend with time, a single-step model of liquid-vapor separation of the deep Aquifer 1 waters is proposed here, based on oxygen-hydrogen isotopes and Cl concentrations. Following this model, the deep Aquifer 1 fluid, during its ascent towards the surface, would produce a vapor phase due to adiabatic expansion occurring with pressure decrease. The single-step model considers the boiling when the two phases, vapor and liquid, remain in contact and in equilibrium until the vapor phase separates at a certain depth and temperature (Truesdell et al. 1977; Giggenbach 1978).

The composition of the liquid (δ_w) and vapor phase (δ_s) of the SP water is related to the isotopic composition of the deep original Aquifer 1 fluid before phase separation (δ_o) by the following set of relations:

$$\delta_s \cdot y_s + \delta_w \cdot (1 - y_s) = \delta_o \quad (4.3)$$

$$\delta_w = \delta_o + \varepsilon . \, y_s \quad (4.4)$$

$$\delta_s = \delta_o - \varepsilon . (1 - y_s) \quad (4.5)$$

with

$$\varepsilon = 1000 \ln \alpha \approx 1000\,(\alpha - 1) \approx \delta_w - \delta_s \quad (4.6)$$

where α is the fractionation factor controlling the distribution of deuterium and oxygen-18 between the liquid and vapor phase at a certain temperature. Values of α used for calculations are those of Majoube (1971) and Horita and Wesolowski (1994). At 100 °C, ε_{18O} is ~5.5 ‰ and ε_D ~ 26 ‰.

y_s is the vapor fraction calculated as:

$$y_s = (H_o - H_w)/(H_s - H_w) \quad (4.7)$$

where H_o is the enthalpy of the deep Aquifer 1 fluid, while H_w and H_s are the enthalpy of the liquid and vapor at the temperature of separation, respectively. Considering a temperature at depth of Aquifer 1 of 220 °C, which is a minimum according to Na/K geothermometers, and a temperature of separation of 100 °C it gives a vapor fraction of 25 % ($y_s = 0.25$). The corresponding enrichment in the residual fluid in δ^{18}O would be +1.3 ‰ and in δD of +7 ‰.

Assuming that the isotopic composition of the AC-AT springs, showing the maximum oxygen shift, constitutes a good proxy for the deep Aquifer 1 isotopic composition ($\delta_o^{18}O = -2.4$ ‰, $\delta_o D = -20$ ‰), the residual fluid separated from its vapor phase at 100 °C would have an isotopic composition in δ^{18}O of −1.1 and δD of −13 (Fig. 4.8). These values approach the lowest ones of SP waters ($\delta^{18}O = -0.1$ ‰; $\delta D = -12$ ‰). However, as mentioned by Peiffer et al. (2011) the AC-AT springs are mixed with AR waters that are characterized by a negative shift in δ^{18}O ($\delta^{18}O = -5.5$ to -4.7). Therefore, the deep waters from Aquifer 1 should present a larger positive shift in δ^{18}O than the one shown by AC-AT spring waters. Hence, the estimated isotopic composition for the Aquifer 1 water, separated from its vapor phase, probably corresponds even better to δ^{18}O and δD values of SP waters. The fumarolic vapor that separates from this water near the surface (t ~100 °C) would have δ^{18}O value of ~−4 ‰ and δD of ~ −32 ‰ which is similar to the average δD-δ^{18}O values of fumarolic condensates from the crater (Taran et al. 1998; Rouwet 2006).

The assumption that the isotopic composition of AC-AT springs fairly reflects the Aquifer 1 water

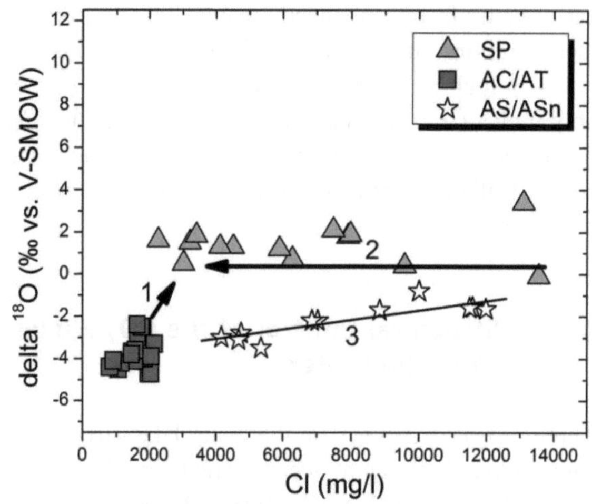

Fig. 4.8 δ^{18}O-Cl diagram for El Chichón thermal waters. *1* Single-step phase (liquid-vapor) separation of the deep Aquifer 1 waters. *2* Dilution of the initial SP reservoir by the deep Aquifer 1 waters. *3* Dilution of the deep Aquifer 2 brines by meteoric waters

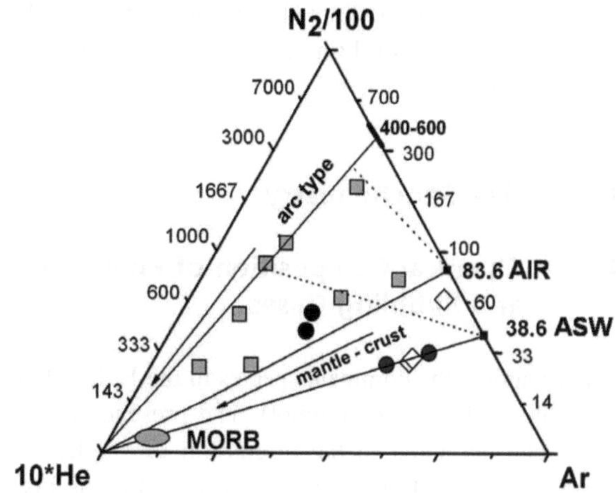

Fig. 4.9 N_2/100-10*He–Ar triangular diagram for El Chichón fumarolic gases (*grey squares*), crater lake bubbling gases (*circles*) and Agua Caliente bubbling gases (*white diamonds*) (modified from Rouwet 2006)

composition is based on the observation that these springs' isotopic composition plot on, or near, the LNWL (Fig. 4.4), therefore ruling out adiabatic boiling as the main process occurring between Aquifer 1 and AC-AT springs outlet. Instead, conductive cooling is probably an important process, given the distal positions of these springs on the flanks of the volcano.

Considering an average Cl concentration of Soap Pool waters since 2004 of ~3,000 mg/L, the deep Aquifer 1 fluid before boiling (vapor fraction 25 %) would have a Cl concentration of ~2,200 mg/L, similar to the Cl content of the AC-AT springs (Table 4.1).

The continuous phase separation model (Truesdell et al. 1977; Giggenbach 1978), which takes into account the separation of the vapor and its removal as soon as it is formed, likely does not apply to El Chichón volcano. In fact, under the same conditions used for the single-step separation model, it causes less enrichment in δ^{18}O and δD of the residual fluid ($\Delta^{18}O < 1$, $\Delta D < 5$), which fails in explaining the specific δ^{18}O values of SP springs.

In addition to phase separation at depth, the evaporation (under equilibrium or kinetic conditions) of the separated fluid at the surface could produce significant enrichments in δ^{18}O and δD like those observed in the SP springs ($\delta^{18}O = -0.1$ to 3.4 ‰; $\delta D = -12$ to 5 ‰; Giggenbach 1978; Taran and Rouwet 2008). Water-rock interaction inside the superficial SP aquifer could also lead to some positive shift in oxygen isotope.

The decreasing trend in Cl concentration observed between 1995 and 2011 probably corresponds to a dilution of the superficial SP volcanic aquifer initially filled with the "buried 1982 crater lake" (Rouwet et al. 2004, 2008). According to this model of phase separation, the dilution in Cl observed, was mainly caused by the continued upflow of water from Aquifer 1. The residual water, separated from the vapor phase, is probably connected by a complex geometry to the SP springs superficial reservoir. It is suggested that the initial 1982 lake water had a value in δ^{18}O similar to the Aquifer 1 residual water (~−2 ‰) so that the SP δ^{18}O values were not affected by the dilution. It seems that since the year 2004, the residual 1982 lake has been completely replaced by waters from the Aquifer 1.

The isotopic composition of the AS and ASn springs in Fig. 4.8 show a positive shift in oxygen-18 up to 3.5 ‰, larger than the one observed for the AC-AT springs. This shift is related to the Cl content of the waters and was interpreted by Taran et al. (2008) as the dilution by meteoric waters of an end member with a high Cl content and positive δ^{18}O. Oil-field brines, located in the Late Jurassic to Late Cretaceous sedimentary basins from southeast of Mexico, have δ^{18}O values between +5.7 and +9.9 ‰, and Cl concentrations up to 25,000 mg/L (Méndez Ortiz 2007). Such values represent an additional evidence for the sedimentary origin and oil-brine resemblance of the deep Aquifer 2 water.

4.7 Gas Geochemistry

4.7.1 Chemical Composition of Fumaroles and Bubbling Gases

Water vapor is the dominant species in the El Chichón fumaroles (841–992 mmol/mol). The presence of the steam cap, as noted in the discussion of single-step steam separation from Aquifer 1 (Sect. 4.6), is also suggested by the chemical composition of the fumarolic exhalations. The concentration of the highly reactive and redox-sensitive species H_2 (Giggenbach 1987) in the fumarolic emissions during the period 1995–2005 varies from 0.1 to 0.3 mol% (Taran et al. 1998; Rouwet 2006; Mazot et al. 2011). The H_2/H_2O ratio in fumaroles (expressed as $R_H = \log(X_{H_2}/X_{H_2O})$) is a measure of the oxidation state and the vapor fraction of the boiling temperature hydrothermal system. Despite the discontinuity of fumarole sampling, the R_H value seems to have increased towards the rock-buffer value of −2.81(FeO-$FeO_{1.5}$ buffer of Giggenbach 1987) during the period 1995–2005, suggesting an increase in the vapor fraction from the boiling aquifer (Aquifer 1).

CO_2 is the dominant species in the dry gas phase of the fumarolic and crater-lake bubbling gases. The acidic gas species HCl, HF and SO_2 are virtually absent; variations in HCl and S_t are probably due to the direct incorporation of liquid droplets of the Cl-SO_4 enriched shallow aquifer (Taran et al. 1998). Total sulfur is thus practically all accounted by H_2S, in accordance with the steam-heated pool nature of the crater-lake and numerous boiling pools inside the SP and fumarolic field. Methane and CO occur in low concentrations (Tassi et al. 2003). The Agua Caliente bubbling gases are also CO_2-dominant though contain higher amounts of atmospheric species (especially N_2) (Taran et al. 1998; Mazot et al. 2011).

The N_2-He-Ar triangular diagram (Fig. 4.9) shows N_2/Ar ratios for fumarolic gases typical for arc-type volcanic gases (108-563) (Giggenbach 1992), with a trend towards higher He contents. Higher He contents can result from enhanced mantle and/or crustal incorporation. The January 1983 fumaroles had N_2/Ar ratios as high as 2,100 (Casadevall et al. 1984) probably indicative of the parental gas for the El Chichón magmatic system. Bubbling gases (crater lake and Agua Caliente spring) show higher atmospheric contamination.

4.7.2 Structural Control of the CO_2 Fluxes from the Crater

Three CO_2 flux surveys on the crater-lake surface and surrounding lake shore were carried out during the period March 2007-April 2008 (Mazot and Taran 2009; Mazot et al. 2011). The CO_2 fluxes averaged values of 1,100 g m^{-2} d^{-1}, or 144 ± 5.9 t d^{-1}, and did not show significant variations during the period of the surveys. Such CO_2 fluxes are typical for diffuse degassing at volcanoes in quiescence (Chiodini et al. 1998).

CO_2 degassing from the crater-lake occurs through two different mechanisms: (1) bubbling degassing, (F_{CO_2} = 465–6,700 g m^{-2} d^{-1}), and (2) diffusion through the water-air interface at the lake surface (F_{CO_2} = 167 g m^{-2} d^{-1}). Clear lineaments of bubbling degassing could be observed and correspond to the highest CO_2 fluxes, showing that degassing in the El Chichón crater is structurally controlled. Three linear zones of enhanced flux are distinguished: (1) the E-W trending structure is probably part of the San Juán sinistral strike-slip fault system (García-Palomo et al. 2004; Macías et al. 2010; Chap. 1), (2) the NE-SW trending structure is likely related to the Chapultenango half-graben system (García-Palomo et al. 2004), and (3) the NW-SE trending structure probably reflects the major regional extension controlling volcanism at El Chichón (Macías et al. 2010; Chap. 1). The latter hypothesis is confirmed by (1) the Chichón-Catedral Fault and the presence of the recently discovered volcanic edifice Catedral, (2) the linear emplacement of the main lava domes and the active crater (Layer et al. 2009) and (3) the unique occurrence of bubbling degassing outside the active crater, at the foot of the NW dome (AS-ASn springs) and in the SE canyon at the Agua Caliente spring.

4.7.3 He and C Isotope Systematics Related to Regional Tectonics

Similar to CO_2 in an acidic medium, the noble gas He is inert and of low solubility when rising from a magma through an active volcano-hydrothermal system: the deep marker of He remains preserved until liberated at the earth surface. These characteristics of CO_2 and He gases convert the He and C isotope systematics into an ideal method to trace deep magmatic degassing at volcanoes and decipher their role within a broad tectonic mark.

The $\delta^{13}C_{CO2}$ values for fumarolic gases in the active crater range from −8.1 to −2.5 ‰ versus V-PDB (Vienna Pee Dee Belemnite), from −4.9 to −1.9 ‰ for bubbling gases at the crater lake, and from −5.5 to −3.6 ‰ for bubbling gases at the Agua Caliente spring (Mazot et al. 2011). Excluding the lowest $\delta^{13}C_{CO_2}$ values (<−7.7 ‰, before 1999), which can be due to a contribution of thermogenic carbon originating from organic material buried by the 1982 eruption, the observed $\delta^{13}C_{CO_2}$ range suggest a mantle source for CO_2.

The air-corrected $^3He/^4He$ ratios for fumarolic gases vary from 6.15 R_A to 8.09 R_A (with R_A being the $^3He/^4He$ ratio of air = 1.40×10^{-6}), from 5.62 R_A to 7.73 R_A for bubbling gases in the crater lake, from 4.98 R_A to 5.56 R_A for Agua Caliente bubbling gases (Mazot et al. 2011), and 2.16 R_A for Agua Salada bubbling gases at the NW dome (Peiffer 2011). The highest $^3He/^4He$ ratios fall within the range for MORB-derived gases (8 ± 1 R_A, Marty and Jambon 1987). The highest $^3He/^4He$ ratio measured at a fumarole in 1998 (8.09 R_A) is the highest reported for a Mexican volcano. The observed $^3He/^4He$ range at El Chichón (7.4 ± 0.7 R_A) is generally higher than $^3He/^4He$ ratios at neighboring volcanic chains (Central American Volcanic Arc: $^3He/^4He$ = 5.97 ± 1.44 R_A, Hilton et al. 2002; Chap. 7). The lower $^3He/^4He$ ratios for the Agua Caliente and Agua Salada bubbling gases could result from mixing with crustal He (R/R_A = 0.01) due to the larger distance from the central degassing vent.

An additional tool, the $CO_2/^3He$ ratio, is often used to trace the origin of the emitted gases (Marty and Jambon 1987). The El Chichón gases (fumaroles, crater lake and Agua Caliente bubbling gases) show a wider range in $CO_2/^3He$ (0.5 to 5.3 × 10^{10}) than generally observed at volcanic arcs (0.4 to 2.6 × 10^{10}). Gases with higher $^3He/^4He$ ratios tend to have lower $CO_2/^3He$ ratios, approaching MORB-like values (2–3 × 10^9, Marty and Jambon 1987). The calculated MORB-proportions (25.3 ± 15.1 %) in El Chichón fumarolic gases, based on the three-component mixing model by Sano and Marty (1995), are significantly higher than at other arc volcanoes (12.8 %, Mazot et al. 2011). The $\delta^{13}C_{CO_2}$ values, $^3He/^4He$ and $CO_2/^3He$ ratios, and high MORB-proportions demonstrate that degassing at El Chichón is more mantle-derived with respect to most arc volcanoes, and that degassing thus probably occurs in a non-classic arc-type setting resulting from melt generation at greater depths. These observations are in agreement with the regional tectonic regime (Manea and Manea 2006).

All the results and interpretations presented in this chapter were possible thanks to more than 15 years of discontinuous chemical monitoring of the El Chichón waters and gases. The volcano-hydrothermal system has remained in a quiet and stable state since the 1982 eruption. However, in case of renewed volcanic-magmatic activity of the volcano, such as the growth of a new extrusive dome, as happened at many volcanoes after a similar type of eruption (Mt Pele in the Antilles; Shiveluch and Bezymyanny in Russia, Mt St Helens in USA, Pinatubo in the Pilippines, and many others), somé precursory signals could be registered in the spring chemistry and dynamics. Nevertheless, monitoring of the flank springs is complicated due to their remote position in deep and quite inaccessible canyons. Chemical monitoring of the springs can be replaced by a monthly monitoring of the river Río Magdalena as stated by Taran et al. (2008) and Taran and Peiffer (2009) (Fig. 4.1). After the same authors, this river constitutes the only drainage of the volcano-hydrothermal system discharge and therefore, could be an ideal point for monitoring the volcanic-magmatic activity at El Chicón.

References

Andrews BJ, Gardner JE, Housh TB (2008) Repeated recharge, assimilation and hybridization in magmas erupted from El Chichón as recorded by plagioclase and amphibole phenocrysts. J Volcanol Geoth Res 175:415–426

Armienta MA, De la Cruz-Reyna S (1995) Some hydrochemical fluctuations observed in Mexico related to volcanic activity. Appl Geochem 10:215–227

Atlas de Agua de la Republica Mexicana (1976) Secretaria de recursos hidraulicos, México
Birkle P, Martínez BG, Milland CP, Eglington B (2009) Origin and evolution of formation water at the Jujo-Tecominoacán oil reservoir, Gulf of Mexico. Part 1: chemical evolution and water–rock interaction. Appl Geochem 24:543–554
Canul RF, Rocha VL (1981) Informe geológico de la zona-geotérmica de "El Chichonal". Chiapas Com Fed de Electr, Morelia, México. Informe pp 32–81, 38 p
Casadevall TJ, De la Cruz-Reyna S, Rose WI, Bagley S, Finnegan DL, Zoller WH (1984) Crater lake and post-eruption hydrothermal activity, El Chichón Volcano, Mexico. J Volcanol Geoth Res 23:169–191
Chiodini G, Cioni R, Guidi M, Raco B, Marini L (1998) Soil CO_2 flux measurements in volcanic and geothermal areas. Appl Geochem 13:543–552
Davis S, Bentley H (1982) Dating groundwater, a short review. In: Currie L (ed) Nuclear and chemical dating techniques: interpreting the environmental record. Am Chem Soc Symp Ser 176:187–222
Espíndola JM, Macías JL, Tilling RI, Sheridan MF (2000) Volcanic history of El Chichón Volcano (Chiapas, Mexico) during the Holocene, and its impact on human activity. Bull Volcanol 62:90–104
Fournier RO (1981) Application of water chemistry to geothermal exploration and reservoir engineering. In: Rybach L, Muffler LJP (eds) Geothermal Systems. Wiley, Principles and Case Histories, pp 109–140
Fournier RO (1989) Geochemistry and dynamics of the Yellowstone National Park hydrothermal system. Ann Rev Earth Planet Sci 17:13–53
García-Palomo A, Macías JL, Espíndola JM (2004) Strike-slip faults and K-alkaline volcanism at El Chichón volcano, southeastern Mexico. J Volcanol Geoth Res 136:247–268
Giggenbach WF (1978) The isotopic composition of waters from the El Tatio geothermal field, Northern Chile. Geochim Cosmochim Acta 42:979–988
Giggenbach WF (1981) Geothermal mineral equilibria. Geochim Cosmochim Acta 45:393–410
Giggenbach WF (1987) Redox processes governing the chemistry of fumarolic gas discharges from White Island, New Zealand. Appl Geochem 2:143–161
Giggenbach WF (1988) Geothermal solute equilibria. Derivation of Na–K–Mg–Cageoindicators. Geochim Cosmochim Acta 52:2749–2765
Giggenbach WF (1992) The composition of gases in geothermal and volcanic systems as a function of tectonic setting. In: Proceedings international symposium water-rock interaction. Balkema vol 7, pp 873–878
Henley RW, Truesdell AH, Barton PB, Whitney JA (1984) Fluid-mineral equilibria in hydrothermal systems. Rev Econ Geol 1:267
Hilton DR, Fischer TP, Marty B (2002) Noble gases and volatile recycling at subduction zones. In: Porcelli D, Ballentine CJ, Wieler R (eds) Noble gases in cosmochemistry and geochemistry. Mineralogical Society of America, Washington, pp 319–370
Horita J, Wesolowski DJ (1994) Liquid-vapor fractionation of oxygen and hydrogen isotopes of water from the freezing to the critical temperature. Geochim Cosmochim Acta 58:3425–2437
Inbar M, Reyes Enriquez A, Graniel JH (2001) Morphological changes and erosion processes following the 1982 eruption of El Chichón volcano, Chiapas, Mexico. Geomorphologie: relief, processes, environment. N3: 175–184
Jutzeler M, Varley N, Roach M (2011) Geophysical characterization of hydrothermal systems and intrusive bodies, El Chichón volcano (Mexico). J Geophys Res 116. doi:10.1029/2010JB007992
Layer PW, García-Palomo A, Jones D, Macías JL, Arce JL, Mora JC (2009) El Chichón volcanic complex, Chiapas, México: stages of evolution based on field mapping and 40Ar/39Ar geochronology. Geofis Int 48:33–54
Lewis AJ, Palmer MA, Sturchio NC, Kemp AJ (1997) The rare earth element geochemistry of acid-sulphate and acid-sulphate-chloride geothermal systems from Yellowstone National Park, Wyoming, USA. Geochim Cosmochim Acta 61:695–706
Luhr JF, Carmichael ISE, Varekamp JC (1984) The 1982 eruptions of El Chichón volcano, Chiapas, Mexico: mineralogy and petrology of the anhydrite-bearing pumices. J Volcanol Geoth Res 23:69–108
Macías JL (1994) Violent short–lived eruptions from small–size volcanoes: El Chichón, Mexico (1982) and Shtyubel', Russia (1907). PhD thesis. Buffalo, State University of New York at Buffalo, 193 pp
Macías JL, Arce JL, Garduño-Monroy VH, Rouwet D, Taran Y (2010) Estudio de prospección geotérmica para evaluar el potencial del volcán Chichónal, Chiapas. Convenio n° 9400047770 IGF-UNAM-CFE
Majoube M (1971) Fractionnement en oxygene-18 et en deuterium entre l'eau et sa vapeur. J Chim Phys 197: 1423–1436
Manea VC, Manea M (2006) The origin of modern Chiapanecan volcanic arc in southern Mexico inferred from thermal models. In: Rose WI, Bluth GJS, Carr MJ, Ewert JW, Patino LC, Vallance JW (eds), ch2: Volcanic Hazards in Central America. GSA Special Paper, vol 412, pp 27–38
Marty B, Jambon A (1987) C/^{3}He in volatile fluxes from the solid Earth: implications for carbon geodynamics. Earth Planet Sci Lett 83:16–26
Mazot A, Taran YA (2009) CO_2 flux from the crater lake of El Chichón volcano (México). Geofis Int 48:73–83
Mazot A, Rouwet D, Taran Y, Inguaggiato S, Varley N (2011) CO_2 and He degassing at El Chichón volcano, Chiapas, Mexico: gas flux, origin and relationship with local and regional tectonics. In: Inguaggiato S, Shinohara H, Fischer T (eds) Geochemistry of volcanic fluids: a special issue in honor of Yuri A. Taran Bull Volcanol vol 73, pp 423–442
Méndez Ortiz BA (2007) Geoquímica e isotopía de aguas de formación (salmueras petroleras) de campos mesozoicos de la cuenca del sureste de México: implicación en su origen, evolución e interacción agua-roca en yacimientos petroleros. PhD Thesis, Universidad Nacional Autonoma de Mexico-CGEO, 224 pp
Michard A (1989) Rare earth element systematics in hydrothermal fluids. Geochim Cosmochim Acta 53:745–750
Molina-Berbeyer R (1974) Informe preliminar geoquímico de los fluidos geotérmicos del volcán del Chichonál, Chiapas. Unpublished report. Comisión Federal de Electricidad, México. Informe 27–74:24

Morton-Bermea O, Armienta MA, Ramos S (2010) Rare-earth element distribution in water from El Chichón Volcano Crater Lake, Chiapas Mexico. Geofis Int 49:43–54

Müllerried FKG (1933) El Chichón, único volcán en actividad descubierto en el estado de Chiapas. Memorias de la Sociedad Científica Antonio Alzate 54:411–416

Peiffer L (2011) Caracterización geoquímica y potencial geotérmico de los acuíferos del volcán El Chichón, Chiapas, México. PhD Thesis, Universidad Nacional Autónoma de México, 182 pp

Peiffer L, Taran Y, Lounejeva E, Solis-Pichardo G, Rouwet D, Bernard-Romero R (2011) Tracing thermal aquifers of El Chichón volcano-hydrothermal system (México) with $^{87}Sr/^{86}Sr$, Ca/Sr and REE. J Volcanol Geoth Res 205:55–66

Peterman ZE, Hedge CE, Tourtelot HA (1970) Isotopic composition of Sr in seawater throughout Phanerozoic time. Geochim Cosmochim Acta 34:105–120

Rouwet D (2006) Estudio geoquímico comparativo de los sistemas hidrotermales de los volcanes activos en Chiapas: El Chichón and Tacaná. PhD Thesis, Universidad Nacional Autónoma de México, 226 pp

Rouwet D (2011) A photographic method for detailing the morphology of the floor of a dynamic crater lake: the El Chichón case (Chiapas, Mexico). Limnology 12(3):225–233

Rouwet D, Taran Y, Varley NR (2004) Dynamics and mass balance of El Chichón crater lake, Mexico. Geofis Int 43:427–434

Rouwet D, Taran Y, Inguaggiato S, Varley N, Santiago SJA (2008) Hydrochemical dynamics of the "lake–spring" system in the crater of El Chichón volcano (Chiapas, Mexico). J Volcanol Geoth Res 178:237–248

Rouwet D, Bellomo S, Brusca L, Inguaggiato S, Jutzeler M, Mora R, Mazot A, Bernard R, Cassidy M, Taran Y (2009) Major and trace element geochemistry of El Chichón volcano-hydrothermal system (Chiapas, Mexico) in 2006–2007: implications for future geochemical monitoring. Geofis Int 48:55–72

Rowe GL, Brantley SL, Fernández JF, Borgia A (1995) The chemical and hydrologic structure of Poás volcano, Costa Rica. J Volcanol Geoth Res 64:233–267

Sano Y, Marty B (1995) Origin of carbon in fumarolic gases from island arcs. Chem Geol 119:265–274

Sruoga P, Rubinstein N, Hinterwimmer G (2004) Porosity and permeability in volcanic rocks: a case study on the Serie Tobifera, South Patagonia, Argentina. J Volcanol Geoth Res 132:31–43

Stimac JA, Goff F, Counce D, Larocque ACL, Hilton DR, Morgenstern U (2004) The crater lake and hydrothermal system of Mount Pinatubo, Philippines: evolution in the decade after the eruption. Bull Volcanol 66:149–167

Takano B, Suzuki K, Sugimori K, Ohba T, Fazlullin SM, Bernard A, Sumarti S, Sukhyar R, Hirabayashi M (2004) Bathymetric and geochemical investigation of KawahIjen Crater Lake, East Java, Indonesia. J Volcanol Geoth Res 135:299–329

Taran YA, Peiffer L (2009) Hydrology, hydrochemistry and geothermal potential of El Chichón volcano–hydrothermal system, Mexico. Geothermics 38:370–378

Taran Y, Rouwet D (2008) Estimating thermal inflow to El Chichón crater lake using the energy-budget, chemical and isotope balance approaches. J Volcanol Geoth Res 175:472–481

Taran YA, Varley NR (1999) New data about El Chichón crater lake. IAVCEI meeting, Jakarta

Taran Y, Fischer TP, Pokrovsky B, Sano Y, Armienta MA, Macías JL (1998) Geochemistry of the volcano–hydrothermal system of El Chichón Volcano, Chiapas, Mexico. Bull Volcanol 59:436–449

Taran Y, Rouwet D, Inguaggiato S, Aiuppa A (2008) Major and trace element geochemistry of neutral and acidic thermal springs at El Chichón volcano, Mexico. Implications for monitoring of the volcanic activity. J Volcanol Geoth Res 178:224–236

Tassi F, Vaselli O, Capaccioni B, Macías JL, Nencetti A, Montegrossi G, Magro G (2003) Chemical composition of fumarolic gases and spring discharges from El Chichón volcano, Mexico: causes and implications of the changes detected over the period 1998–2000. J Volcanol Geoth Res 123:105–121

Templos LA, Munguia Bracamontes F, Barrera VM (1981) Observaciones geoquímicas en la zona geotérmica del Chichonál, Chiapas, México. Unpublished report. Comisión Federal de Electricidad, Mexico. Informe 33–81:32 pp

Truesdell AH, Nathenson M, Rye RO (1977) The effects of subsurface boiling and dilution on the isotopic compositions of Yellowstone thermal waters. J Geophys Res 82:3694–3703

Varekamp JC, Kreulen R (2000) The stable isotope geochemistry of volcanic lakes, with examples from Indonesia. J Volcanol Geoth Res 97:309–327

Varekamp JC, Pasternack GB, Rowe GL (2000) Volcanic lake systematics II. Chemical constraints. J Volcanic Geoth Res 97:161–180

Wohletz K, Heiken G (1992) Volcanology and Geothermal Energy. University of California Press, Berkeley, CA, p 432

Wood SA (2006) Rare earth element systematics of acidic geothermal waters from the Taupo Volcanic Zone, New Zealand. J Geochem Explor 89:424–427

Comparison of the Seismicity Before and After the 1982 El Chichon Eruption

5

D. Legrand, J.M. Espíndola, Z. Jiménez, T. Scolamacchia, C. Valdés-González, S.K. Singh, J. Lermo, Z. Spica, and R.W. Valenzuela

Abstract

The seismicity recorded at El Chichón Volcano shows significant differences before and after the 1982 eruptive episodes. The analysis of the seismicity was performed using two well-known laws in seismology: the Gutenberg-Richter law, which describes the frequency-magnitude distribution of seismicity and the Omori law, which applies to the temporal decay of the number of earthquakes. Results of the analysis suggest that large quantities of fluids (hydrothermal fluids, water, and/or magma) were involved in the physical processes that generated the precursory seismicity one month before the onset of the eruption; only shallow (2–8 km depth) hybrid earthquakes were located and long-period and volcanic tremors were recorded. The volcanic signature of the seismicity is better described by two slopes instead of only one as is usually the case for tectonic events. Furthermore, the seismicity did not follow the Omori law. In contrast, for one month after the last eruptive phases of the 1982 eruptive events, the seismicity showed a more 'tectonic' signature, as evidenced by the occurrence of five large earthquakes (magnitudes ~ 3.8) on 4 April 1982. These followed both the Gutenberg-Richter law (with a single slope) and the Omori law. The tectonic signature is confirmed further by the single-slope observed in the distribution of the seismicity recorded during the period 1985–1990. Such behavior suggests the absence of abundant fluid at depth after the last plinian event of 4 April 1982, with a slow return to a regular tectonic response of the volcanic system. Seismic analysis of the 1982 eruptive sequence illustrates clear volcano-tectonic feedback interactions.

5.1 Introduction

Seismic monitoring is an efficient—and arguably the most diagnostic—tool to forecast the possible re-activation of a volcano and monitor an eruption. A permanent seismological network, located at about 50 km from the El Chichón volcano, was operating before the 1982 volcanic activity, and a portable network was installed around the volcano during the activity. The seismicity recorded by the two networks allowed for a better understanding of the operative processes before

D. Legrand (✉) · J.M. Espíndola · Z. Jiménez · C. Valdés-González · S.K. Singh · Z. Spica · R.W. Valenzuela
Instituto de Geofísica, Universidad Nacional Autónoma de México, Av. Universidad 3000, 04510 México D.F, Mexico
e-mail: denis@geofisica.unam.mx

T. Scolamacchia
Department. Geo Und Umweltwissenschaften, Ludwig Maximilians Universität, Theresienstr. 41-III, 80333 Munich, Germany

J. Lermo
Instituto de Ingeniería, Universidad Nacional Autónoma de México, Av. Universidad 3000, 04510 México D.F, Mexico

T. Scolamacchia and J.L. Macías (eds.), *Active Volcanoes of Chiapas (Mexico): El Chichón and Tacaná*, Active Volcanoes of the World, DOI 10.1007/978-3-642-25890-9_5, © Springer-Verlag Berlin Heidelberg 2015

and after the 1982 eruptions. Seismic analysis of those data show that the 1982 El Chichón eruptive sequence provides a text-book case of volcano-tectonic interaction, as illustrated by a significant change in the nature of the seismicity with respect to time.

One month before the first eruptive event of 29 March 1982, only hybrid events (Hs) at shallow depths (from 0 to 5– 8 km), long period events (LPs) and volcanic tremors occurred, while volcano-tectonic events (VTs) were absent. From one month to a few years following the last, and largest, eruptive event of 4 April 1982 at 1110 UT, the opposite trend is observed. Only VTs are present and these are located at greater depths (down to 20 km). This fundamental change in the nature of seismicity is clearly evidenced by: 1) the Gutenberg-Richter law that shows a break in the slope before the first eruptive event and a single slope after the last, large eruptive event; and 2) the Omori law, which is not followed before the first eruptive event but is satisfied after the final one (see inset box 1). This observation is interpreted to mainly reflect a change from a volcanic character of the seismicity before the eruption, driven by the presence of high quantities of fluids (magma and/or water), to a tectonic regime afterwards.

Inset Box 1: Scaling Laws of Seismic Swarms
Seismic swarms are large concentrations of earthquakes in space and time, higher than the background seismicity (Mogi 1962, 1963). They can be of tectonic or volcanic origin, or a combination of both. A tectonic earthquake swarm is often an aftershock sequence following a 'main-shock'. It starts abruptly with a large magnitude earthquake (i.e. main-shock) followed by aftershocks which are, generally, of smaller magnitudes. Such swarms may last a few months depending on the magnitude and on the focal mechanism of the main-shock. In contrast, a volcanic earthquake swarm often starts gradually, without the occurrence of a comparatively large earthquake at the beginning. The largest earthquakes may occur in the middle of the seismic swarm, or close to its end (Yamasato et al. 1991; Sapin et al. 1996). The number of earthquakes increases with time until it reaches a paroxysm and decreases afterwards (Mogi 1962, 1963). A volcanic swarm can last from a few days to a few months, or even several years (Bullen and Bolt 1947, 1993; Ukawa and Tsukahara 1996; Kunugi et al. 2000).

Two basic laws of seismology can help in discriminating tectonic from volcanic swarms. The Omori law (Omori 1895), one of the first relationships established in seismology, shows that the number of aftershocks (*N*) decreases with time (*t*) as a power law. The temporal derivative of *N*(*t*) follows the relation: $\frac{dN}{dt} = \frac{K}{(t+c)^p}$, where *K*, *c* and *p* are constants. *K* represents the number of aftershocks of the first time interval (i.e. hour, day or year). The exponent *p* represents the decay rate of aftershocks with respect to time. Larger *p* values indicate a faster decrease of the earthquake sequence. The constant *c* is used to avoid a discontinuity at $t = 0$ (Omori 1895; Utsu 1961). An aftershock sequence that follows a large tectonic earthquake fits this law; in contrast, inter-seismic-tectonic and volcanic-earthquake swarms do not.

A second basic law in seismology, known as the Gutenberg-Richter law, considers the distribution of the magnitude of earthquakes (Ishimoto and Iida 1939; Gutenberg and Richter 1944). It describes the number (*N*) of earthquakes of magnitude greater than or equal to a magnitude (*M*) during the time window (*t*) as: $\log_{10}[N(t) \geq M] = a(t) - bM$, where *a*(*t*) is the number of earthquakes of magnitude ≥ 0 during the time window (*t*), and *b* is a constant over the time interval (*t*). The *b*-value describes the number of small earthquakes with respect to large ones (i.e. the bigger is *b*, the larger is the number of small earthquakes). Note that the *b*-value is the slope of a line fitted to the plot of the logarithm of the cumulative frequency of events of a given magnitude versus that magnitude. Aftershock sequences, inter-seismic earthquake sequences, and volcanic swarms follow the Gutenberg-Richter law. The *b*-value is commonly calculated following Aki (1965).

Both, the Omori and the Gutenberg-Richter laws are different for tectonic and volcanic swarms. A tectonic swarm differs from a volcanic swarm in both time and magnitude distributions, due to the existence of an external force that modifies the temporal evolution of the seismicity as described below.

A tectonic aftershock sequence represents the response of an elastic medium (the earth), to a single main-shock of short duration (few to few tens of seconds) compared with the duration of the complete sequence (few tens of days to a few months). Hence, the number of aftershocks decreases with time. In contrast, a volcanic swarm is generated by movement of fluids (magma, water and/or gas), acting as external forces, which can increase and/or decrease with time. Therefore, a volcanic swarm may last a very long time (a few days, months, or years), with an increase or decrease of the number of earthquakes during its course. For this reason, the simple study of the temporal evolution of the number of earthquakes (the Omori law) helps in discriminating the tectonic or volcanic origin of a swarm.

Aftershock sequences, inter-seismic tectonic earthquake sequences, and volcanic swarms follow the Gutenberg-Richter law, but yield different *b*-values. In general, the *b*-value is smaller for an aftershock sequence of tectonic origin than for a volcanic swarm, which means that an aftershock sequence shows a greater number of larger events than a volcanic swarm, characterized by more small earthquakes than large ones. Typical *b*-values of tectonic earthquake sequences vary between 0.6 and 1.0, whereas volcanic swarms have *b*-values much greater than 1 (e.g. Wiemer and McNutt 1997; Wyss et al. 1997, 2001). As an example, $b = 1$ indicates the occurrence of ten times more earthquakes of $M \geq 5$ than earthquakes of $M \geq 6$. Likewise, there occur a hundred times more earthquakes of $M \geq 4$ than earthquakes of $M \geq 6$, and so on. It follows that the magnitude of earthquakes is not randomly distributed: only a few large earthquakes occur, whereas many small ones are produced, and their relative numbers follow a non-linear power law. Therefore, the Gutenberg-Richter law suggests the existence of a self-organization of the distribution of the earthquake magnitudes (Bak and Tang 1989), and a self-organization in the spatial distribution of earthquakes, a fact that is not intuitive a priori. As an example of such self organization is the equation $D = 2b$, where D is the fractal dimension of the spatial distribution of earthquakes (Aki 1981). This equation indicates that the spatial distribution of earthquakes depends directly on their magnitude. It can be shown that this relation depends on magnitudes. For large earthquakes, $D = b$ which means that these earthquakes have the tendency to be distributed over a line. On the other hand, for small earthquakes, as is often the case for volcanoes, $D = 3b$, which means that small earthquakes tend to be distributed over a volume (Legrand 2002). For intermediate-size earthquakes, $D = 2b$, which means that earthquakes are distributed over a plane (i.e. the fault plane). Note that the magnitude ranges of small, intermediate and large earthquakes are not absolute values but depend on the earthquake sequence.

When fluids are present, as it occurs in most volcanic or geothermal environments, the frequency-magnitude distribution shows a very specific behavior, namely: a break in the slope of the distribution, that is, two different *b*-values. This peculiar behavior has been observed in volcanoes (Caplan-Auerbach and Duennebier 2001; Legrand et al. 2004; Kundu et al. 2012), geothermal fields, and tectonic earthquake sequences where fluids are involved (Neunhöfer and Hemmann 2005).

Therefore, many swarms considered as purely 'volcanic' have in fact both, a volcanic and tectonic origin, generated through a feedback process (Legrand et al. 2002; Roman 2005; Roman and Cashman 2006; Roman and Heron 2007; Roman et al. 2008; Legrand et al. 2011; Kundu et al. 2012; Vargas-Bracamontes and Neuberg 2012) as discussed below. Faults in volcanic regions represent preferential paths for magma movement. Many interactions and feedback between the tectonic and the volcanic process are possible when faults intersect the volcanic edifice. Magma movement may reactivate a pre-existing fault, crossing the volcano edifice at local, or regional, distances. Faults reactivated during an earthquake with a normal focal mechanism, and/or a focal mechanism with an extensional component (i.e. due to the opening of the two blocks located at each side of the fault), may facilitate magma migration. Usually this phenomenon is not unidirectional (i.e. one phenomenon triggering the other as an earthquake that facilitates magma migration or, alternatively, magma movement that generates an earthquake). Therefore the movement of magma and earthquakes over faults is a complex, dynamical, feedback process. Both, magma movement and earthquakes will change the local stress tensor, favoring the creation of new faults, or re-activating pre-existing ones at local distances. Such feedback interactions between tectonism and volcanism may be illustrated in pull-apart systems, as in the case of El Chichón (García-Palomo et al. 2004).

5.2 Tectonic and Seismic Setting of El Chichón

As discussed in Chap. 1, the regional tectonics of the area is compressional, related to the subduction of the oceanic Cocos plate beneath the continental North American and Caribbean plates. At a local level, several faults exist close to the volcano, some of them cutting the edifice itself. El Chichón is one of the few volcanoes in México located within a local strike-slip fault system (the directions of the principal stress axes of the stress tensor are σ_1 N70°E, σ_2 vertical, and σ_3 N20°W) associated with left-lateral strike-slip faults oriented NNW-SSE and E-W (García-Palomo et al. 2004; Macías et al. 2010, Chap. 1). This set of faults is compatible with an extensional regime, with en échelon faults, and occurs within a pull-apart basin system. Normal faults related to this last system have orientations ranging from NW-SE to almost N-S (Guzmán-Speziale et al. 1989; Guzmán-Speziale and Meneses-Rocha 2000; Bursik 2009, Chap. 1). The E-W left-lateral strike-slip San Juan Fault system (SJF) (García-Palomo et al. 2004), and the NNW-SSE Chichón-Catedral fault (Macías et al. 2010) were the loci of old magmatic and volcanic activity (see Chaps. 1 and 3). Additional faults in the N45°E direction (Chapultenango Fault System), dipping NW with normal faulting, generate a half-graben on top of which lies the volcanic edifice (Macías et al. 1997; García-Palomo et al. 2004). A right-lateral strike-slip conjugate fault system exists in the N-S direction, almost perpendicular to the E-W left-lateral fault system (Duffield et al. 1984; García-Palomo et al. 2004, Chap. 1). All these left-lateral (E-W) and right-lateral (N-S) strike-slip faults and normal faults (in a N45°E direction) are compatible with the regional stress-tensor described above.

A seismic swarm occurring in the 1920s motivated the visit of noted geologist Friedrich K.G. Müllerried to El Chichón in 1928 and in 1930 (Müllerried 1932, 1933). A second seismic swarm took place in 1967–1968 along with an increase of the superficial thermal activity (Gonzalez-Salazar 1973; Molina-Berbeyer 1974). Yet none of those two seismic swarms culminated in an eruption (Tilling 2009). During fieldworks in 1964 and in the 1980s, geoscientists felt local earthquakes, which they associated to the volcano (Medina 1982; Tilling 2009). During their fieldwork in late 1980 and early 1981, Canul and Rocha (1981) heard rumbling sounds from the subsurface, '*being stronger and more frequent in the crater*', and felt local earthquakes related to the volcano, suggesting that the volcano could erupt soon. People living close to the volcano felt further local seismic events during the late 1981-early 1982 and noticed an increase in the fumarolic activity three months before the first eruption (Tilling 2009, see Chap. 3 for more details).

5.3 Seismological Networks and seismicity

Although the 1982 eruptive sequence lasted only one week, seismic activity continued for about 4 weeks after the eruption (Medina 1982; Havskov et al. 1983; Medina et al. 1990, 1992; Yokoyama et al. 1992; Jiménez et al. 1999; Espíndola et al. 2002, 2006; Tilling 2009). Two independent seismic networks recorded this activity: one was operating before the 1982 (permanent CFE network); a second one (temporary network) was installed on 30 March 1982 after the first eruptive event, when UNAM scientists arrived in the field (De la Cruz-Reyna and Martín del Pozzo, 2009, see Chap. 8).

5.3.1 The permanent CFE network

A permanent network of 14 vertical-component, short period (1 Hz) seismometers, installed by Federal Power Commission (CFE) in collaboration with the Instituto de Ingeniería of UNAM to monitor the seismicity related to the Chicoasen dam, was in operation at the time of the 1982 eruption (Fig. 5.1). The first station (CSN), was installed in March 1976, three more stations (CZC, OZC, PEN) were installed in May–June 1979. The ten additional ones were installed between December 1979 and January 1981 (Table 5.1, Fig. 5.1). The complete network was fully operational by January 1981, and consisted of six telemetry stations (i.e. data were transmitted analogically to Tuxtla Gutiérrez where they were continuously recorded on helicorders), and three autonomous smoked-paper stations. The station closest to the volcano (TPN) was at distance of about 27 km, had a low gain, and did not correctly record the signals at all frequencies. Station CR3 was located about 62 km S–SE of the volcano, had a high gain and was recording continuously. CR3 recorded the diversity of volcanic signals better than TPN, the station nearest to the volcano. Thus, CR3 has been used previously to study the local seismicity before and during the 1982 eruptive activity sequence (Haskov et al. 1983; Medina et al. 1990; Yokoyama et al. 1992; Jiménez et al. 1999). Additionally, the signals recorded at this station during the volcanic activity, were not clipped, in contrast to the stations close to the volcano. Consequently, station CR3 was used to estimate seismic magnitudes, especially the largest ones, because the signals were not saturated (Havskov et al. 1983). Therefore, the data from this station allow a precise study of the Gutenberg-Richter and Omori laws.

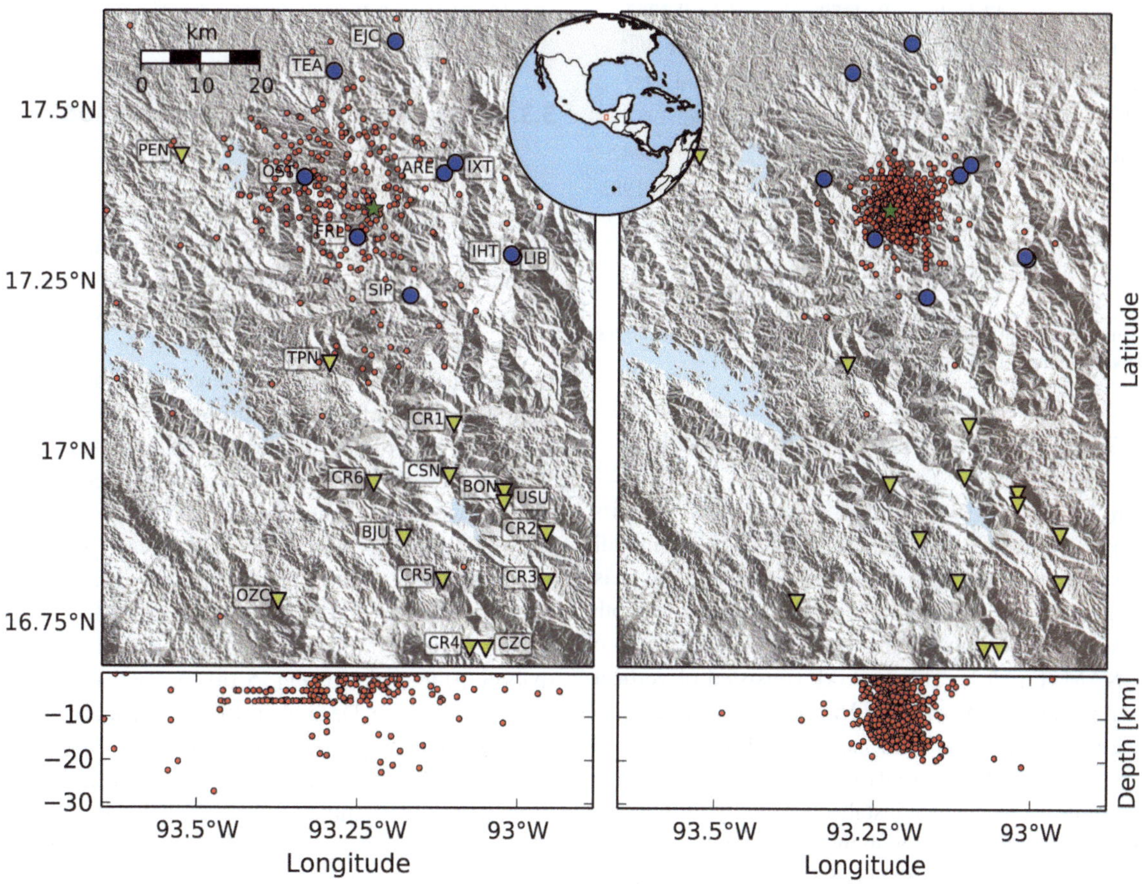

Fig. 5.1 Seismological stations (*inverted yellow triangles* for the CFE stations and *blue circles* for UNAM portable seismic network following the 1982 eruptions), referred in Table 1. *Green star* shows the 1982 crater. **a** *Red dots* show the pre-eruption seismicity from 1 January 1980 to 26 March 1982. **b** *Red dots* show the post-eruption seismicity from 1 April 1982 to 27 April 1982

Table 5.1 List of coordinates, and date of installation of the two seismological networks

Name of the station	Latitude (North)	Longitude (West)	Date of installation
BJU	16.8790	93.1790	April 1980
BON	16.9460	93.0220	April 1980
CR1	17.0450	93.1012	April 1980
CR2	16.8850	92.9550	April 1980?
CR3	16.8140	92.9540	January 1981
CR4	16.7172	93.0748	January 1981
CR5	16.8160	93.1180	April 1980
CR6	16.9582	93.2260	April 1980
CSN	16.9690	93.1080	March 1976
CZC	16.7170	93.0500	May 1979
PEN	17.4360	93.5280	June 1979?
OZC	16.7850	93.3730	May 1979
TPN	17.1330	93.2950	December 1979
USU	16.9300	93.0220	April 1980
ARE	17.4120	93.1180	1st April 1982
EJC	17.6045	93.1955	2nd April 1982
IHT	17.2905	93.0083	9th April 1982
IXT	17.4272	93.1008	1st April 1982
FRL	17.3180	93.2530	25th April 1982
LIB	17.2940	93.0120	30th March 1982
OST	17.4055	93.3347	1st April 1982
SIP	17.2330	93.1698	2nd April 1982
TEA	17.5612	93.2905	8th April 1982

CFE permanent seismological network (*inversed triangles*). UNAM temporal seismological network (*circles*)

5.3.2 The Temporary UNAM Network

On 30 March 1982, almost 2 days after the first explosive event (29 March 1982, 0515 UT), an additional temporary network of nine short-period seismometers was installed around the volcano by both, Instituto de Geofísica and Instituto de Ingeniería of UNAM (Fig. 5.1). This network consisted of vertical-component, short-period (1 Hz) seismometers coupled with continuous, smoked paper MEQ analog recorders. Given the high activity of the volcano at this time, these instruments were installed relatively far from the volcano (see Chap. 8), with the nearest station at about 8 km from the summit (Fig. 5.1).

5.3.3 Classification of the 1982 Seismic Signals

During a volcanic eruption, it is important to characterize the seismicity (via a classification of the signals) in order to understand and follow the physical process of an eruption. Jiménez et al. (1999) and Espíndola et al. (2006) classified the seismic activity related to the 1982 eruption using the four categories defined by Lahr et al. (1994) and Chouet (1996): (1) high-frequency (1–2 Hz) volcano-tectonic events (VTs), characterized by clear P and S waves; (2) low-frequency (between 0.2 and 0.5 Hz) long-period events (LPs), characterized by emergent P waves and almost no S waves; (3) hybrid events (Hs) characterized by a combination of low and high frequency waves, between 0.2 and 2 Hz, with emergent P waves, and distinguishable S waves; and (4) low-frequency (between 0.2 and 0.5 Hz) volcanic tremors. The same classification is followed in this chapter.

5.3.4 Seismicity Before 29 March 1982, 0515 UT

The permanent seismological network of CFE (Fig. 5.1, Table 5.1) recorded the seismicity of the region for a few years before the 1982 eruption (Fig. 5.1a). An anomalous seismicity was first detected in late 1980, which further increased in late 1981. Before January 1980 only a few earthquakes were recorded, with only four occurring in 1979 (Jiménez et al. 1999). From 1 January 1980 to 28 February 1982, 77 events (average of less than three events/month), with local magnitudes between 2.8 and 3.8, occurred within 15 km from the volcano (Jiménez et al. 1999, Fig. 5.1a). The number of events (up to 170) increased from 1 March to 28 March 1982. Even if their locations were poor, their foci were probably shallow (0 to 5–6 km, Fig. 5.1a). A few earthquakes recorded between December 1981 and January 1982 were also felt by people (Espíndola et al. 2002), and accompanied by an increase in fumarolic activity. An earthquake of magnitude 2.9 on 26 February 1982 at 1806 UT marked the start of sustained seismic activity (Jiménez et al. 1999; Espíndola et al. 2006, Table 5.2).

Table 5.2 Chronology of the seismological activity registered during the 1982 eruptions

Date	Hour of the eruptions	Seismological records	Descriptions of eruptive activity and interpretations offered by previous works	VEI
29 March	0515 (2315 March 28, local time)[b, e]	Both isolated and continuous tremors from March 28 at 2300 to March 29 at 0328 that stopped abruptly about 2 h before the eruption	Plinian column with a maximum altitude of 17.5 km. Destroyed partially (1/4–1/3) the summit dome (see also Chap. 3). First eyewitness accounts at 0532. Lasted 5 h[c].	Initially estimated to VEI = 4[a]
		Complete seismic quiescence after 0328 for 1 h and 10 min		
		Tremor of strong amplitude started at 0438 and last until 0530, the signal was clipped out during the first 30 min. A high amplitude tremor marked the start of the eruption, and stopped around 0600[b, h, i]		Re-estimated to VEI = 5[j]
30 March	1500[c, h]	Complete seismic quiescence for the first 12 h of the day. Small amplitude tremor overlapped with LPs swarms at 1315[a]. Seismic activity is small[a, b]	Small eruption[e, f]	
30 March	2100[c, h, i]	Small seismic activity, lasted about 4 h[h, i]	Small eruption with a plume of about 2–3 km altitude[e, f]	
31 March	1930[h]	Tremor started at 1315 on March 30 reaches a maximum amplitude at 0200 on March 31[h]. Small LPs swarms between 1340 and 1600 overlapping small amplitude tremors[h]	Small eruption with plume of 2–3 km of altitude[c, d]	
01 April		Small LPs swarms between 1340 and 1600[h]		
02 April	1700[e, i] (satellite)	Small amplitude tremor started on March 31 continued until April 2 at 0400[a] Small LP swarm from 1339 to 1600[a].	Small mushroom-shaped plume, which rose up to 3.5 km in 30 min[c, e, i]	
		Small seismic activity[h, i] No seismic signal correlated with satellite observations at 1700[h]		
03 April	0000–0030[c] (satellite)	Absence of seismicity[b]	Small eruption. Plume dispersed to the E[c]	
03 April	0830–0900[c] (satellite)	Tremor and overlapping LPs from 0830 to 0915 (with strong amplitude during 24 min)[h, i]	Plinian event A2[c, f, g]	
	0840[b]	Seismic quiescence from 0915 to 0936[h]		
03 April	1003[b]	Tremors of varying amplitude from 0936 to 1045[h]	Minor eruption[b]	
03 April	1500[c] (satellite)	No LPs registered before satellite records[h]	Small eruption plume dispersed toward SW and SE[c, e]	
		Strong seismic activity between 1400 and 1600[h, i]. Alternating periods of tremor and quiescence between 1218 and 1545, and between 1630 and 1810–2020		
04 April	0130–0200[h]	Volcano-Tectonic earthquakes from 0130 to 0212 (with strong amplitude for 25 min), and from 1630 to 1810[h, i]	Plinian event (fall B)[c, d, e, f, g, h]	VEI = 5
	0139[b]		Lasted about 4–7 h (See chap. 3)	
	0135[c, e] (eyewitness)			

(continued)

Table 5.2 (continued)

Date	Hour of the eruptions	Seismological records	Descriptions of eruptive activity and interpretations offered by previous works	VEI
04 April	1110[b]	Intense tremor started at 1050 until 1310 saturated all seismic stations within 70 km^2, and recorded by many seismometers up to 720 km (Mexico city)[b,h]	Major phreatoplinian eruption (fall C)[b, c, d, e, f]	VEI = 5
	1122[c, e] (eyewitness)			
	1130[c] (satellite)	Seismic activity started again at 1106 (end not determined)[i]		

All times are UT (Universal Time = local time + 6 h)

[a] Newhall and Self 1982

[b] Haskov et al. 1983

[c] Sigurdsson et al. 1984

[d] Matson 1984

[e] Carey and Sigurdsson 1986

[f] Macías et al. 1997

[g] Macías et al. 1998

[h] Jiménez et al. 1999

[i] Espíndola et al. 2002

[j] Tilling 2009

An intense seismic swarm started on 1 March 1982, 28 days before the first eruptive event. The largest earthquake (local magnitude 4.0) occurred on 6 March 1982 at 1324 UT, 5 days after the beginning of the seismic swarm (Table 5.3, Fig. 5.2). This earthquake prompted some villagers to get away from the volcano (Espíndola et al. 2002). Most of this pre-eruptive seismicity was composed mainly of hybrid events (Hs), and lasted until 28 March 1982 (Figs. 5.1a, and5.2). The depth of such events was shallow, between 0 and 5–6 km (Havskov et al. 1983; Jiménez et al. 1999; Espíndola et al. 2006, Fig. 5.1a). Sequences of LPs started on March 23, reaching a peak number of 66 within 4 h (from 0906 to 1316 UT) on March 27. LPs of small magnitude started on March 28, at 0400 UT, 25 h before the onset of the eruption (Fig. 5.2). The same day, at 1237 UT, an earthquake of magnitude 3.8 occurred (Medina et al. 1992; Yokoyama et al. 1992; Jiménez et al. 1999; Espíndola et al. 2006, Table 5.3, Fig. 5.2), followed by ten smaller events within 10 min. The last hybrid event occurred on March 28, at 1458 UT. It was followed by an incessant overlapping of LP events that began at 1800 UT, and turned into a continuous volcanic tremor of varying amplitude at 2300 UT, which suddenly stopped on March 29 at 0328 UT (Jiménez et al. 1999; Espíndola et al. 2002), 2 h before the first eruption (Table 5.2). This seismic quiescence of about 70 min was again followed by a tremor of larger amplitude during 30 min, and followed by the first eruption of March 29 at 0515 UT. The tremor was sustained for 30 min after the beginning of the eruption.

Table 5.3 List of the largest earthquakes recorded at El Chichón before the first eruption on 29 March 1982 at 0515 UT (data from Yokoyama et al. 1992; Jiménez et al. 1999; Espíndola et al. 2002)

Date	Hour (UT)	Magnitude
26 November 1981	1241	3.2
04 December 1981	2018	2.4
13 January 1982	1719	3.1
26 February 1982	1806	2.9 (Start of the seismic swarm)
04 March 1982		30 earthquakes within 24 h
06 March 1982	1324	4.0 (Strongest event recorded during the crisis)
28 March 1982	1237	3.8

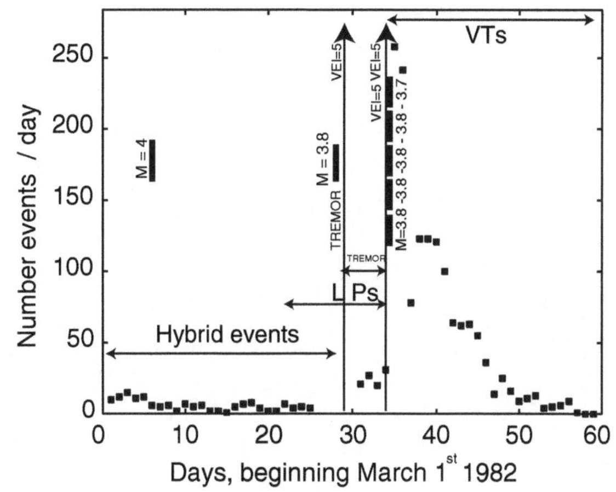

Fig. 5.2 Number of earthquakes/day from 1 March to 27 April 1982 (*black dots*). *Vertical bars* correspond to the largest earthquakes with corresponding magnitude. *Vertical arrows* correspond to the largest eruptions, with the corresponding VEI (the three largest eruptive events of 29 March 1982 at 0515 UT and 4 April 1982 at 0130 and 1110 UT). The *horizontal arrows* correspond to the main periods of hybrid, long period, volcano-tectonic events and tremors

5.3.5 Seismicity Between 29 March and 30 April 1982

The seismic activity that occurred between March 29, 0515 UT and April 4, 1110 UT consisted of LP events and volcanic tremors, except for the first 12 h of March 30, which were almost seismically quiet.

Three seismic swarms, composed of 23, 26 and 65 LP events occurred on March 30 (at 1315 UT), March 31 (at 1340 UT) and April 2 (at 1340 UT) respectively. Considering the coincidence of these swarms, all starting at about the same time of the day (1340 UT) and ending at about 1600 UT, with the maximum earth tidal strain, Jiménez et al. (1999) and Espíndola et al. (2006) attributed these events to the overpressurization of the magmatic system induced by the earth tide, and the occurrence of strong degassing.

After the eruptive events occurring on April 4 at 0130–0200 and 1110 UT (i.e. Plinian events B and C, respectively), and until April 30, the seismicity alternated between periods of quiescence and of intense activity. It consisted of VTs with depths down to about 25 km already during the events that produced the plinian fallout B, much deeper than the Hs and LPs hypocenters (0–6 km), which occurred before the first eruptive event on March 29 (Havskov et al. 1983; Medina et al. 1990; Yokoyama et al. 1992; Jiménez et al. 1999; Espíndola et al. 2006). Five strong earthquakes (M = 3.8; 3.8; 3.8; 3.8; 3.7) occurred on April 4 at 1353; 1634; 1730; 1829 and at 1937 UT, respectively, after the last major phreatoplinian event of April 4 at 1110 UT (Fig. 5.2).

The velocity model (Jiménez et al. 1999) used to locate these events was deduced from geological data published by Canul and Rocha (1981). The duration magnitude was calculated with the equation $Md = -0.87 + 1.86 \log_{10}(T)$, where T is the total earthquake duration in seconds. It was calibrated using regional earthquakes in the state of Oaxaca, México (González-Ruiz 1980) (Fig. 5.1).

5.3.6 Seismicity Registered a Few Years After the 1982 Eruptive Events

Six earthquakes, with magnitudes smaller than 3.0, were reported within 8 km from the volcano in 1984 by Lermo et al. (1989a). Following this report, the CFE installed a new network, and 17 short period seismometers were operating by August 1985 (Fig. 5.3). From August 1985 to 1990, the seismicity was recorded and located (Lermo et al. 1989a, b; Díaz de León and Lermo 1990; Lermo et al. 1990; Díaz de León 1991, Fig. 5.3). The duration magnitude used by Lermo et al. (1989a) was obtained from the equation $Md = -1.59 + 2.4 \log_{10}(T) + 0.00046\ D$, where T is the total event duration in seconds, and D is the epicentral distance in km. Considering that the epicentral distance (D) is about 10 km, this term was discarded and the local magnitude Md of Lermo et al. (1989a) was transformed into $Md = -0.87 + 1.86 \log_{10}(T)$ corresponding to the magnitude used by Jiménez et al. (1999) in order to have a uniform definition of the magnitude.

5.4 Analysis of the Seismicity

The Omori and Gutenberg-Richter laws (see inset box 1) are useful to show the different origin of the seismicity recorded before and after the 1982 eruptive events.

To perform this analysis, the seismicity was divided into three periods: (a) 1 month before the first eruption of March 29, (b) 1 month after the Plinian event at 1110 UT April 4, and (c) a period of 5 years between

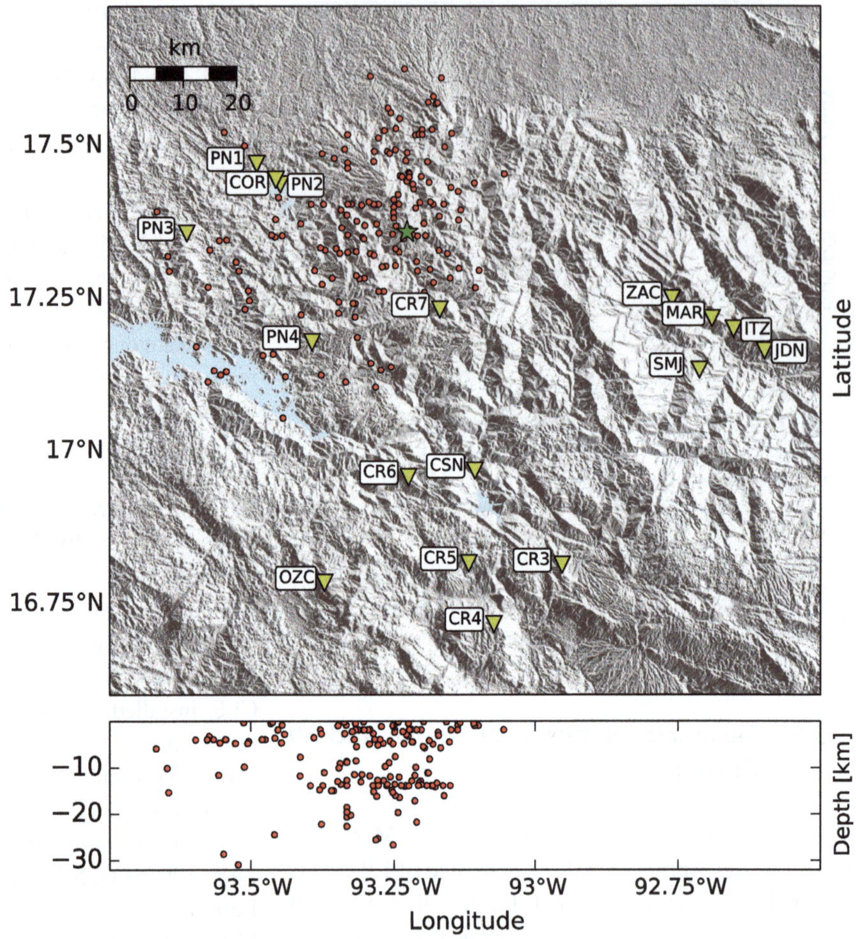

Fig. 5.3 Seismicity from 22 September 1985 to 22 December 1990 located by the CFE seismic network (*inverted yellow triangles*). *Green star* shows the 1982 crater

1985 and 1990. A real absence of seismicity, or a lack of data, prevent the analysis of others periods. As an example, between January 1980 and February 1982 the number of earthquakes registered was too small to be useful for a statistical analysis. On the other hand, between March 26 and March 31 the seismicity was almost continuous, or absent, preventing the analysis. Being statistical laws, the Gutenberg and Omori laws can be applied only to large time spans (i.e. several weeks), and it is not possible to associate them to a specific volcanic event. No detail on the nature of eruptions can be therefore inferred using these analyses.

A total of 1938 earthquakes that occurred between 1980 and 1990 were analyzed: 77 between 1 January 1980 and 28 February 1982, 152 from March 1 to March 26 1982, 1533 from April 1 to 27 April 1982, and 176 from 1985 to 1990.

5.4.1 Omori Law

The number of events/day for the seismic sequences that occurred from March 1 to March 26, and from April 1 to April 27 (1685 earthquakes in total) is shown in Fig. 5.2. Recalling the names used to describe the major Plinian phases of the eruption (see Chap. 3), eruption A refers to the event on 29 March, 0515 UT, eruption B to the one on April 4, 0130 UT, and eruption C to the one on April 4, 1110 UT. Before the onset of eruption A, the number of earthquakes was almost constant (Fig. 5.2) and considering that the number of events/day does not decrease as a power law of time, the Omori law is not fulfilled (Fig. 5.4). The strongest earthquakes (vertical bars in Fig. 5.4) did not occur at the beginning of the sequence but were continuously distributed over time, as normally observed for volcanic swarms. In contrast, after the

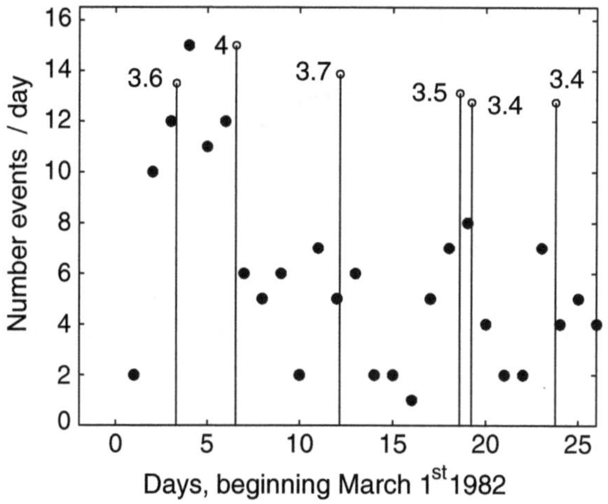

Fig. 5.4 Number of earthquakes per day before the first eruption (from 1 March 1982 to 27 March 1982)

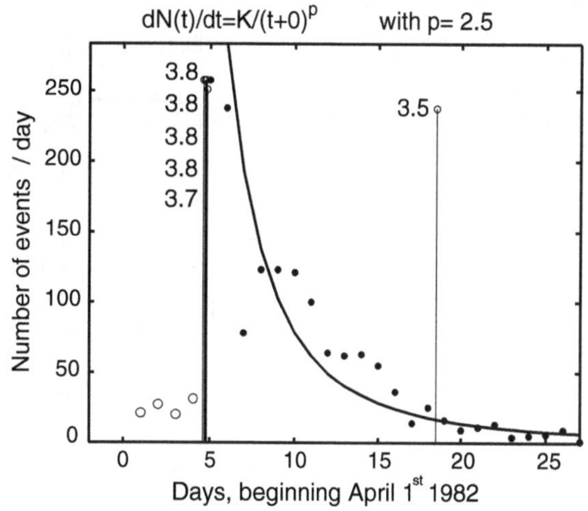

Fig. 5.5 Number of earthquakes per day from 1 April 1982 to 27 April 1982

eruption C, and for the following month, the number of earthquakes decreased as a power law with respect to time, satisfying the Omori law (solid dots in Fig. 5. 5), as is often observed for tectonic swarms. The five strongest magnitude earthquakes occurred at the beginning of this post-eruptive sequence on April 4 between 1353 and 1937 UT (vertical bars in Fig. 5.5), as is often observed for tectonic swarms. For this sequence, the Omori law shows a high p-value of 2.5, indicating that the number of earthquakes decreases very quickly with respect to time.

5.4.2 Gutenberg-Richter (GR) Law

The same time periods used for the Omori law were also analyzed for the Gutenberg-Richter law. The plot of cumulative magnitude of earthquakes versus magnitude recorded before March 29 at 0515 UT (Fig. 5.6a, b; data set Fig. 5.4) shows two different slopes. For a tectonic swarm, the data follows a single slope. To confirm that the double-slope is real and not an artifact, three hypotheses were tested: (1) a single slope was used to fit only the smallest magnitudes, (2) a single slope was used to fit only the largest magnitudes (3) two slopes were used to fit the entire data set.

For the first hypothesis (Fig. 5.6a) a single linear slope was considered, corresponding to a b-value $b_1 = 1.1 \pm 0.2$ calculated between $Mc_1 = 2.4$, the magnitude of completeness (magnitude for which the seismic catalogue is complete), and magnitude 3.2. For the second hypothesis, only the largest magnitudes, between $Mc_2 = 3.2$ and 3.9, were considered and the b-value is $b_2 = 2.2 \pm 0.2$ (Fig. 5.6a). For the third case, all magnitudes between $Mc_1 = 2.4$ and 3.9 with two slopes b_1 and b_2 were considered.

The results show that for the first hypothesis, the earthquakes with magnitudes greater than 3.2 deviate significantly from the linear part corresponding to the slope $b_1 = 1.1$, indicating that the number of large events is in deficit with respect to the number expected with slope $b_1 = 1.1$. Such deviation could be due to the absence of large events in the catalogue, even if the probability of missing large events ($M > 3.2$) is very low. Given that some stations were located close enough to the hypocenters (i.e. less than 30–40 km), the detection of all events of magnitude greater than 2.4 (then including those of magnitude greater than 3.2) is guaranteed.

Furthermore, if the slope $b_2 = 2.2$ were valid for magnitudes between 2.4 and 3.2, a number of 562 earthquakes of magnitude $M > 2.5$ (point B on Fig. 5.6a) should have been recorded. Yet, only 112 earthquakes of magnitude $M \geq 2.5$ were recorded (point A on Fig. 5.6a). Failing to record those 450 earthquakes in one month is unlikely. Thus, the third hypothesis is the most likely: there was a real deficit of large earthquakes with magnitudes greater than 3.2. It is therefore necessary to fit a second slope for these magnitudes (i.e. b_2 in Fig. 5.6a), leading to the observation of two real slopes: one for magnitudes

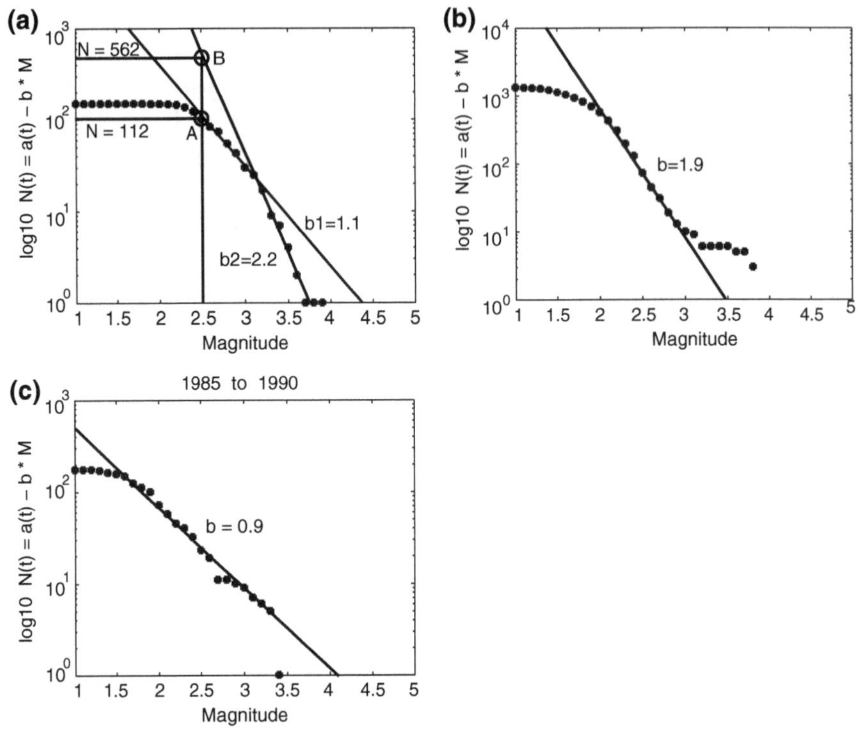

Fig. 5.6 Gutenberg-Richter law for the seismicity. **a** From 1 March 1982 to 27 March 1982. **b** From 1 April 1982 to 27 April 198. **c** From 1985 to 1990

from 2.4 to 3.2 with a b-value $b_1 = 1.1 \pm 0.2$, and another one for magnitudes from 3.2 to 3.8 with a b-value $b_2 = 2.2 \pm 0.2$. The errors estimated for b_1 and b_2 are small enough to guarantee that b_1 and b_2 are significantly different.

The Gutenberg-Richter law of earthquakes from April 1 to April 27, after the first eruptive event A, is shown in Fig. 5.6b and corresponds to the data set of Fig. 5.5. In contrast to Fig. 5.6a, a single slope is observed for magnitudes between 2.0 and 3.2 with a b-value of 1.9 ± 0.2. This value falls between the b_1 and b_2 values found for the pre-eruptive sequence suggesting that the seismicity slowly returned to a 'tectonic' regime.

The number of earthquakes registered after May 1982 is too small to estimate a continuous temporal evolution of b-values. Nevertheless, the seismicity from 1985 to 1990 allowed the estimation of the post-eruptive b-value. The magnitude distribution of the post-eruptive period (1985–1990) is shown in Fig. 5.6c and corresponds to the data set of Fig. 5.3. As in Fig. 5.6b, only one slope is observed for magnitudes from 1.6 to 3.3 and $b = 0.88 \pm 0.15$. In this case, the seismicity follows the Gutenberg-Richter law, showing a typical 'tectonic' signature, with no significant amount of fluids intervening to generate this seismicity.

5.5 Discussion

The characteristics of the seismicity (tectonic vs. volcanic nature of the events, distributions in space and magnitudes) before and after the 1982 eruptive sequence show great differences. Most of the pre-eruptive events (from January 1980 to February 1982 and, at an increased level, from late February 1982 until the first eruptive event) are almost exclusively hybrid events with magnitudes greater than 2.0 and depths between 0 and 5 km, with a few LP events and tremors. Deep events up to 20 km, were very rare (Havskov et al. 1983; Medina et al. 1992; Jiménez et al. 1999; Espíndola et al. 2006). Hybrid events were restricted to the first 2 km 2 days before the first eruption A (Havskov et al. 1983). This observation indicates the presence of fluids (magma and/or water) in the pre-eruptive physical process (Havskov et al. 1983; Espíndola et al. 2006). The absence of seismicity between 7 and 13 km depth before eruption A suggest the presence of a magma chamber at those depths (Jiménez et al. 1999). Such depths are consistent with those of 7–8 km inferred by Luhr (1990), based on isotopic values of coexisting minerals, and composition of the matrix glass, which indicate pre-eruptive temperatures for the 1982 trachyandesitic melt of 800 ± 50 °C, and equilibration at about $P_{tot} = P_{vap} = 2$ kb. The occurrence of volcanic tremors indicates the ascent of fluids. Thus, the seismicity between the first eruption A, and the last two plinian eruptions B and C, consisting mainly of LPs and tremors, indicates that fluids were involved until the last phases of the eruption.

The occurrence of the three LP swarms (Table 5.2), on March 30 (at 1315 UT), on March 31 (at 1340 UT) and on April 2 (at 1340 UT) suggests a maximum overpressure of the magmatic system and strong degassing.

After the last plinian eruption C, the seismicity was almost entirely composed of deeper VT events (down to 20 km). Such events occurred in a short time span (a few tens of days), without the occurrence of any LP events. The presence of deeper VTs, without any LPs and Hs, points to a decrease in fluid content in the physical process that generated such seismicity, which is more likely of tectonic origin (Havskov et al. 1983; Espíndola et al. 2006), as suggested also by their locations close to the regional and local faults, confirming their tectonic origin (Jiménez et al. 1999).

The presence of fluids generated a specific seismicity pattern (see inset box 2), in particular on the Gutenberg-Richter law. The presence of a large amount of fluids before the eruptive events A1, B and C and the absence of fluids few years after the last eruption C, is confirmed by the temporal and magnitude analysis of the seismicity. Before the first eruptive event A1, the Omori law is not satisfied, a condition that is common for volcanic seismic swarms and the frequency-magnitude distribution shows a break in the slope, condition also commonly observed for volcanic swarms, especially when high quantities of fluids are involved. Such occurrence is consistent with the existence of a large amount of hydrothermal fluids and water. In contrast, the post-eruptive seismic events follow the Omori law, as common for swarms of tectonic aftershocks, and the magnitude distribution shows only one slope, with a smaller *b*-value, suggesting that fewer fluids were involved, as discussed below. A few years later (from 1985 to 1990), the Gutenberg-Richter distribution shows a single slope with a *b*-value of 0.88, which is much smaller than the one observed during, or just after, the eruption. This *b*-value is close to 1 and suggests a tectonic origin of the seismicity near the volcano.

Inset Box 2: Influence of Water on Seismicity

It is well established that a very small (<1 MPa) increase in the pore fluid pressure decreases the effective normal stress on fractures triggering 'induced' earthquakes more easily than in those situations where fluids are absent (Hubbert and Rubey 1959; Healy et al. 1968; Pearson 1981; Pine and Batchelor 1984; Fehler 1989; Talebi 1998; Guha 2000). Hence, the presence of fluids (water and/or magma) increases the number of small earthquakes with respect to large earthquakes, leading to high *b*-values (usually much greater than 1.0), as is often observed in geothermal sites (Wyss 1973; Cornet et al. 1997; Cuenot et al. 2008; Dorbath et al. 2009) and volcanoes (Mc Nutt 1986; Wyss et al. 1997, 2001; Wiemer and Benoit 1996; Wiemer and McNutt 1997; Murru et al. 1999; Del Pezzo et al. 2003; Legrand et al. 2004; Kundu et al. 2012). A summary of such effect is presented below.

The critical shear stress τ_c required to cause slip on a fault is given by:

$$\tau_c = \tau_0 + \mu\sigma_n$$

where τ_0 is the inherent shear strength of the material (cohesion), μ is the friction coefficient, and σ_n is the normal stress acting on the fault. For most rocks, μ varies between 0.6 and 1.0. When the rock is saturated in water, the effective normal stress is reduced, and the shear stress becomes:

$$\tau_c = \tau_0 + \mu(\sigma_n - P)$$

where P is the pore pressure.

The presence of fluids can reduce the stress on the faults, inducing seismic slip more easily than in their absence. Hence, the main factors that control the level of induced seismicity are the ambient tectonic stress, fluid injection, and the orientation of pre-existing faults.

The presence of a break in the slope of the magnitude-distribution shows that the spatial distribution of earthquakes is not self-similar at all scales. Such a break is related to a critical size of local faults having a different rupture process.

Medina et al. (1992), Jiménez et al. (1999), and Espíndola et al. (2006) proposed that different kinds of pre-eruptive seismicity (Hs, LPs, tremors) attest to the interaction between rising magma and aquifers. The volcano hosts a shallow rain-fed aquifer at depths less than 280 m, due to high precipitation (average of about 4–5 m/year). A second, deeper aquifer exists about 2 km below sea level, i.e. 3 km below the crater of the volcano (see Chap. 4). Hence, El Chichón volcano hosts a well-developed hydrothermal system (Taran and Rouwet 2008; Taran et al. 2008; see Chap. 4). The presence of these fluids may explain part of the seismic signals observed (Hs and LPs). The importance of water involved in the 1982 eruption is also clearly revealed by stratigraphic data (see Chap. 3).

Zobin and Jiménez (2008) proposed that the absence of a new lava dome, 30 years after the 1982 eruption, would be due to the '*rapid emission of gas and water from the magmatic and hydrothermal system beneath the volcano*', which increases the magma viscosity, decreasing the probability of its extrusion. An alternative hypothesis, consistent with the absence of a spatial seismic gap between 7 and 13 km, observed in the post-eruptive seismicity (Jiménez et al. 1999; Espíndola et al. 2006), could be that most of the magma was ejected in 1982, and the magma chamber is not completely recharged. In fact, Andrews et al. (2008) proposed time intervals for the magma chamber recharge between 50 and 500 years, which are similar to the intervals between 100 and 500 years proposed by Espíndola et al. (2000).

Havskov et al. (1983), Medina et al. (1990) and Yokoyama et al. (1992) clearly noted that the Omori law is not followed before the first eruptive event (March 29 at 0515 UT), but it is followed after April 1. According to Yokoyama et al. (1992) the seismicity after April 1 corresponds to a classical aftershock sequence, with a p-value of 1.29. A similar conclusion is reached here, even though with a larger p-value of 2.5. For comparison, the post-eruptive VT sequence observed at Pinatubo in 1991 followed the Omori law (Mori et al. 1996) with a p-value of 2.42. This value is also higher than the typical value <1.0 commonly observed for tectonic aftershock sequences, and is very similar to the one obtained at El Chichón. These high p-values, which suggest a very short time window for the occurrence of aftershocks following a large earthquake, are classically observed when fluids are involved, for example in geothermal areas.

Medina et al. (1992) found that b-values increased from 1.10 to 1.39 before the first eruption on 29 March 1982, and then further increased to 1.71 between this event, and the last one on 4 April 1982 at 1110 UT. After the last Plinian event C, they found that b varied from 1.21 to 0.93. Yokoyama et al. (1992) found $b = 2.70 \pm 0.2$ for the post-eruption seismicity. This value is greater than the value of 1.90 ± 0.2, found here, which may be due to the use of the single station (CSN) located at a greater distance with respect to the local network, and to the different magnitude definition they used. Yokoyama et al. (1992) mentioned that this b-value is typical of LP events, which implicitly suggests the presence of fluids, and is consistent with the present interpretation.

The 6-day time window (March 29–April 4) between the first and the last Plinian event C can be therefore considered as a bifurcation period between two different regimes. The first one, before the eruptive sequence, is mainly 'volcanic', while the second one, which started after the Plinian event C and lasted until 1990, is mainly 'tectonic'. A preliminary conclusion is that the pre-eruptive 'volcanic' seismicity triggered the five large 'tectonic' earthquakes of April 4. Nevertheless, the temporal separation of the 'volcanic' from the 'tectonic' response is not sharp. In fact, the volcanic signature of the seismicity continued for 1 month after the last Plinian eruption C, as revealed by the persistence of high b-and p-values. This behavior is consistent with a 'volcanic' response. In the time period from 1985 to 1990, the 'tectonic' signature returned to a 'normal' regime (i.e. b-values <1). Hence, the overall seismicity reveals a continuous feed-back interaction between the volcanic and tectonic contributions, with a predominantly volcanic signature before, and a predominantly tectonic signature after, the 1982 eruptive sequence, suggesting that the magma migrated towards the surface following a pre-existing fault or creating new ones compatible with the local stress tensor. The absence of a break in the magnitude-distribution one month until few years after the last large eruptive event C shows a decrease in the amount of fluids (magma and/or water) that generated the seismicity.

The presence of large amounts of water is indicated by the abundant phreatomagmatic eruptions in stratigraphic observations (see Chap. 3), by the break in the slope observed in the Gutenberg-Richter magnitude-distribution, and the high b-value (2.2). The return to a common single *b*-value after April 4 indicates that fluids were much less involved in the physical process of generation of the seismicity in the months and years following 4 April 1982. Medina et al. (1992), Jiménez et al. (1999), and Espíndola et al. (2006) pointed out that all the earthquakes that occurred after the April 4 (eruption C) are VT, and are located to the north, on the trace of the San Juan fault, one of the most important faults in the region (Chap. 1). Furthermore, the seismicity is deeper (about 10–20 km), after April 4 clearly related to active faults. This observation is in agreement with the observation of a single slope of the Gutenberg-Richter law a few years (1985–1990) after the eruption, showing the tectonic signature of this seismicity.

Moreover, the chemical composition of magmas of El Chichón is adakitic-like (Ignacio et al. 2003), suggesting that a high quantity of fluids was involved in their generation. Manea and Manea (2008) suggested that these fluids could derive from the strong dehydration of oceanic sediments at depths between 40 and 80 km, and from the dehydration of the serpentinized oceanic lithosphere beneath the Tehuantepec Ridge at depths of 180–200 km, consistent with slab depths beneath El Chichón. Luhr et al. (1984) indicated that the trachyandesitic melt erupted in 1982 was water-rich (4–10 wt% H_2O). The large amount of primary anhydrite found in its product, and magma water contents, were related to the location of the subducted Tehuantepec ridge (Luhr 2008; another interpretation postulates, though, that El Chichón magmas could be related to a rift system associated to the Motagua-Polochic fault system—see Chap. 2). Such high magmatic water content before the eruption is also compatible with the high b-values of the Gutenberg-Richter law and its associated break observed before the eruption. An attenuation study based on the coda of seismic waves (Zuñiga and Díaz 1994) indicates the existence of a high attenuation volume (a magma body and/or a fracture zone) in an elongated E–W direction, at depths between the subsurface and 25 km. If interpreted as a fracture zone, this volume would correspond to the San Juan Fault, which trends in the same direction. A magma chamber has also been suggested on the basis of the absence of seismicity between 7 and 13 km depth (Jiménez et al. 1999). The attenuation below El Chichón is much higher than the one observed below the Trans Mexican Volcanic Belt and the Central American Volcanic Arc (Zuñiga and Díaz 1994).

Hence, before and during the initial eruptive events, the seismicity pattern (type of earthquakes, depth, and scaling laws) can be explained by the ascent of magma which entered in contact with ground water of a shallow aquifer. In contrast, most of the seismic signatures after the last Plinian event indicate the absence of fluids. This is compatible with the fact that more than 30 years after the 1982 eruptions, no new lava dome has formed.

References

Aki K (1965) Maximum likelihood estimate of b in the formula log (N) = a-bM and its confidence limits. Bull Earthq Res Inst Tokyo Univ 43:237–239

Aki K (1981) A probabilistic synthesis of precursory phenomena. In: Simpson DW, Richards PG (eds) Earthquake prediction: an international review, maurice ewing series 4. American Geophysical Union, Washington, pp 566–574

Andrews BJ, Gardner JE, Housh T (2008) Repeated recharge, assimilation, and hybridization in magmas erupted from El Chichón as recorded by plagioclase and amphibole phenocrysts. J Volcanol Geoth Res 175:415–426

Bak P, Tang C (1989) Earthquakes as a self-organized critical phenomenon. J Geophys Res 9:15635–15637

Bullen K, Bolt B (1947) An introduction to the theory of seismology. Cambridge University Press, Cambridge

Bullen K, Bolt B (1993) An introduction to the theory of seismology, 4th revised edition. Cambridge University Press, Cambridge

Bursik M (2009) A general model for tectonic control of magmatism: examples from long valley caldera (USA) and El Chichón (México). Geofis Int 48(1):171–183

Canul R, Rocha V (1981) Informe Geologico de la Zona Geotermica de "El Chichónal", Chiapas, Mexico. Comision Federal de Electricidad. Rep 32–81, 38 pp (Unpublished)

Caplan-Auerbach J, Duennebier F (2001) Seismicity and velocity structure of loihi seamount from the 1996 earthquake swarm. Bull Seis Soc Am 91:178–190

Carey S, Sigurdsson H (1986) The 1982 eruptions of El Chichon volcano, Mexico (2): observations and numerical modelling of tephra-fall distribution. Bull Volcanol 48:127–141

Chouet B (1996) Long-period volcano seismicity: its source and use in eruption forecasting. Nature 380:309–316

Cornet F, Helm J, Poitrenaud H, Etchecopar A (1997) Seismic and aseismic slips induced by large-scale fluid injections. Pure Appl Geophys 150:563–583

Cuénot N, Dorbath C, Dorbath L (2008) Analysis of the microseismicity induced by fluid injection in the hot dry rock site of Soultzsous-Forêts (Alsace, France): implications for the characterization of the geothermal reservoir properties. Pure Appl Geophys 165:797–828

De la Cruz Reyna S, Martin Del Pozzo A (2009) The 1982 eruption of El Chichón volcano, Mexico: eyewitness of the disaster. Geof Int 48(1):21–31

Del Pezzo E, Bianco F, Saccorotti G (2003) Duration magnitude uncertainty due to seismic noise: inference on the temporal pattern of G-R b-value at Mt. Vesuvius, Italy. Bull Seism Soc Am 93:1847–1857

Díaz de León J (1991) Actividad sísmica en el área del volcán Chichonal durante 1990. CFE, Proyecto 1719

Díaz de León J, Lermo J (1990) Actividad sísmica en el área del volcán Chichonal durante 1989. CFE, Proyecto 0729

Dorbath L, Cuénot N, Genter A, Frogneux M (2009) Seismic response of the fractured and faulted granite of Soultz-sous-Forêts (France) to 5 km deep massive water injections. Geophys J Int 177:653–675

Duffield W, Tilling R, Cañul R (1984) Geology of E1 Chich6n Volcano, Chiapas,Mexico. J Volcanol Geoth Res 20:117–132

Espíndola JM, Macías JL, Godínez ML, Jiménez Z (2002) La erupción de 1982 del volcán Chichonal, Chiapas, México. 37–65, In: Lugo HJ, Inbar M (eds) Desastres naturales en América Latina, Fondo de Cultura Económica, México, 2002, pp 501

Espíndola JM, Macías JL, Tilling RI, Sheridan MF (2000) Volcanic history of El Chichón Volcano (Chiapas, Mexico) during the Holocene, and its impact on human activity. Bull Volcanol 62:90–104.

Espíndola JM, Zamora-Camacho A, Jiménez Z (2006) Some aspects of the seismicity associated with the 1982 eruption of El Chichon Volcano, Chiapas, Mexico. J Volcanol Geoth Res 157:367–374

Fehler M (1989) Stress control of seismicity pattern observed during hydraulic fracturing experiments in the Fenton Hill hot dry rock geothermal energy site, New Mexico. Int J Rock Mech Min Sci Geomech 26:211–219

García-Palomo A, Macías JL, Espíndola JM (2004) Strike-slip faults and K-alkaline volcanism at El Chichón volcano, southeastern Mexico. J Volcanol Geoth Res 136:247–268

González-Ruiz LC (1980) Study of the aftershocks (November 29 to December 17, 1978, m 3.0) of the November 29, 1978 Oaxaca earthquake, calibration of magnitudes, BSc thesis, UNAM, Mexico. 68 pp (in Spanish)

Gonzalez-Salazar A (1973) Preliminary report on the geothermal zone of the Volcan E1 Chichónal, Chiapas. Department Geothermal Resources, Federal Energy Commission (unpublished report in Spanish) 4 pp

Guha S (2000) Induced earthquakes. Kluwer Academic Publishers, Dordrecht, 314 pp

Gutenberg B, Richter C (1944) Frequency of earthquakes in California. Bull Seismol Soc Am 34:185–188

Guzmán-Speziale M, Meneses-Rocha J (2000) The North American-Caribbean plate boundary west of the Motagua Polochic fault system: a fault jog in southeastern Mexico. J South Am Earth Sci 13:459–468

Guzmán-Speziale M, Pennington W, Matumoto T (1989) The triple junction of the North America, Cocos and caribbean plates: seismicity and tectonics. Tectonics 8:981–997

Havskov J, De la Cruz-Reyna S, Singh K, Medina F, Gutierrez C (1983) Seismic activity related to the March–April, 1982 eruptions of El Chichon Volcano, Chiapas, Mexico. Geophys Res Lett 10:269–293

Healy J, Rubey W, Griggs D, Raleigh C (1968) The Denver earthquake. Science 161:1301–1310

Hubbert M, Rubey W (1959) Mechanics of fluid-filled porous solids and its application to overthrust faulting. Geol Soc Am Bull 70:115–166

Ignacio C, Castiñeiras P, Márquez A, Oyarzun R, Lillo J, López Y (2003) El Chichón Volcano (Chiapas Volcanic Belt, Mexico) transitional calc-alkaline to adakitic-like magmatism: petrologic and tectonic implications. Int Geol Rev 45:1020–1028

Ishimoto M, Iida K (1939) Observations sur les seismes enregistres par le microsismographe construit dernièrement (1). Bull Earthq Res Inst Univ Tokyo 17:443–478

Jiménez Z, Espíndola VH, Espíndola JM (1999) Evolution of the seismic activity from the 1982 eruption of El Chichón Volcano, Chiapas, Mexico. Bull Volcanol 61:411–422

Kundu B, Legrand D, Gahalaut K, Gahalaut V, Mahesh P, Kamesh Raju K, Catherine J, Ambikapthy A, Chadha R (2012) The 2005 volcano-tectonic swarm in the Andaman Sea: triggered by the 2004 great Sumatra-Andaman earthquake. Tectonics 31, TC5009. doi:10.1029/2012TC003138

Kunugi T, Fukao Y, Ohno M (2000) Underdamped responses of a well to nearby swarm earthquakes off the coast of Ito city, central Japan, 1995. J Geophys Res 105:7805–7818

Lahr J, Chouet B, Stephens C, Power J, Page R (1994) Earthquake classification, location, and error analysis in a volcanic environment: implications for the magmatic system of the 1989–1990 eruptions at Redoubt Volcano, Alaska. J Volcanol Geoth Res 62:137–151

Legrand D (2002) Fractal dimensions of small, intermediate and large earthquakes. Bull Seismol Soc Am 92:3318–3320

Legrand D, Calahorrano A, Guillier B, Rivera L, Ruiz M, Villagómez D, Yepes H (2002) Stress tensor analysis of the 1998–1999 tectonic swarm of northern Quito related to the volcanic swarm of Guagua Pichincha Volcano, Ecuador. Tectonophysics 344:15–36

Legrand D, Villagómez D, Yepes H, Calahorrano A (2004) Multifractal dimension and b value analysis of the 1998–1999 Quito swarm related to Guagua Pichincha volcano activity, J Geophys Res 109. doi: 10.1029/2003JB002572

Legrand D, Barrientos S, Bataille K, Cembrano J, Pavez A (2011) The fluid-driven tectonic swarm of Aysen Fjord, Chile (2007) associated with two earthquakes (Mw=6.1 andMw=6.2) within the Liquiñe-Ofqui Fault Zone. Continental Shelf Research 31:154–161

Lermo J, Coyoli R, Díaz de León J, Gutiérrez C, Nava E (1989a) Actividad sísmica en el area del volcán Chichonal durante el periodo 1985 to 1987 observada en la red sísmica de la C. H. Peñitas. CFE, Proyecto 8719

Lermo J, Cesati G, Coyoli R, Aguirre J (1989b) Análisis de la sismicidad local en la zona de la central hidroeléctrica de Peñitas durante 1987 y 1988. CFE, Proyecto 9731

Lermo J, Díaz de León J, Javier C (1990) Análisis de la sismicidad local en la zona de la central hidroeléctrica de Peñitas durante 1989. CFE, Proyecto 0729

Luhr J (1990) Experimental phase relations of water-and-sulfur saturated arc magmas and the 1982 eruption of El Chichón volcano. J Petrol 31:1071–1114

Luhr J (2008) Primary igneous anhydrite: progress since its recognition in the 1982 El Chichón trachyandesite. J Volcanol Geoth Res 175:394–407

Luhr J, Carmichael I, Varekamp J (1984) The 1982 eruptions of El Chich6n Volcano, Chiapas, Mexico: mineralogy and petrology of the anhydrite-bearing pumices. J Volcanol Geoth Res 23:69–108

Macías JL, Espíndola JM, Taran Y, García PA (1997) Explosive volcanic activity during the last 3500 years at El Chichon Volcano, Mexico. In: IAVCEI, General Assembly, Puerto Vallarta, Mexico. Field Trip Guide

Macías JL, Espíndola JM, Bursik M, Sheridan MF (1998) Development of lithic-breccias in the 1982 pyroclastic flow deposits of El Chichón Volcano, Mexico. J Volcanol Geoth Res 83:173–196

Macías JL, Arce JL, Garduño-Monroy VH, Rouwet D, Taran Y, Layer P, Jiménez A, Álvarez R (2010) Estudio de prospección geotérmica para evaluar el potencial del volcán Chichonal, Chiapas. Unpublished Report no. 9400047770 IGF-UNAM-CFE

Manea M, Manea V (2008) On the origin of El Chichón volcano and subduction of tehuantepec ridge: a geodynamical perspective. J Volcanol Geoth Res 175:459–471

Matson M (1984) The 1982 El Chichon volcano eruptions - a satellite perspective. J Volcanol Geotherm res 23:1–10

McNutt S (1986) Observations and analysis of B-type earthquakes, explosions, and volcanic tremor at Pavlof Volcano, Alaska. Bull Seismol Soc Am 76:153–175

Medina F (1982) El Volcán Chichón. GEOS. Bull Union Geofis Mex 2:4–19

Medina F, Gonzalez-Moran T, Gonzalez L (1990) Gravity and seismicity analyses of the El Chichon Volcano, Chiapas, Mexico. Pageoph 133:149–165

Medina F, Gonzalez L, Gutierrez C, Aguilera R, Espíndola JM (1992) Analysis of the seismic activity related to the 1982 eruptions of the El Chichon Volcano, Mexico. In: Gasparini P, Aki K (eds) Volcanic seismology, vol 3. Proceedings in volcanology. Springer, Berlin, pp 97–108

Mogi K (1962) Magnitude-frequency relation for elastic shocks accompanying fractures of various materials and some related problems in earthquakes. Bull Earth Res Inst Tokyo Univ 40:831–853

Mogi K (1963) Some discussions on aftershocks, foreshocks, and earthquake swarms—the fracture of a semi-infinite body caused by an inner stress origin and its relation to the earthquake phenomena, 3. Bull Earth quake Res Inst Univ Tokyo 41:615–658

Molina-Berbeyer R (1974) Preliminary report on the geochemistry of geothermal fluids from the Volcan Del Chich6nal, Chiapas. Department Geothermal Resources, Federal Energy Commission, (unpublished report in Spanish), 23 pp

Mori J, White R, Harlow D, Okubo P, Power J, Hoblitt R, Laguerta E, Lanuza A, Bautista B (1996) Volcanic earthquakes following the 1991 climactic eruption of Mount Pinatubo: strong seismicity during a waning eruption. In: Newhall CG, Punongbayan, RS (eds) Fire and Mud University of Washington Press, Washington, pp 339–350

Müllerried F (1932) Der Chichón, ein bischer unbegannter tätiger Vulkan im nordlichen Chiapas. Zeit. Vulkanologie, XIV, Mexiko, pp 191–209

Müllerried F (1933) "El Chichón", volcán en actividad, descubierto en el estado de Chiapas. Mem. Rev. Acad. Nac. Cienc., 54, 413–416. (English translation in: Volcano News, 1982; 7: 5–6)

Murru M, Montuori C, Wyss M, Privitera E (1999) The location of magma chambers at Mt. Etna, Italy, mapped by b-values. Geophys Res Lett 26:2553–2556

Neunhöfer H, Hemmann A (2005) Earthquake swarms in the Vogtland/Western Bohemia region: spatial distribution and magnitude–frequency distribution as an indication of the genesis of swarms? J Geodyn 39:361–385

Newhall CG, Self S (1982) The Volcanic Explosivity Index (VEI): an estimate of explosive magnitude for historical volcanism. J Geophys Res 87:1231–1237

Omori F (1895) On the aftershocks of earthquakes. Coll Sci Imper Univ Tokyo 7:111–200

Pearson C (1981) The relationship between microseismicity and high pore pressures during hydraulic stimulation experiments in low permeability granitic rocks. J Geophys Res 86:7855–7864

Pine R, Batchelor A (1984) Downward migration of shearing in jointed rock during hydraulic injections. Int J Rock Mech Min Sci 21:249–263

Roman D (2005) Numerical models of volcanotectonic earthquake triggering on non-ideally oriented faults. Geophys Res Lett 32, L02304. doi.org/10.1029/2004GL021549

Roman D, Cashman K (2006) The origin of volcano-tectonic earthquake swarms. Geology 34(6):457–460

Roman D, Heron P (2007) Effect of regional tectonic setting on local fault response to episodes of volcanic activity. Geophys Res Lett 34, L13310. doi.org/10.1029/2007GL030222

Roman D, De Angelis S, Lachtman JL, White R (2008) Patterns of volcanotectonic seismicity and stress during the ongoing eruption of Soufrière Hills Volcano, Montserrat (1995–2007). J Volcanol Geoth Res 173:230–244

Sapin M, Hirn A, Lépine JC, Nercessian A (1996) Stress, failure and fluid flow deduced from earthquakes accompanying eruptions at Piton de la Fournaise volcano. J Volcanol Geoth Res 70:145–167

Sigurdsson H, Carey SN, Espíndola JM (1984) The 1982 eruptions of El Chichonvolcano, Mexico: stratigraphy of pyroclastic deposits. J Volcanol Geotherm Res 23:11–37

Talebi S (1998) Seismicity associated with mines, reservoirs and fluid injection.In: Birkhauser, Basel. Talebi S (eds) Pageoph. Springer, Berlin, 720 pp

Taran Y, Rouwet D (2008) Estimating thermal inflow to El Chichón crater lake using the energy-budget, chemical and isotope balance approaches. J Volcanol Geoth Res 175:472–481

Taran Y, Rouwet D, Inguaggiato S, Aiuppa A (2008) Major and trace element geochemistry of neutral and acidic thermal springs at El Chichón volcano, Mexico. Implications for monitoring of the volcanic activity. J Volcanol Geoth Res 178:224–236

Tilling R (2009) El Chichón's "surprise" eruption in 1982: lessons for reducing volcano risk. Geofis Int 48(1):3–19

Ukawa M, Tsukahara H (1996) Earthquake swarms and dike intrusions off the east coast of Izu Peninsula, central Japan. Tectonophysics 253:285–303

Utsu T (1961) A statistical study of the occurrence of aftershocks. Geophys Mag 30:521–605

Vargas-Bracamontes D, Neuberg J (2012) Interaction between regional and magma-induced stresses and their impact on volcano-tectonic seismicity. J Volcanol Geoth Res. doi.org/10.1016/j.jvolgeores.2012.06.025

Wiemer S, Benoit J (1996) Mapping the b-value anomaly at 100 km depth in the Alaska and New Zealand subduction zones. Geophys Res Lett 23:1557–1560

Wiemer S, McNutt S (1997) Variations in the frequency-magnitude distribution with depth in two volcanic areas: Mount St. Helens, Washington, and Mt. Spurr, Alaska. Geophys Res Lett 24:189–192

Wyss M (1973) Towards a physical understanding of the earthquake frequency distribution. Geophys J R Astron Soc 31:341–359

Wyss M, Shimazaki K, Wiemer S (1997) Mapping active magma chambers by b values beneath the off-Ito volcano, Japan. J Geophys Res 102:20413–20422

Wyss M, Klein F, Nagamine K, Wiemer S (2001) Anomalously high b-values in the South Flank of Kilauea volcano, Hawaii: evidence for the distribution of magma below Kilauea's East rift zone. J Volcanol Geoth Res 106:23–37

Yamasato H, Yokota T, Kashiwabara S (1991) Earthquake swarm and volcanic tremors off Eastern Izu peninsula in 1989—Spectral investigation and characteristics of waveforms. J Phys Earth 39:79–92

Yokoyama I, De la Cruz-Reyna S, Espíndola JM (1992) Energy partition in the 1982 eruption of El Chichon Volcano, Chiapas, Mexico. J Volcanol Geoth Res 51:1–21

Zobin V, Jiménez Z (2008) Some regularity in the process of re-awakening of andesite and dacite volcanoes: specific features of the 1982 El Chichón volcano, México reactivation. J Volcanol Geoth Res 175:482–487

Zúñiga R, Díaz L (1994) Coda attenuation in the area of El Chichón volcano, Chiapas, Mexico. Tectonophysics 234:247–258

Eruptive History of the Tacaná Volcanic Complex

6

J.L. Macías, J.L. Arce, P.W. Layer, R. Saucedo, and J.C. Mora

Abstract

Tacaná is the northernmost volcano of the Central American Volcanic Arc, and one of the four volcanic structures of the Tacaná Volcanic Complex (TVC), from oldest to youngest: Chichuj, Tacaná, and San Antonio volcanoes, and Las Ardillas dome. Geologic and radiometric data show that volcanic activity of the TVC began around 225 ka with the construction of Chichuj volcano within the 2 Ma old San Rafael Caldera. The edifice of Tacaná began its construction west of Chichuj volcano around 50 ka. San Antonio volcano, and Las Ardillas Dome formed southwest of Tacaná volcano during Late Pleistocene. Effusive and explosive eruptive activity has alternated from all eruptive centers of the complex. Flank collapses of Chichuj, Tacaná, and San Antonio edifices have generated debris-avalanches. At least four plinian-subplinian events—two of which rank ~5 on the Volcanic Explosivity Index (VEI)—and nine other smaller explosive eruptions occurred at Tacaná during the Holocene, the most recent one around 150 year BP. The 1949 and 1986 phreatic explosions from Tacaná attracted scientific and public attention to the complex. At present, Tacaná represents the second most dangerous volcano in Mexico after Popocatepetl.

J.L. Macías (✉)
Instituto de Geofísica, Unidad Michoacán, Universidad Nacional Autónoma de México, Campus-Morelia, 58090 Morelia, Michoacán, Mexico
e-mail: macias@geofisica.unam.mx

J.L. Arce
Insituto de Geología, Universidad Nacional Autónoma de México, Cd. Universitaria, 04510 Coyoacán, Mexico, D.F., Mexico
e-mail: jlarce@geologia.unam.mx

P.W. Layer
College of Natural Science and Mathematics and Geophysical Institute, University of Alaska Fairbanks, Fairbanks, AK 99775, USA
e-mail: player@gi.alaska.edu

R. Saucedo
Instituto de Geología/Fac. Ingeniería UASLP, Dr. M. Nava no 5, Zona Universitaria, 78240 San Luis Potosí, Mexico

J.C. Mora
Instituto de Geofísica, Universidad Nacional Autónoma de México, Cd. Universitaria, Coyoacán, Mexico D.F., 04510, Mexico
e-mail: jcmora@geofisica.unam.mx

6.1 Previous Studies

The summit of Tacaná volcano (15° 08′N, 92° 09′W) straddles the State of Chiapas, México, and the San Marcos Department, Guatemala (Fig. 6.1a). It delineates the international boundary between the two countries, and represents the second highest peak in Central America after Tajumulco volcano (4,220 m) in Guatemala. It consists of four NE-trending volcanic edifices (oldest to youngest): Chichuj (3,800 m above sea level), Tacaná (4,060 m), Las Ardillas dome (3,782 m), and San Antonio (3,700 m) (Figs. 6.1b and 6.2a).

The volcano was first mentioned as Soconusco (Humboldt 1862), then Istak (Dollfus and Monserrat 1867), and finally Tacaná (Sapper 1896, 1899; Böse 1902, 1903). Bergeat (1894) made the first petrographic description of Tacaná rocks, which he described as augite-bearing andesites. Detailed studies of the volcano began after a series of regional earthquakes

T. Scolamacchia and J.L. Macías (eds.), *Active Volcanoes of Chiapas (Mexico): El Chichón and Tacaná*,
Active Volcanoes of the World, DOI 10.1007/978-3-642-25890-9_6, © Springer-Verlag Berlin Heidelberg 2015

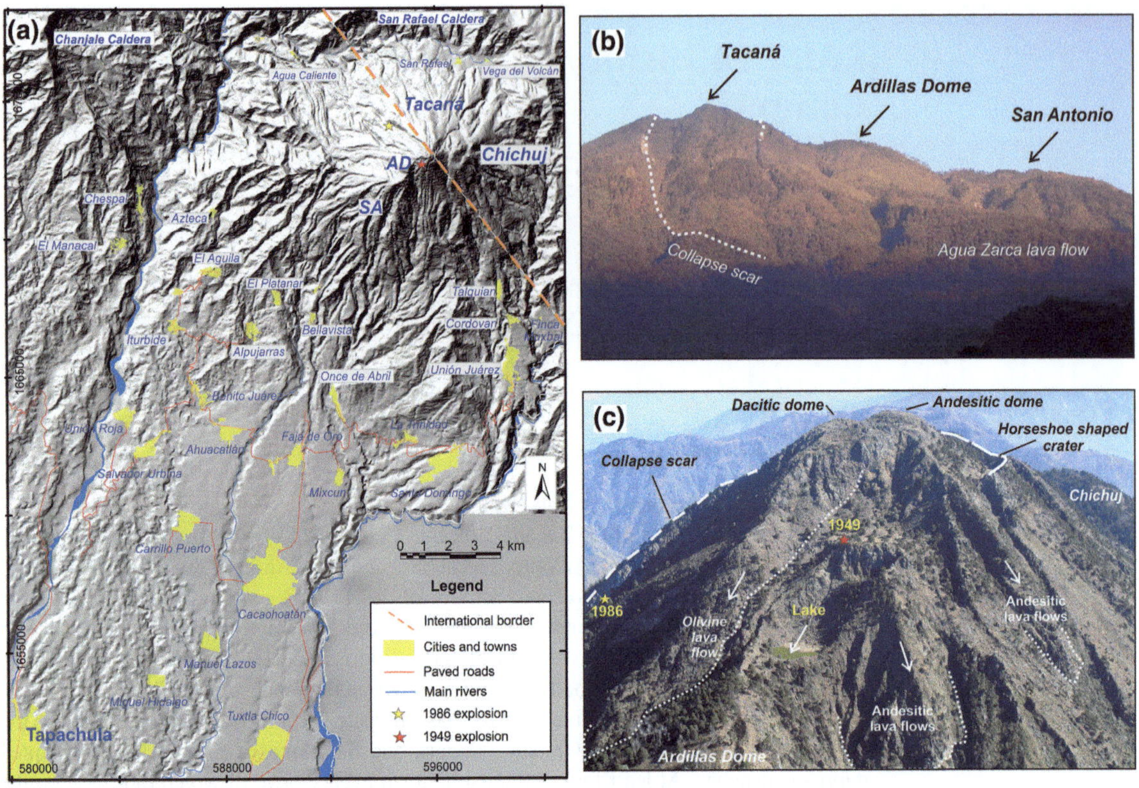

Fig. 6.1 **a** Shaded relief model of the Tacaná Volcanic Complex (TVC) showing the City of Tapachula, main towns and access roads. Abbreviations are: *AD* Ardillas Dome, and *SA* San Antonio. **b** Panoramic view of the Tacaná Volcanic Complex looking from the northwest. **c** Aerial view to the northeast of the summit of Tacaná volcano showing the large *horseshoe-shaped* crater, summit domes, the locations of the 1949 and 1986 phreatic explosions and the *green-colored* lake. Hills in the background are the northern slopes of the San Rafael Caldera facing Guatemalan territory

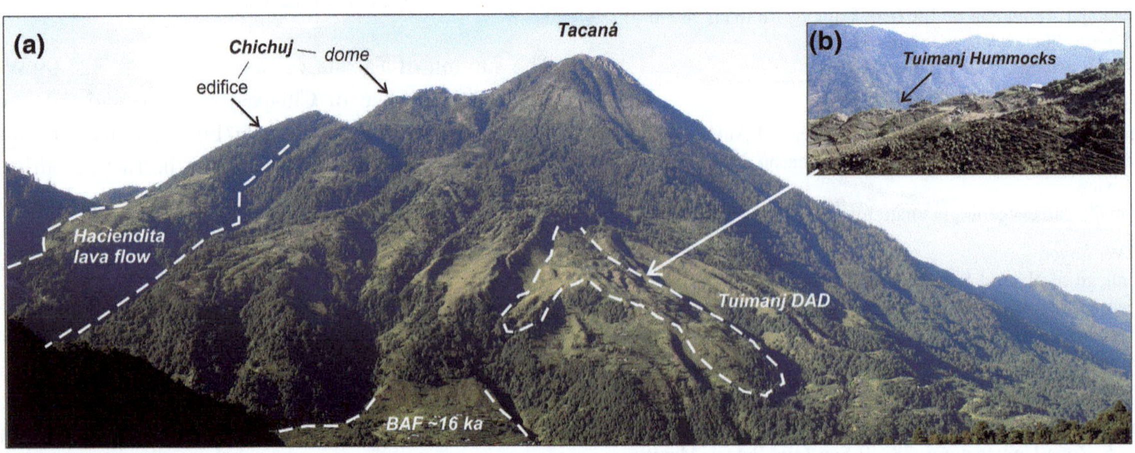

Fig. 6.2 **a** View of the Tacaná Volcanic Complex from the north showing Chichuj volcano and summit dome, the Haciendita lava flow, the Tuimanj debris-avalanche deposit (DAD) and the ~16 ka block-and-ash flow fan (BAF) in the vicinity of the San Rafael village. The inset shows the hummocky topography of the Tuimanj debris avalanche at the hamlet of the same name (**b**)

occurred on September 22, 1902, and October 24, 1902 eruption of Santa Maria volcano in Guatemala, prompting several geologists to visit the region. Indeed Böse (1902, 1903) made the first detailed study of Tacaná volcano, indicating that its base was located at an elevation of 2,200 m on a granitic basement. He also concluded that Tacaná consisted of three "terraces", eventually representing craters, located at elevations of 3,448, 3,655 and 3,872 m (the uppermost one on top of a lava dome). The rock samples collected "around the summit", were later classified by Ordóñez as hypersthene-hornblende andesites (Fig. 6.1c). Böse (1902, 1903) also described an elliptical explosion crater ($\sim$50 m wide and 5 m deep) located SW of the main summit, which discharged sulfur waters and fumes. Sapper (1897) reported the formation of fissures with sporadic emission of fumes on the flanks of Tacaná after the earthquake occurred on January 12, 1855 (Mooser et al. 1958). Böse (1902, 1903) and Waibel (1933) considered Tacaná as a dormant volcano but not extinct. For the following few decades, the volcano remained unstudied.

In 1949, Tacaná reawakened with a small phreatic explosion (Müllerried 1951), after which it was included in the Catalogue of Active Volcanoes of the World (Mooser et al. 1958) (Fig. 6.1c). During the 1980s, the *Comisión Federal de Electricidad* (National Power Company) began a series of studies to evaluate the geothermal potential of the volcano (De la Cruz and Hernández 1985; Saucedo and Esquivias 1988), and presented the first geological map, and a composite stratigraphic column of the volcano, lying on granitic basement and Tertiary andesites. Another eruptive episode in 1986, reminded both the authorities and scientists about the potential hazards from Tacaná volcano (De la Cruz-Reyna et al. 1989) (Fig. 6.1c). After the 1986 crisis, several studies began. de Cserna et al. (1988) presented a photogeological map of the volcano, defined 14 stratigraphic units, and concluded that Tacaná was a stratovolcano composed of three NE–SW aligned volcanoes. Older pyroclastic deposits exposed on the flanks and the apron of the volcano were described and dated (Espíndola et al. 1989, 1993). Mercado and Rose (1992) published a general hazard map of Tacaná together with the chemistry of some calc-alkaline andesitic rocks. An increasing number of studies have been carried out during the last decade. These studies focused on the stratigraphic record of the volcano (Macías et al. 2000, 2010, 2011; Borjas 2006; Ordaz 2006; García-Palomo et al. 2006; Macías 2007; Campa 2009; Rangel 2009; Limón 2011; Arce et al. 2012), the chemical evolution of eruptive products (Mora et al. 2004; Arce et al. 2014; Chap. 2), and the volcaniclastic sedimentation (Macías et al. 2000; Murcia 2008; Murcia and Macías 2009a, b). The results of these last studies indicate that the city of Tapachula, Chiapas ($\sim$350,000 inhabitants), 31 km SW from the summit (Fig. 6.1a), was built on top of lahar deposits that originated from one of the centers of the TVC (Macías et al. 2000; Murcia 2008; Murcia and Macías 2009a, b, 2014). In this chapter, we review previous information and present new geologic mapping (Fig. 6.3), stratigraphic data (Fig. 6.4), and radiometric analyses ($^{40}Ar/^{39}Ar$ in Table 6.1, and ^{14}C in Table 6.2), collected during the past decade, which have aided in deciphering the evolution of the Tacaná Volcanic Complex.

6.2 Geomorphology and Hydrologic Network

The TVC stands in high relief with respect to the elevations of the surrounding terrain: higher by $\sim$3,000 m in Mexico, and $\sim$1,300 m in northern Guatemala (García-Palomo et al. 2006) (Fig. 6.2a). The southern portion of the TVC is characterized by large coalesced pyroclastic and debris fans that extend to the Pacific coastline (Macías et al. 2000), and that pinch out against the walls of the San Rafael caldera to the north. The main rivers that drain the area are the Coatán and Suchiate (the latter marks the international border between Mexico and Guatemala), which flow into the Pacific Ocean. The San Rafael River drains the northern apron of Tacaná and flows 12 km to the west to debouch into the Coatán River (Fig. 6.1a). From this point the Coatán River flows for another 60 km to the Pacific Ocean. The Coatán and Suchiate rivers bound a large alluvial fan and smaller fans on top of which is situated the City of Tapachula (Macías et al. 2000; Murcia 2008; Murcia and Macías 2009a, b, 2014).

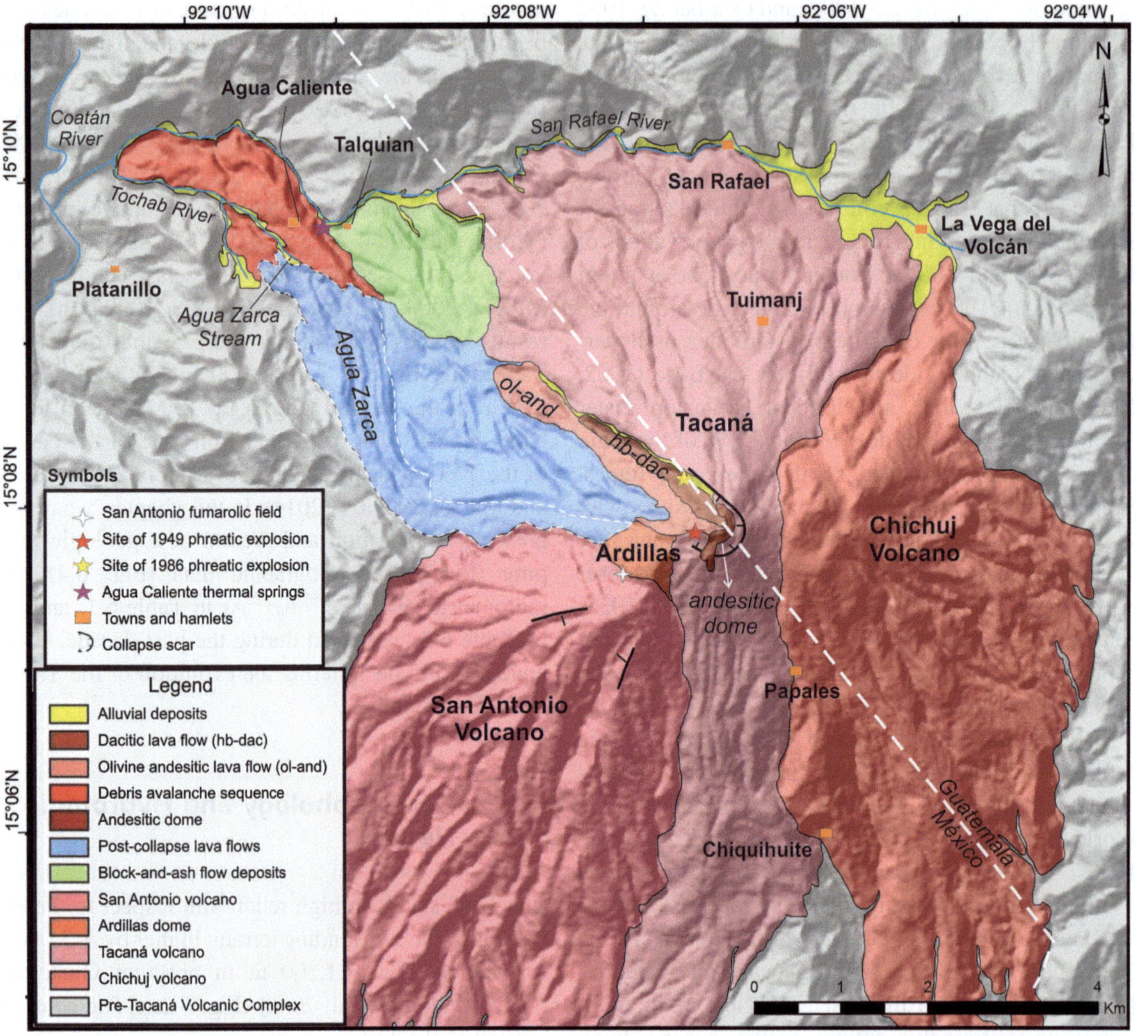

Fig. 6.3 Simplified geologic map of the TVC and surrounding areas after Macías et al. (2010). The map shows the distribution of the deposits associated with Chichuj, Tacaná, and San Antonio volcanoes and the ~15 ka Las Ardillas dome. Tacaná volcano has an open horseshoe-shaped crater produced by the collapse of its northwestern flank that generated the Agua Caliente debris- avalanche deposits. Post-collapse activity was marked by the emplacement of andesitic lava flows (i.e., Agua Zarca) into the amphitheater crater (see text for details)

6.3 The Tacaná Volcanic Complex (Late Pleistocene-Recent)

The Tacaná Volcanic Complex (TVC) lies on Mesozoic metamorphic rocks that have been affected by two episodes of intrusion of granites, granodiorites, and tonalites, 29–35 and 13–20 Ma (Mujica 1987; García-Palomo et al. 2006; Chap. 1) (Fig. 6.3). These metamorphic and intrusive rocks represent the regional basement and are overlain by volcanic deposits of the San Rafael (2 Ma), Chanjale (1 Ma), and Sibinal (unknown age) calderas (García-Palomo et al. 2006). The TVC began its formation during late Pleistocene, ~225 ka ($^{40}Ar/^{39}Ar$ date, Table 6.1) inside San Rafael caldera, through the construction of four NE-oriented volcanic edifices (oldest to youngest): Chichuj, Tacaná, Las Ardillas dome, and San Antonio (Figs. 6.1 and 6.2). The stratigraphic sequence of the volcanic complex is shown in Fig. 6.4 and described below.

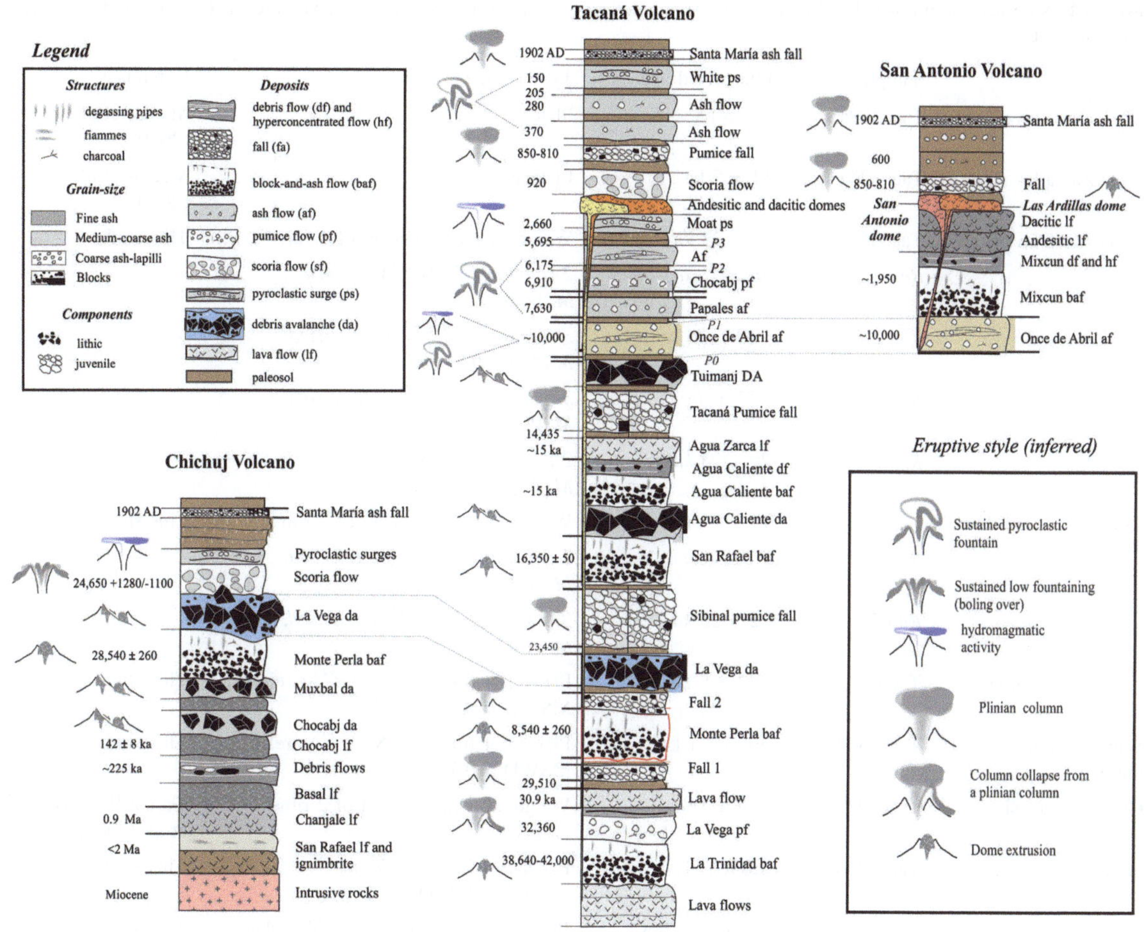

Fig. 6.4 Composite stratigraphic columns of Chichuj, Tacaná, and San Antonio volcanoes and the Las Ardillas dome, forming the Tacaná Volcanic Complex. The columns display a complex set of deposits indicating lava effusion and dome construction followed by collapses of Chichuj, Tacaná and San Antonio edifices. Plinian to subplinian events and associated column collapses generated pyroclastic deposits concentrated at Tacaná volcano

6.3.1 Chichuj Stratovolcano (~225 ka)

Chichuj (3,800) has a semi-conical shape with a 1.1 km wide summit crater and an inner lava dome (Fig. 6.2a). The edifice is dissected to the west by a collapse scar that hosts the younger Tacaná stratovolcano (see below). The early stages of activity at Chichuj were characterized by the eruption of andesitic lava flows that extend as far as 4 km from the present crater rim. This effusive activity was followed by the extrusion of lava domes, which were subsequently destroyed by explosive eruptions that generated pyroclastic density currents (Fisher et al. 1987). This explosive activity was followed by effusive activity, which emplaced several lava flows.

The oldest deposits dated at Chichuj so far are a series of partly indurated block-and-ash flow deposits (BAF), and volcaniclastic deposits, exposed south and southwest of the present Chichuj summit. These deposits, >100 m thick, formed the base of a thick fan in the Muxbal coffee plantation (Finca Muxbal) and appear as isolated outcrops northwest of the town of Union Juarez (Fig. 6.1a). The deposits consist of several light-pink to gray flow units. Each unit is massive, matrix-supported, monolithologic, consisting of blocks of dacite embedded in a medium ash matrix of the same composition. A ~20 m-thick, lithified debris flow deposit composed of angular to subangular clasts in a medium to coarse-grained altered sand matrix outcrops near Santa Maria La Vega. The clasts

Table 6.1 Summary of published and new $^{40}Ar/^{39}Ar$ age determinations for rocks of the Tacaná Volcanic Compex and underlying basement rocks

Unit	Sample	Interpreted age	Type of age	Reference
Agua Zarca lava flow	Tac0309a	9 ± 5 ka	Plateau	Macías et al. (2010)
Agua Zarca lava flow	Tac0324a	11 ± 5 ka	Plateau	Macías et al. (2010)
Agua Zarca lava flow	Tac0361c	13 ± 23 ka	Weighted mean plateau of 4 runs	This study
	Run 1	*86 ± 51 ka*	*2 fractions, 74 % ^{39}Ar released, MSWD = 0.2*	
	Run 2	*−22 ± 36 ka*	*3 fractions, 93 % ^{39}Ar released, MSWD = 0.7*	–
	Run 3	*−37 ± 48 ka*	*3 fractions, 58 % ^{39}Ar released, MSWD = 0.1*	–
	Run 4	*29 ± 28 ka*	*2 fractions, 72 % ^{39}Ar released, MSWD = 0.0*	–
Agua Zarca lava flow	Tac0334c	17 ± 23 ka	Weighted mean plateau of 2 runs	This study
	Run 1	*7 ± 39 ka*	*2 fractions, 39 % ^{39}Ar released, MSWD = 0.6*	–
	Run 2	*22 ± 28 ka*	*3 fractions, 81 % ^{39}Ar released, MSWD = 0.1*	–
Chocabj lava flow	Tac0342c	142 ± 8 ka	Plateau: 4 fractions, 72 %^{39}Ar release, MSWD = 0.2	This study
White rhyolitic pumice	Taca0564a	399 ± 31 ka	Plateau: 4 fractions, 82 %^{39}Ar release, MSWD = 0.7	This study
Chanjale andesite	Tac0333	815 ± 16 ka	Weighted mean plateau of 2 runs	García-Palomo et al. (2006)
Green ignimbrite	Tac0328a	1,035 ± 164	Plateau: 6 fractions, 77 % ^{39}Ar released, MSWD = 2.2	This study
Azteca lava flow	Tac0363c	1,340 ± 11 ka	Plateau: 6 fractions, 91 % ^{39}Ar release, MSWD = 0.7	This study
Azteca ignimbrite	Tac0362c	1,650 ± 15 ka	Plateau: 3 fractions, 83 % ^{39}Ar release, MSWD = 0.9	This study
Tochab volcaniclastic deposits	Tac0322a Bi	1,688 ± 17	Plateau: 10 fractions, 99 % ^{39}Ar release, MSWD = 0.5	This study
Dome-lavas	Tac0313a	1,697 ± 32 ka	Plateau: 9 fractions, 77 % ^{39}Ar release, MSWD = 3.1	This study
Dacitic lava flows	Tac0313a Bi	1,707 ± 20 ka	Plateau: 11 fractions, 99 % ^{39}Ar release, MSWD = 0.9	This study
Dome-lavas	Tac0346c	1,768 ± 17 ka	Plateau: 5 fractions, 76 % ^{39}Ar release, MSWD = 1.5	This study
Tochab BAF	Tac0321a	1,769 ± 23 ka	Plateau: 7 fractions, 64 % ^{39}Ar release, MSWD = 1.6	This study
San Rafael lava flow	Tac0349c	1,872 ± 24 ka	Weighted mean plateau of two runs	García-Palomo et al. (2006)
Tochab andesitic lava flow	Tac0323a	1,995 ± 82 ka	Weighted mean plateau of 2 runs	García-Palomo et al. (2006)

(continued)

Table 6.1 (continued)

Unit	Sample	Interpreted age	Type of age	Reference
Granodoritic xenolith	Taca0564b1 Bi	3.77 ± 0.14 Ma	Weighted mean plateau of 3 runs	This study
	Run 1	*3.99 ± 0.16 Ma*	*Plateau: 7 fractions, 95 % ^{39}Ar release, MSWD = 1.2*	–
	Run 2	*3.73 ± 0.25 Ma*	*Plateau: 5 fractions, 63 % ^{39}Ar release, MSWD = 1.2*	–
	Run 3	*3.54 ± 0.17 Ma*	*Plateau: 5 fractions, 79 % ^{39}Ar release, MSWD = 0.7*	–
Granodoritic xenolith	Taca0564b2 Bi	6.84 ± 0.08 Ma	Weighted mean plateau of 2 runs	This study
	Run 1	*6.81 ± 0.12 Ma*	*Plateau: 6 fractions, 85 % ^{39}Ar release, MSWD = 0.2*	This study
	Run 2	*6.87 ± 0.11 Ma*	*Plateau: 5 fractions, 79 % ^{39}Ar release, MSWD = 2.2*	This study
Sibinal granite	Sibinal Granite Bi	12.65 ± 0.08 Ma	Plateau: 7 fractions, 99 % ^{39}Ar release, MSWD = 0.5	This study
Schist with Bi	Tac0358	13.3 ± 0.1 Ma	Plateau: 9 fractions, 91 % ^{39}Ar release, MSWD = 1.6	This study
Granodiorite	Tac0364 Bi	13.3 ± 0.2 Ma	Plateau	García-Palomo et al. (2006)
Qtz-Bi intrusive granite	Tg104 Bi	13.88 ± 0.08 Ma*	Plateau: 8 fractions, 73 % ^{39}Ar release, MSWD = 1.1	This study
Granodiorite	Tac0359c Bi	13.9 ± 0.1 Ma	Plateau	García-Palomo et al. (2006)
Granite	Tac0364c Bi	29.4 ± 0.2 Ma*	Plateau	García-Palomo et al. (2006)

All samples are whole rock analyses except where noted: Bi = biotite, Standard TCR-2 with an age of 27.87 Ma was used to calculate the irradiation parameter except for * where mmhb-1 with an age of 513.9 Ma was used. Ages reported at the 1-sigma level. Interpreted ages are single run plateau ages or weighted averages of two or more plateau ages from multiple runs
Italics refer to individual runs that were averaged to produce a weighted mean age for new data. Analytical methods and detailed data tables for all samples are provided in the data appendix

consist of plagioclase, amphibole, and quartz in a glassy groundmass. A block from this deposit yielded an age of ~225 ka, which can be considered up to now the maximum age of the volcano.

The effusion of several andesitic lava flows (García-Palomo et al. 2006) on the eastern flanks of the volcano followed. The oldest flow (*Chocabj lava flow*) was erupted around 142 ± 8 ka (Table 6.1). It is gray in color with a mineralogical assemblage of plagioclase + amphibole + pyroxene and xenoliths in a fine gray glassy groundmass. The following lava flows, were characterized by basal breccias (amphibole + pyroxene + plagioclase ≫ quartz) that flowed towards the south (*Talquian lava flows*). Dark-gray porphyritic andesites flowed to the northeast (*Haciendita lava flow*) (Fig. 6.2a). These lava flows contributed to the construction of Chichuj stratovolcano.

The construction of Chichuj was disrupted by at least three flank collapses. The remains of a collapse scar to the west attest to the largest collapse, which left no deposits due to the following burial by products of the younger Tacaná volcano (García-Palomo et al. 2006). Two debris avalanche deposits to the southeast, and northeast attest to minor collapses, which occurred in these directions from the summit.

The massive debris-avalanche deposit to the SE (Muxbal), exposed at the base of the Muxbal gully (Espíndola et al. 1993), consists of meter-size jigsaw blocks in a matrix of coarse-grained ash, with yellow to orange hydrothermally altered areas (Fig. 6.5a). The blocks consist of an assemblage of plagioclase + amphibole + quartz embedded in a fine-grained groundmass. The absence of other pyroclastic deposits on top of this debris-avalanche deposit suggests that

Table 6.2 Summary of published and new radiocarbon dates of the Tacaná Volcanic Complex

Unit	Sample number	Lab No.	Location		Sample	$d^{13}C$ (%)	Conventional age yr. BP	Calibrated age	Calibrated age	Observations	References
			North	West				Range 1σ AD/BC	Range 2σ		
White Fall-Surge deposit	TAC 9716	9580	15° 04′ 24″	92° 03′ 49″	Paleosol	−24.9	105 ± 40	AD 1952–1953	AD 1951–1954	Paleosol below 1902	Mora et al. (2004)
	TAC 9737E	AU	na	na	Paleosol	−25	140 ± 50	AD 1951–1952	AD 1902–1953	Paleosol with pottery and charcoal below 1902	Mora et al. (2004)
	TAC0814b	14862	15° 06′ 21″	92° 06′ 27.2″	Charcoal	−23.3	150 ± 40	AD 1950–1952	AD 1666–1784 1795–1892 1908–1949	White pyroclastic surge deposit at section## at the Tacaná moat	This study
	TAC0814a	14867	15° 06′ 21″	92° 06′ 27.2″	Paleosol	−25.6	205 ± 85	AD 1916–1952	AD 1616–1954	Paleosol between white surge deposit and <800 year fallout at the moat of Tacana	This study
Papales Ash flow deposit 2	TAC 0445a	13733	15° 07′ 05″	92° 06′ 05″	Charcoal	−24.9	280 ± 60	AD 1499–1501, 1512–1601, 1616–1666, 1784–1795	AD 1454–1682, 1737–1757, 1761–1803, 1936–1951*	Ash flow deposit	García-Palomo et al. (2006)
Yellow Pyroclastic flow deposit	TAC0837a	14864	15° 09′ 10″	92° 06′ 23.1″	Charcoal	−25.6	370 +80/−75	AD 1558–1631	AD 1417–1664	Yellow pyroclastoc flow deposit at the northern flank of Tacaná	This study
San Antonio	TAC9865b	10003	15° 07′ 28″	92° 07′ 09″	Charcoal	−25.1	600 ± 50	AD 1304–1365, 1384–1402	AD 1288–1417	Charcoal in paleosol from an ash flow atop San Antonio	Macías et al. (2000)
Pumice fall	TAC 0448a	13734	15° 07′ 43.3″	92° 06′ 24.9″	Charcoal	−24.1	810 +110/−105	AD 1048–1086, 1123–1138, 1150–1285	AD 1016–1325, 1344–1394	Paleosol 5	This study
	TAC 0453	13737	15° 07′ 51.2″	92° 06′ 26.1″	Paleosol	−25.2	850 ± 40	AD 1247–1251	AD 1148–1266	Paleosol 5	This study

(continued)

Table 6.2 (continued)

Unit	Sample number	Lab No.	Location		Sample	$d^{13}C$ (%)	Conventional age yr. BP	Calibrated age	Calibrated age	Observations	References
			North	West				Range 1σ AD/BC	Range 2σ		
Scoria flow deposit	TAC0306a	12888	15° 09′ 21″	92° 09′ 04″	Charcoal	−29.1	920 ± 35 AMS	AD 1043–1104,1118–1158	AD 1027–1187, 1199–1206	BAFD	Macías et al. (2010)
Mixun BAF	TAC 9721	9581	15° 01′ 12″	92° 08′ 19″	Charcoal	−24.2	1,825 ± 140	AD 54–358, 364–381	BC 152–139,BC 112–AD 537	MFD	Macías et al. (2000)
	TAC0803	14860	15° 07′ 50.6″	92° 06′ 03″	Charcoal	−26.3	1,830 ± 50	AD 126–243	AD 278–328	Brown ash flow Papales-Top	Limón (2011)
	TAC 9712-1	9779			Charcoal	−26.3	1,935 ± 105	BC 45–AD 182, BC 186–214	BC 196–AD 265, AD 273–335	MFD	Macías et al. (2000)
	TAC 9737-d				Charcoal	−25	1,950 ± 50	BC 17–15, BC 0–AD 89, AD 102–123	BC 81–53, AD 174, 192	MFD	Macías et al. (2000)
	TAC 65c-1		15° 07′ 28″	92° 07′ 09″	Charcoal	–	1,980 ± 40	BC 37–29, 22–11, BC 2–AD 63	BC 87–78, 55–91, BC99	Charcoal, surge deposit, top of SAV	Macías et al. (2000)
	TAC 9865d	10004	15° 07′ 28″	92° 07′ 09″	Paleosol	−24.5	2,015 ± 45	AD 36–52	AD 118–74	Paleosol on top of SAV	Macías et al. (2000)
	TAC 0443a	13731	15° 07′ 03.3″	92° 06′ 01″	Charcoal	−24.5	2,070 ± 70	BC 176–AD 1	BC 354–291, BC 23–AD 74	AFD3	García-Palomo et al. (2006)
	TAC 9739	9786	15° 07′ 28″	92° 07′ 09″	Charcoal	−27.6	2,370 +280/−203	BC 804–159, 134–116	BC 1186–1151, 1150–1129	MFD	Macías et al. (2000)
Moat Pyroclastic surge deposit	TAC 0451	13736	15° 08′ 3.5″	92° 06′ 27.4″	Paleosol	−23.7	2,660 +55/−50	BC 895–871, 850–793	969–961, 932–76	Paleosol 4	This study

(continued)

Table 6.2 (continued)

Unit	Sample number	Lab No.	Location		Sample	$d^{13}C$ (%)	Conventional age yr. BP	Calibrated age	Calibrated age	Observations	References
			North	West				Range 1σ AD/BC	Range 2σ		
Pyroclastic surge/flow deposit	TAC0301a P3	12886	15° 09′ 19″	92° 08′ 02″	Paleosol P3	−25.3	5,695 +190/ −185	BC 4728–4347	BC 4131–4069	Paleosol 3 east of Agua Caliente	Macías et al. (2010)
	TAC 9730	9583	na	na	Charcoal	−25	5,860 ± 125	BC 4877–4871, 4848–4549	BC 5046–4450	BAF above Papales	Mora et al. (2004)
	TAC0301a P2	12885	15° 09′ 19″	92° 08′ 02″	Paleosol P2	−25.7	6,175 ± 70	BC 5217–5033	BC 5302–4952	Paleosol 2 east of Agua Caliente	Macías et al. (2010)
Chocabj Pumice flow deposit	TAC9891	10442	15° 06′ 30″	92° 04′ 40″	Charcoal	−25.9	6,910 ± 95	BC 5896–5713	BC 5983–5937, 5934–5644	Pumice flow deposit (Chiqui 8) at the M-G border	García-Palomo et al. (2006)
Papales ash flow deposit 1	TAC 0444a	13732	15° 07′ 03.6″	92° 06′ 03.4″	Charcoal	−22.7	7,630 ± 115	BC 6603–6392	BC 6695–6230	Ash flow deposit between Papales and the top	García-Palomo et al. (2006)
Once de Abril ash flow deposit	TAC9737C-1	–	15° 03′ 25″	92° 08′ 42″	Charcoal	−25	9,960 ± 50	BC 9648–9607, 9524–9496, 9458–9315	BC 9670–9295	Ash flow deposit around Once de Abril	Macías et al. (2000)
	TAC0837b	14863	15° 09′ 10″	92° 06′ 23.1″	Charcoal	−26.6	10,050 +260/ −250	BC 10089–9289	BC 8901–8850	PF with pumice northern flank	Limón (2011)
	TAC03-B13	12893	na	na	na	−25.2	10,500 ± 115	BC 10742–10398, 10368–10339, 10333–10289	BC 10841–10135	BAFD	Mora et al. (2004)
	TAC0301a	12884	15° 09′ 19″	92° 08′ 02″	Charcoal	−23.2	10,610 +330/ −315	BC 10945–10094	BC 11175–9447	Paleosol with embedded charcoal	Macías et al. (2010)
	TAC0299	12882	15° 09′ 28.4	92° 08′ 28″	Charcoal	−25	10,615 ± 55 AMS	BC 10838–10687, 10481–10476	BC 10863–10620, 10545–10450	BAFD east of Agua Caliente	Mora et al. (2004)
	TAC9737B-1	–	15° 03′ 25″	92° 08′ 42″	Charcoal	−25	10,960 ± 50	BC 10979–10916	BC 11044–10887	Ash flow deposit around Once de Abril	Macías et al. (2000)

(continued)

Table 6.2 (continued)

Unit	Sample number	Lab No.	Location		Sample	$d^{13}C$ (%)	Conventional age yr. BP	Calibrated age	Calibrated age	Observations	References
			North	West				Range 1σ AD/BC	Range 2σ		
Tacaná Pumice	TACA0427	na	15° 13′ 15″	92° 03′ 26″	Paleosol	−20.9	14,435 +155/−150	BC 15380–15298	BC 16028–15136	Paleosol below the Tacana fallot	Macías et al. (2011)
San Rafael BAF	TAC9752				Charcoal	16,350 ± 50	BC 17,591–17,478	BC 17642–17444			–
Sibinal Pumice	TACA0411B	13739	15° 09′ 23″	92° 02′ 03″	Paleosol	−24.8	23,540 +255/−495	BC 26768–25951	BC 27425–25643	Paleosol below Sibinal Pumice	Macías et al. (2011)
La Vega DAD	TACA0415A	13741	15° 09′ 02″	92° 05′ 03″	Charcoal	−24.3	24,650 +1,280/−1,100	BC 28641–26357	BC 29391–24770	Pyroclastic Surge above DAD	Macías et al. (2011)
Pumice fall 1	TACA041A	13738	15° 09′ 23″	92° 02′ 03″	Paleosol	−23.8	29,510 +620/−575	BC 32740–31482	BC 33207–30530	Paleosol below the Pomez fall 1	Macías et al. (2011)
Monte Perla BAF	TAC9704	na	15° 02′ 34″	92° 05′ 11″	Charcoal	na	28,000	nd	nd	BAFD at Monte Perla south of Tacana	Espíndola et al. (1989)
	TAC9332	6923	15° 02′ 34″	92° 05′ 11″	Charcoal	−25.3	>30,845	nd	nd	BAFD at Monte Perla	Espíndola et al. (1993)
	TAC9714	na	15° 05′ 03″	92° 04′ 35″	Charcoal	na	28,540 ± 260	BC 31400–30543	BC 31727–29910	BAFD at Muxbal southeast of Tacana	[4d]Mora et al. (2004)
La Vega pyroclastic flow	TACA0415B	13742	15° 09′ 02″	92° 05′ 03″	Charcoal	−23	32,360 +560/−525	BC 36633–34219	BC 34023–33532	Pumice flow deposit	This study
La Trinidad BAF	TAC93TRI	6924	15° 02′ 12″	92° 06′ 51″	Charcoal	−25	38,630 +5,100/–	BC 47734–34202	BC 34037–33464	BAF at La Trinidad	Espíndola et al. (1993)
	Trinidad BAF	na	15° 02′ 12″	92° 06′ 51″	Charcoal	na	40,000	nd	nd	BAFD at La Trinidad	Espíndola et al. (1989)

na not available and *nd* not determined

Fig. 6.5 **a** Outcrop showing the Muxbal debris-avalanche deposit (MDAD, *right*) covered by Tacaná block-and-ash flow deposits (BAFD) at the entrance of the Muxbal coffee plantation in the vicinity of Cordovan village (Fig. 6.1). **b** Exposure of a shattered block belonging to La Vega debris-avalanche deposit that is covered by a scoria flow and stratified pyroclastic surge deposits with disseminated charcoal (*insets*). **c** Exposure of the La Vega pyroclastic flow deposit made of massive to stratified layers rich in block to lapilli-sized pumice in a coarse ash matrix containing charcoal. The deposit is covered by a series of paleosols and reworked beds

the collapse was not triggered by volcanic activity, but was instead related to the hydrothermal alteration of the edifice. This debris-avalanche deposit is covered by a sequence of lacustrine and fluvial deposits, and a block-and-ash flow deposit (*Muxbal BAF*). The Muxbal BAF, dated at 28,540 ± 260 year BP, was attributed to volcanic activity occurred at Tacaná volcano (Mora et al. 2004).

The debris-avalanche deposit related to a NE flank collapse of Chichuj is exposed in the vicinity of the La Vega del Volcan village (Macías et al. 2011). It consists of megablocks of up to 4 m in diameter with jigsaw-fit textures (Fig. 6.5b). The blocks are banded gray to red porphyritic andesites with a mineralogical assemblage of amphibole + pyroxene + plagioclase. This deposit is covered by a massive scoria deposit, which in turn is overlain by cross-stratified to massive beds of fine lapilli to coarse ash with soft-state sedimentary deformation. These scoria beds are attributed to the emplacement of a pyroclastic density current produced by a lower fountaining of pyroclasts without the development of an eruption column (so-called "boiling-over" activity) followed by the emplacement of pyroclastic surges (cross-stratified ash beds) derived from hydromagmatic activity. Radiocarbon ages of 24,650 +1,280/−1,100 year BP were obtained on charcoal in the matrix of the cross-stratified beds (Insets in Fig. 6.5b).

The activity at Chichuj probably continued with the extrusion of a summit dome, and subsequent explosions lead to the generation of pyroclastic density currents toward the east of the present Chichuj crater, as indicated by a deposit exposed around Buenavista village consisting of scoria fragments, and minor lithic clasts, embedded in a coarse- to fine-grained ash matrix. The scoria fragments are poorly vesiculated and have the same mineralogy as the lithic fragments consisting of plagioclase + olivine + pyroxene glomerocrysts all set in a glassy groundmass. The exact age of this deposit is unknown, but its relative age was inferred from field relationships with adjacent lava flows.

6.3.2 Tacaná Volcano (Late Pleistocene-Recent)

This edifice was built during the past 45 ka with the effusion of several andesitic lava flows and domes ranging in composition from andesites to dacites (Macías 2007; Limón 2011). During this time period, Tacaná has also produced extensive block-and-ash flow deposits. To date, we have identified three pyroclastic flow fans around the volcano (dated respectively at ~42, ~28 and ~16 ka; Macías 2007) interbedded with lava flows, most of which are of unknown age. At least four plinian to subplinian eruptions occurred between 30 and 14 ka (Macías et al. 2011; Arce et al. 2012).

The inception of volcanic activity at Tacaná edifice is still poorly known because of the paucity of exposures, and dateable samples. Apparently, the early stages were characterized by effusion of andesitic lava flows that are mostly covered by younger products. On the southern flanks, between the villages of La Trinidad and Santo Domingo, these lava flows are covered by a block-and-ash flow deposit, *La Trinidad BAF*, which is the oldest deposit dated so far. Charred logs embedded in this deposit yielded ages between ca. 42,000 year BP (Espíndola et al. 1989) and 38,630 +5,100/−3,100 year BP (Espíndola et al. 1993) (Table 6.2).

A second gray block-and-ash flow deposit, *Monte Perla BAF*, (Espíndola et al. 1993) is exposed at two locations south of Tacaná, near the village of Monte Perla (Mora et al. 2004), and the Muxbal Coffee Plantation (Finca Muxbal) (Fig. 6.1a), where it overlies a debris avalanche deposit, attributed to the collapse of the former Chichuj volcano, and lacustrine deposits (García-Palomo et al. 2006). The Monte Perla BAF deposit was dated around 28,540 ± 260 year BP (Table 6.2). A third block-and-ash flow deposit is exposed near the village of San Rafael, on the northern flanks of Tacaná (Fig. 6.2a). This deposit consists of at least four massive beds of subangular andesitic blocks set in a medium to fine ash matrix, separated by thin beds of fine-grained ash. A charcoal sample in this deposit yielded an age of 16,350 ± 50 year BP (Mora et al. 2004).

The presence of block-and-ash flow deposits, led Espíndola et al. (1993) to define Tacaná as a Pelean-type stratovolcano. The pyroclastic flow deposit fans around Tacaná suggest that andesitic domes were extruded at the summit crater in several occasions and destroyed by Vulcanian eruptions.

The present-day Tacaná edifice was built during the past 30.9 ka (dated by $^{40}Ar/^{39}Ar$), during at least four eruptive stages, each of which involved the effusion of andesitic lava flows and domes from the central crater, followed by their subsequent destruction by explosive activity (Limón 2011). This predominantly effusive activity was frequently accompanied by explosive activity as indicated by the stratigraphic record during the past 35 ka (Macías et al. 2010, 2011; Arce et al. 2012). The oldest deposit known consists of massive to stratified beds with pumice and ash fragments that outcrops on the north-northeastern flanks of the volcano (Fig. 6.5c). The apparent lack of associated fall deposits at its base suggests that this deposit was not associated with the collapse of a high eruption column. More likely, it was related to a low fountaining of pyroclasts without the development of an eruption column ("boiling-over" activity). This deposit, in the vicinity of La Vega del Volcán village, contains carbonized trees dated at 32,360 +560/−525 year BP.

After ~3,000 years, the explosive activity at Tacaná consisted of a plinian to subplinian eruption, which generated plumes that drifted to the north-northeast, due to prevailing winds, reaching what is today Guatemala. In fact, several studies (Arce et al. 2008, 2012; Macías et al. 2011) identified at least four fall deposits, separated by paleosols, namely: Pumice fall 1, Pumice fall 2, Ochre Pumice, Sibinal Pumice, and Tacaná Pumice (Fig. 6.6a, b).

The paleosol underlying the lowermost fall deposit (Pumice fall 1) yielded a ^{14}C age of 29,510 +620/−575 year BP. Pumice fall 2 is bracketed between Pumice Fall 1 and the La Vega debris avalanche

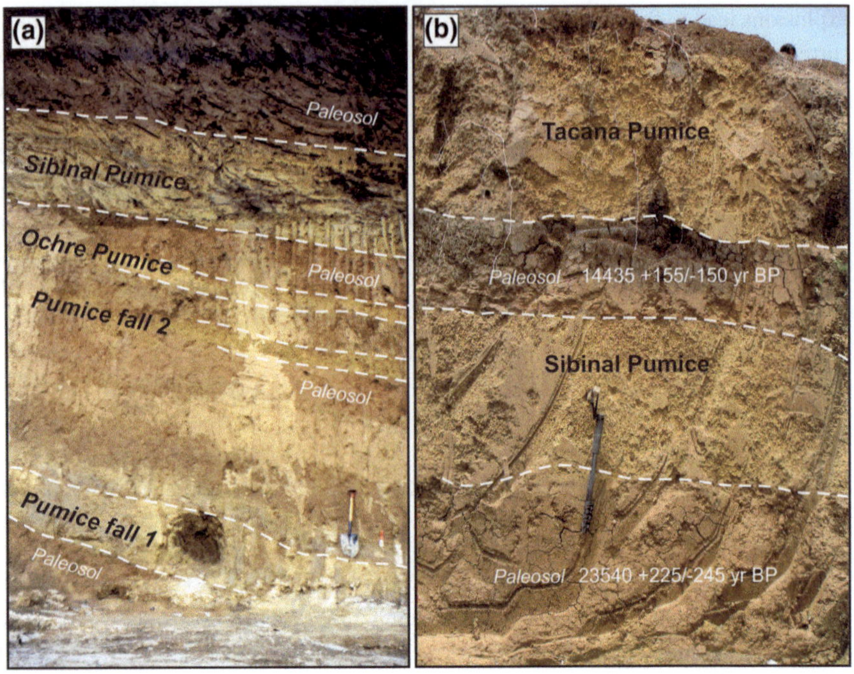

Fig. 6.6 **a** View of the three fall deposits located at ~8 km from Tacaná summit, near the village of Sibinal. The outcrop shows the lower Pumice fall 1, Ochre Pumice, and Pumice fall 2 separated by paleosols, and the thicker Sibinal pumice above, also covered by *thick* paleosols. **b** View of the Tacaná pumice deposit overlying a paleosol dated at ~14 ka. BP and the Sibinal Pumice

deposit (>24,650 +1,280/−1,100 year BP; see description below) (Fig. 6.6a). To date, only two of these deposits have been described in detail: the Sibinal Pumice (Arce et al. 2012), and the Tacaná Pumice (Arce et al. 2008; Macías et al. 2011). The Sibinal Pumice was eventually emplaced from a plinian to subplinian eruption column, 19–21 km high, as: (1) a lower stratified member of at least seven normal graded clast-supported layers of pumice lapilli interbedded with cross-bedded layers, and (2) an upper clast-supported massive member (MM). The Sibinal Pumice overlies a paleosol dated at 23,540 +225/−245 year BP (Arce et al. 2012). The lower member was dispersed to the northeast and involved a minimum volume of 2.9 km^3 (1.1 km^3 DRE) and erupted at a mass discharge rate of 4.7×10^7 kg/s. The upper member, greater in volume 4.6 km^3 (1.9 km^3 DRE), was erupted at a mass discharge rate of 8.1×10^7 kg/s, and dispersed to the north (N22°E). Using the index of Newhall and Self (1982), the eruption that produced the Sibinal pumice would rank a VEI = 5 constituting one of the largest magnitude eruptions of Tacaná. Pumice compositions range from basaltic to andesitic (48–61 wt% SiO_2) with phenocrysts of plagioclase, augite, hypersthene, Fe–Ti oxides, and rare amphibole (Arce et al. 2012).

The Tacaná Pumice is an ochre, massive, clast-supported fall deposit composed of lapilli-sized altered pumice clasts (Fig. 6.6b). The underlying dark-brown paleosol was dated at 14,435 +155/−150 year BP (Table 6.2). A rough estimation of the tephra volume of this event is 8 km^3 (3 km^3 DRE), with a column height calculated at 28 km (Arce et al. 2008). As with the eruption that generate the Sibinal Pumice, this eruption is also estimated to have been VEI = 5.

6.3.3 Sector Collapses of the Volcano

The construction of Tacaná has been interrupted at least twice during its history by the collapse of sectors of the volcano. One of the major collapses occurred around ~15 ka ago emplacing a debris-avalanche deposit (Agua Caliente debris avalanche) toward the northwest. This deposit covers a minimum area of 4 km^2, has a total volume of 0.8 ± 0.5 km^3, and has a coefficient of friction H/L (H = height/L = length) of ~0.35 (Macías et al. 2010). This flank failure apparently also caused the collapse of a growing summit dome, as evidenced by the presence of a block-and-ash flow deposit, on top of the debris-avalanche deposit. The collapse left a 600 m-wide summit amphitheater scar with a 30° opening to the northwest (Figs. 6.1b–c and 6.3). Eventually the emplacement of both the debris-avalanche, and the block-and-ash flow deposits produced a ~200 m-thick natural dam that blocked and diverted the course of the San Rafael River. The subsequent failure of this dam and remobilization of the volcaniclastic material produced lahars that eroded the primary deposits and cascaded into the Coatán River.

A lava flow (Agua Zarca lava flow, Figs. 6.1b and 6.3), which overlies the debris avalanche and the block-and-ash flow deposits, was dated at 10 ± 6 ka by $^{40}Ar/^{39}Ar$ (Macías et al. 2010) providing a minimum age for the collapse event. The lava flow is a dark-gray porphyritic andesite (59.58 wt% SiO_2) with plagioclase and pyroxene phenocrysts in a fine-grained groundmass; it has steep flow fronts and levees composed of meter-size blocks.

A second flank collapse of Tacaná occurred to the north, from the summit crater toward Guatemala. The resulting debris avalanche (Figs. 6.2a, b and 6.3) swept 4 km until reaching the village of Tuimanj (Macías et al. 2010; Limón 2011). The deposit consists of elongated to semi-conical hills (1–20 m high) of light-gray andesitic lava blocks with jigsaw-fit textures. The debris-avalanche deposit overlies the ~16 ka *San Rafael BAF* (Fig. 6.2a) and is covered by a massive, yellowish deposit, consisting of pumice lapilli and accidental lithics with mm-sized charcoal fragments set in a medium to fine-grained ash matrix. Radiocarbon dating of the charcoal yielded ages of 10,050 +260/−250 year BP, which corresponds to an explosive event that followed the collapse of the edifice, named Once de Abril ash-flow sequence (Macías et al. 2010). The Once de Abril sequence is one of the most widespread pyroclastic deposits around the flanks and smoothed fans of Tacaná. It was initially described as a massive, matrix-supported deposit, exposed on the outskirts of the Once de Abril village, 10.5 km southwest of the crater, and dated by ^{14}C at 9,960 ± 50 and 10,960 ± 50 year BP (Macías et al. 2000). Subsequent studies recognized a succession of stratified beds of light-gray, pink lapilli and ash with voids and soft-state deformation structures also in the proximity of Agua Caliente village, 7 km northwest of the crater, in the same stratigraphic position as the massive deposit described at the Once de Abril village (Macías et al. 2010). This sequence is bracketed by poorly developed paleosols (P0 and P1). The lower massive ash bed observed around the volcano is interpreted to have been emplaced by a sustained pyroclastic fountain, while the stratified upper part is interpreted as the product of hydromagmatic activity (i.e. pyroclastic surges).

Disseminated charcoal inside the succession yielded ages of 10,610 +330/−315 year BP (Table 6.2). At other locations, (e.g. near the Tuimanj village), the Once de Abril sequence consists of a block-and-ash flow deposit with dispersed charcoal. ^{14}C dating yielded ages between 10,615 ± 55 year BP (Accelerator Mass Spectrometry) and 10,500 ± 115 year BP (^{14}C Standard).

The absence of fall deposits intercalated within this sequence, and the occurrence of block-and-ash flow deposits, suggests that these widespread deposits were generated by the explosive destruction of a dacitic dome, followed by a sustained low fountaining of pyroclasts (Branney and Kokelaar 2002), that dispersed pyroclastic density currents out from the crater.

6.3.4 Holocene Activity

At least nine pyroclastic sequences, each of which produced by distinct eruptions, have been identified around Tacaná. Based on radiocarbon dating of charcoal in the deposits and paleosols separating them, these eruptive events occurred during the past ~8,000 years.

The oldest deposit outcrops 4 km from the crater on the southern flanks of the volcano near Papales village (*Papales ash flow* 1). It consists of a massive, 1.7 m-thick bed of pumice lapilli and rare lithic clasts with disseminated charcoal, set in a matrix of medium-grained ash. The deposit overlies a dark-brown paleosol. Radiocarbon dating on a charcoal fragment yielded an age of 7,630 ± 115 year BP (Fig. 6.7a).

A 6 m-thick yellow bed of pumice lapilli and blocks in a coarse ash matrix (*Chocabj pumice flow*, Table 6.2 and Fig. 6.4), mantled by 6 m of reworked material in the outskirts of the Chocabj village, outcrops only on the eastern slopes of Chichuj volcano at distances of 4 km from Tacaná. Charcoal inside the deposit was dated at 6,910 ± 95 year BP (García-Palomo et al. 2006).

A following sequence of fine-grained pyroclastic deposits is bracketed by two light-brown paleosols P2 (6,175 ± 70 year BP) and P3 (5,695 +190/−185 year

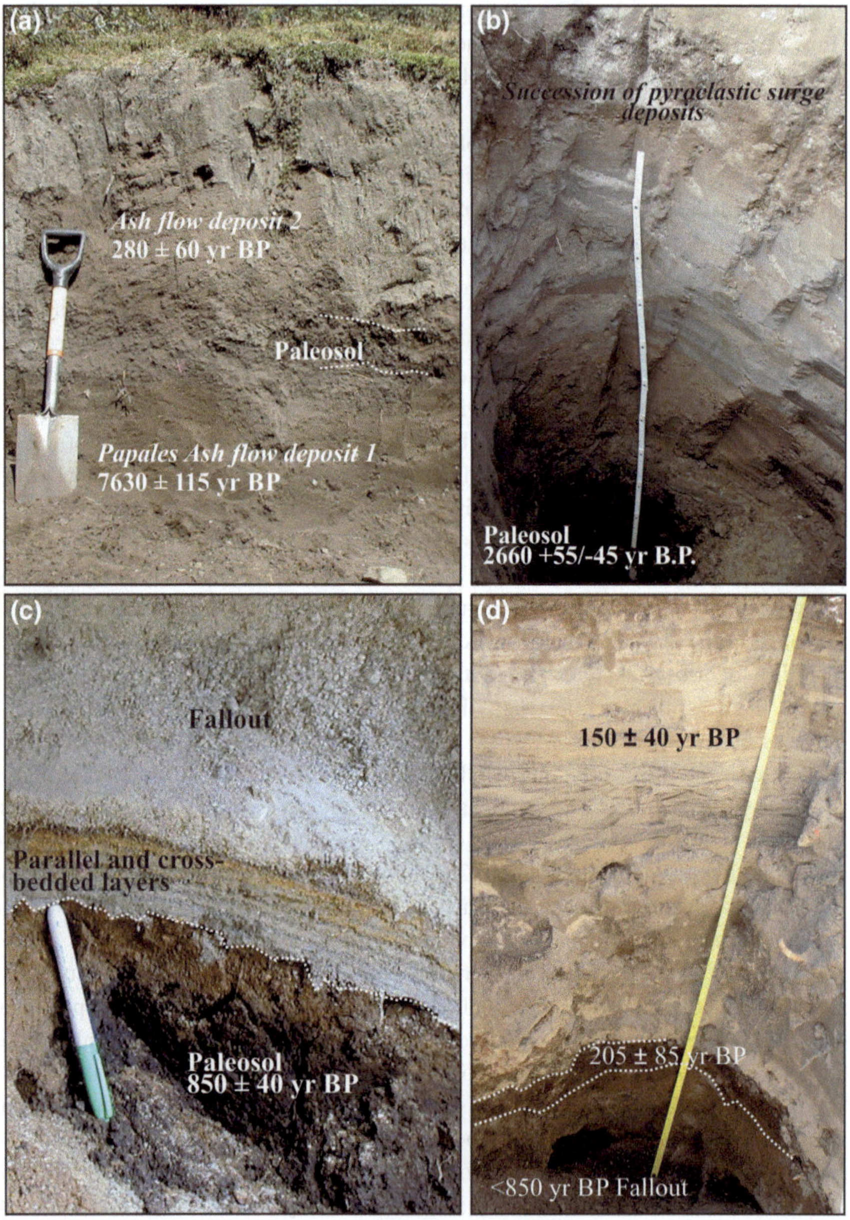

Fig. 6.7 Outcrops of some of the main Holocene deposits at Tacaná Volcano, as described in the stratigraphic column of Fig. 6.3b–d. **a** View of two ash flow deposits with pumice lapilli and charcoal separated by a paleosol in the vicinity of the Papales hamlet (*Papales ash flow deposits 1 and 2*). **b** Detail of the pyroclastic surge succession on *top* of a *dark-brown* paleosol. **c** View of the <850 year BP tephra fall deposit and a succession of basal pyroclastic surge deposits on *top* of a *dark-brown* paleosol dated at 850 ± 40 year BP, the fallout layer is widely dispersed around the Tacaná crater. **d** Exposure of a *white* to *light-gray* fine-grained pyroclastic surge deposit that overlies the <850 year BP tephra fall deposit and a paleosol dated at 205 ± 85 year BP. Charcoal found within this deposit yielded an age of 150 ± 40 year BP, making it the youngest deposit recognized to date at Tacaná

BP). It is exposed on the northwestern slopes of Tacaná close to the village of Agua Caliente, some 7 km from the crater. The sequence consists of a basal bed (32 cm thick) with a discontinuous stratification of white, light-gray to pink, fine-medium ash, rich in crystals with sparse pumice lapilli and lithics, overlain by a gray, massive layer of fine ash and lithic lapilli. Based on the structural and textural characters, this deposit was interpreted as the product of pyroclastic density currents, generated from a sustained pyroclastic fountain (Macías et al. 2010).

The moat area of Tacaná volcano, between the summit andesitic dome and the crater rim, is occupied by a complex sequence of laminated to cross-bedded ash layers. This sequence, consists of multicolored, alternating coarse and fine ash beds, and was interpreted as being deposited from pulsating dilute pyroclastic density currents generated by phreatomagmatic eruptions. The deposit overlies a dark-brown paleosol dated at 2,660 +55/−50 year BP (Fig. 6.7b).

A gray massive matrix-supported deposit, at least 2.6 m thick, outcrops on the northwestern slopes of Tacaná volcano. It has a coarse base of subrounded dark-gray andesitic scoria blocks, with sparse red andesitic accidental blocks, set in a fine to medium-grained ash matrix, grading upward into a massive bed of ash. This deposit was attributed to a sustained low pyroclastic fountain. However, additional and more detailed, studies of the deposit are needed to confirm this hypothesis. Radiocarbon dating yielded ages of 920 ± 35 year BP (Macías et al. 2010).

A sequence composed of two parts drapes the upper flanks of Las Ardillas dome, Tacaná and San Antonio volcanoes (Macías et al. 2010): (1) a basal, stratified sequence of at least ten matrix to clast-supported beds of fine-to-coarse ash, and minor andesitic lithic and pumice lapilli, gray to yellow in color; (2) a 42 cm-thick, massive, clast-supported bed of angular pumice and andesite lapilli lithics. An underlying paleosol dated at 810 +110/−105 year BP, and 850 ± 40 year BP, gives a maximum age for the eruption that produced this sequence. Structural and textural features of this sequence suggest that the lowermost bedded portion was emplaced by pyroclastic density currents, and was followed, almost immediately by the establishment of a low altitude plume that emplaced the pyroclastic fall deposit (Fig. 6.7c).

A thin, massive, yellow bed of fine ash with dispersed andesite lithic clasts is exposed on the northern flanks of Tacaná. It covers the Tuimanj debris avalanche deposit and grades upward into a paleosol. This bed contains abundant charcoal fragments dated at 370 +80/−80 year BP (Limón 2011) suggesting that pyroclastic density currents burned a large amount of vegetation.

The southern flanks of Tacaná are draped by another dark-brown massive deposit of fine ash with dispersed lithic blocks and lapilli, inferred to be emplaced by a pyroclastic density current, which is bracketed between a paleosol, and the modern soil. Charcoal found within the Papales ash flow deposit 2 yielded an age of 280 ± 60 year BP (Fig. 6.7a).

The youngest sequence at Tacaná was observed in a trench dug between the summit crater and the andesitic dome (the highest part of Tacaná volcano). Here, this sequence overlies a paleosol dated at 205 ± 85 year BP (Fig. 6.6d) and is followed by a reworked deposit, and the modern soil. It consists of several white to green, massive, clast-supported ash beds and low-angle cross-stratified multicolored beds with disseminated charcoal (dated at 150 ± 40 year BP, Table 6.2). High-angle (30°), gray, cross-stratified layers of ash beds, occur in the middle part of the sequence. A low-altitude eruptive column deposited the lower part of the sequence (low angle beds) followed by a brief pause, allowing strong winds to remobilize the pyroclastic material and produce the high-angle cross-bedding (the middle part of the sequence). This seems to be the most likely hypothesis to explain these two parts of the sequence. Then the low-altitude eruptive column eventually ended with a pulsating fountain of pyroclasts (to generate pyroclastic surges) that deposited the upper part of the sequence (upper low-angle beds). Most of the Holocene and older sequences are overlain by a dark-brown paleosol that, in places, contain abundant pottery shards and the ash fall layer from the 1902 eruption of Santa Maria volcano in Guatemala (Williams and Self 1983), and modern soils.

Summit domes

At present, the summit, horseshoe-shaped crater of Tacaná, is occupied by two lava domes and associated lava flows (Figs. 6.1b–c and 6.2a). An andesitic lava dome fills the eastern part of the crater, and a hornblende-bearing dacitic lava dome (Fig. 6.3) occupies the western sector that opens to the northwest, forming an amphitheater scar. The prolonged extrusion of the dacitic dome overspilled the northwestern edge of the breached crater forming a 1.5 km-long lava flow. An olivine-bearing lava flow fills the western-northwestern part of the crater with two lava lobes 1.5 and 2.3 km long (Figs. 6.1c and 6.3). According to our stratigraphic reconstruction (Fig. 6.4), the two domes should overlie the sequence of pyroclastic surge deposits emplaced ~2,660 year BP at the moat (see above), and are covered by dispersed pumice fragments from the ~850 year BP fall deposit.

6.3.5 Historical Phreatic Explosions

6.3.5.1 1949 Explosion

On December 22, 1949, a strong earthquake occurred at Tacaná volcano. In the following days, area residents observed white vapor columns rising above the summit and described ash falling on the outskirts of

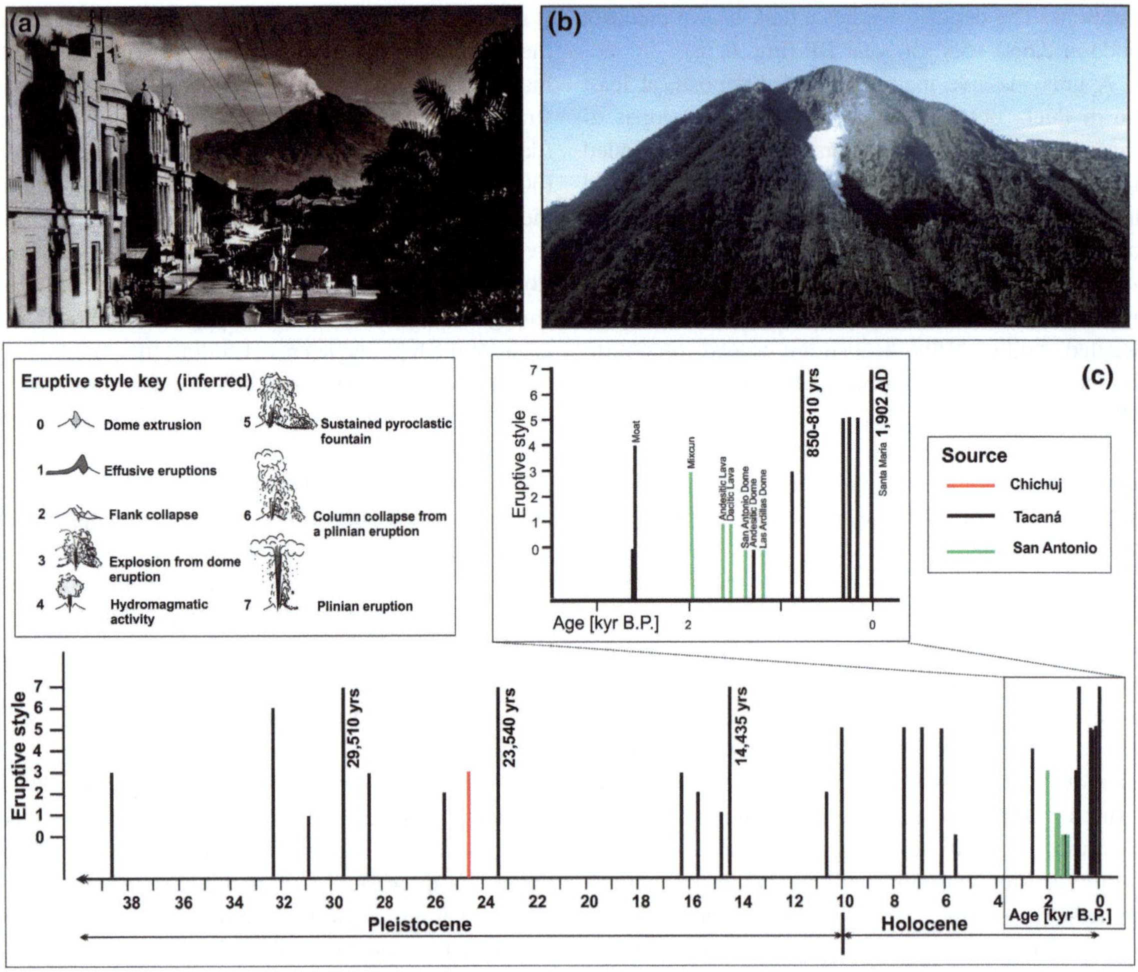

Fig. 6.8 Schematic chart summarizing the eruption types versus time occurred at the Tacaná Volcanic Complex during the past 40,000 year BP. **a** Water vapor column rising above Tacaná volcano in December 1949 as viewed from the main square of the city of Tapachula (Photo Anonymous). **b** View from the northeast of the fumarole produced after the 1986 phreatic explosion of Tacaná. The explosion occurred along the base of a lava cliff left by the Agua Caliente debris avalanche (*photograph* by Servando de la Cruz)

Unión Juarez village (Fig. 6.8a). This event alerted the local authorities, who asked for help from the Institute of Geology of the Universidad Nacional Autónoma de México. On January 7, 1950, geologist Federico Müllerried flew from Mexico City to Tapachula to visit the area. The detailed accounts of the eruption are reported from his observations (Müllerried 1951). Similar to what was reported by Böse (1902, 1903), on January 10th the volcano consisted of three craters, or "steps", located 70, 160, and 230 m below the summit (4,030 m, average height of all altitudes determined by Müllerried in 1951). The main crater of Tacaná was located 70 m below the summit, with 20–30 m-high walls opened to the north-northwest. This crater contained the summit cone of Tacaná. A second crater (160 m below the summit), had a flat morphology, and hosted a dry lake. A third crater (230 m below the summit) contained a small lake. According to the descriptions of Müllerried (1951), the fumarolic activity observed on January 10th, 1949 occurred from 16 small vents, all located to the southwest of the summit (Fig. 6.1c). The area occupied by these vents seems to coincide with the area source of the 19th century eruptions mentioned by Böse (1902, 1903). The vents hosting the fumaroles in 1949 were 2–4 m wide and, when the bottom was visible, had depths of ~4 m. All vents discharged "rose, and white to transparent fumes" visible from the City of Tapachula, and in some cases, sulfur (Müllerried 1951).

6.3.5.2 1986 Eruption

After 35 years of repose, in late December 1985, Tacaná volcano experienced several earthquakes and seismic events (De la Cruz-Reyna et al. 1989). Following these signals, a portable seismic network was installed in January 1986 at the Agua Caliente village northwest of the volcano summit. The largest single earthquake occurred on February 3, 1986, damaging adobe houses at Ixchiguan village located 25 km ENE from the crater in Guatemala, (De la Cruz-Reyna et al. 1989). Afterwards the seismic activity decreased, but persisted at lower levels, for 7 weeks with epicenters located between 15 and 25 km East to ENE of the crater. On April 20th 1986, stronger earthquakes were felt and heard in areas located to the W-SW of the volcano. Such activity increased until May 7th, when an intense earthquake swarm provoked panic among the population living around the volcano. On May 8, when the frequency of earthquakes felt by the population was two per minute, a phreatic eruption occurred at an elevation of 3,600 m above sea level on the northwestern flank of the volcano, almost along the Mexico-Guatemala international border (De la Cruz-Reyna et al. 1989) (Figs. 6.1c and 6.3). The eruption produced a 1 km-high column that left an 8-m wide vent (Fig. 6.8b). Afterwards, the seismic activity declined notably and 2 days later the seismic activity returned to pre-May levels. The plume was rich in water vapor, indicating few or no juvenile magmatic components (Martíni et al. 1987).

Both, the 1949 and 1986 phreatic explosions occurred along the horseshoe-shaped crater generated by the ~10 ka collapse that produced the Agua Caliente debris avalanche deposit (Macías et al. 2010) (Fig. 6.3). The 1949 eruption occurred at the southern tip of the breached summit crater, while the 1986 eruption occurred at the base of the northern wall forming the collapse scar of the horseshoe-shaped crater. These historical eruptions, and the current geothermal manifestations located SE of the village of Agua Caliente (Chap. 7), occurred along the collapse scar, reflecting a weak zone of the volcano, which likely corresponds to a fracture system (Macías et al. 2010) that needs to be better characterized by further studies.

The summit area of Tacaná has been affected by prolonged hydrothermal alteration over geologic time, with the largest and most altered areas located along the northwestern edge of the horseshoe-shaped crater, and on the northwestern flanks of San Antonio volcano.

The springs on the northwestern flanks of Tacaná and the fumaroles on top of San Antonio volcano have been sampled and monitored during the past 10 years (Martíni et al. 1987; Rouwet et al. 2004, 2009; Chap. 7).

6.3.6 San Antonio Volcano

San Antonio volcano is located southwest of Tacaná stratovolcano and Las Ardillas dome (Figs. 6.1b and 6.3). It has a horseshoe-shaped scar open to the southwest. The initiation of the eruptive activity of San Antonio is not well known. Apparently, the early stages consisted of the eruption of lava flows that are not yet dated. The following discussion summarizes the youngest eruption of San Antonio dated so far (~1950 year BP), and its subsequent evolution.

The deposits produced by this eruption are exposed to the southwest of the volcano in the vicinity of the Mixcun village. They consist of several massive flow units of light-gray to pink block-and-ash flow deposits that cover an area of 25 km^2 with a total estimated volume of 0.12 km^3. This eruption was probably similar in style, volume, and magnitude to the 1902 eruption of Mt. Pelée (Bourdier et al. 1989); therefore, we speculate that this eruption may rank a VEI = 4. The channel-filling facies of this deposit consists of light-gray, massive, >10 m-thick, units consisting of two-pyroxene andesite (60–63 wt% SiO_2) blocks supported by a coarse ash matrix. Disseminated charcoal in the deposit at several locations has been dated between 1,825 ± 140 and 2,370 +280/−203 year BP with an average age of 1950 year BP (Macías et al. 2000).

The distribution of the Mixcun BAF deposit towards the southwest of the volcano, and the horseshoe-shaped crater of San Antonio volcano, suggest that a dacitic summit dome was destroyed by collapse (i.e. Pelean-type eruption) in multiple stages, each of which generated pyroclastic density currents. The eruption was followed by lahars that flooded the main ravines south of the Tacaná Volcanic Complex toward the ceremonial center of Izapa. Even if these lahars did not impact Izapa directly, they caused its isolation from nearby population centers, causing its abandonment (Macías et al. 2000).

The 1950 year BP eruption left a horseshoe-shaped crater opened to the SW. Sometime after this main event, the activity at San Antonio continued with the emplacement of several andesitic lava flows (58–61 wt % SiO_2) filling the horseshoe-shaped crater (Macías et al. 2010; Mora et al. 2013), reaching distances of 4 km from the source. The eruption ended with the emission of a ~1 km-long dacitic lava flow (62–64 wt % SiO_2) and a summit dome (64.4 wt% SiO_2). We do not have an age for the occurrence of the andesitic and dacitic lavas of San Antonio that occurred after the 1950 year BP collapse of the volcano. Considering that both dacitic lavas, and the summit dome are covered by the ~850 year BP fall deposit from Tacaná, we infer that the andesitic and dacitic lavas of San Antonio were emplaced between 850 and 1950 year BP.

The ~850 year BP fall deposit is overlain by at least two other, partially eroded, pumice fall deposits from Tacaná volcano that are separated by paleosols. One of these deposits was dated at 600 ± 50 year BP, indicating that no other eruption occurred at San Antonio since that time.

6.3.7 Las Ardillas Dome

Las Ardillas dome is located between San Antonio and Tacaná volcanoes (Figs. 6.1b and 6.3). García-Palomo et al. (2006) interpreted this structure as a central dome and two associated lava flows that encroached the northwestern and southwestern flanks of San Antonio and Tacaná volcanoes, respectively. However, more recent mapping suggests that these lava flows were emplaced from Tacaná volcano. Las Ardillas dome is a porphyritic gray andesite made up of plagioclase and amphibole and abundant dark gray enclaves set in a glassy matrix (62.96 wt% SiO_2). No age determination is available for this summit dome.

6.4 Discussion: Evolution of the Tacaná Volcanic Complex

Volcanic activity in the area which hosts Tacaná volcano began during late Pliocene and continued in the Early Pleistocene with the formation of three calderas named San Rafael (2 Ma), Chanjale (ca. 1 Ma), and Sibinal (unknown age). The ancestral San Rafael caldera possibly began its formation as a stratovolcano, which erupted several lava flows of andesitic composition around 2 Ma, and continued with the extrusion of andesitic domes, and associated lava flows. The domes were subsequently destroyed generating block-and-ash flow deposits around 1.76 Ma. A more complex evolution of these calderas is envisaged, considering field evidences, but to date no detailed studies exist.

The Tacaná Volcanic Complex grew within the remains of the San Rafael Caldera (García-Palomo et al. 2006). The early stages of construction of the first edifice, Chichuj, likely occurred >225 ka with the emission of andesitic to dacitic lava flows, and domes, that were subsequently explosively destroyed, producing widespread block-and ash flow deposits which can be observed today in the vicinity of Muxbal near the border between Mexico and Guatemala, and north of Cordovan village (Fig. 6.1). It is possible that both towns of Union Juarez and Cordovan were built on top of these old pyroclastic deposits. The episodes of lava dome destruction were followed by the generation of debris flows mainly to the south of Tacaná. Chichuj continued to grow with the emplacement of the Chocabj lava flow (~142 ka ago) and the Talquian and La Haciendita lava flows (unknown ages) (Fig. 6.2a). The final stages of construction of Chichuj involved the accumulation of andesitic lavas and their associated breccias on the upper flanks of the volcano. Subsequently, the volcano experienced a collapse to the south, producing a small debris avalanche deposit (i.e. Muxbal DAD, Fig. 6.5), which remained confined in the Muxbal valley.

By projecting the remains of the actual Chichuj crater, and the present day lava dome, it is possible to envisage that the volcano extended for 4 km in diameter with a ~1.2 km wide crater. Chichuj rises from elevations of 1,600 m on the south and 2,500 m on the north, up to 3,800 m (present elevation). Hypothesizing a basal circumference of 8 km and an average altitude of 1.8 km for this structure, we obtained an approximated original volume for this edifice of ~30 km^3. Half of it was destroyed sometime during its evolution by a sector collapse to the west (García-Palomo et al. 2006). The deposits of such collapse (i.e. debris avalanches), are probably buried by younger edifices, and their eruptive products. It is quite probable that this major event in the eruptive evolution of the TVC allowed the inception of new volcanic activity, responsible for the construction of Tacaná edifice, 1 km to the southwest of the main summit of Chichuj. The volcanic activity at Chichuj continued with the extrusion of a lava dome, which was partially destroyed producing the Buenavista pyroclastic density current to the east. The northern flank of Chichuj volcano collapsed around 24,650 +1,280/−1,100 year BP generating the La Vega debris-avalanche deposit (Fig. 6.8c). The eruption continued with a boiling over event that emplaced a scoria flow to the northeast of the source, and was followed by hydromagmatic activity dispersing pyroclastic surges.

As mentioned above, Tacaná began its construction inside the amphitheater left by the second collapse of Chichuj. Tacaná formed a semi-conical edifice by the recurrent production of andesitic lava flows and summit domes. These domes were partly destroyed by gravitational collapse, or more probably, by Vulcanian events, dispersing pyroclastic density currents that emplaced block-and-ash flow deposits around the main cone. The oldest block-and-ash flow products of this activity near La Trinidad village have been dated ~42 ka. The oldest explosive products of Tacaná dated at 32,360 year BP, are pumiceous pyroclastic flow deposits found on the northern flanks of the volcano near La Vega village (Fig. 6.8).

Therefore, the construction of Tacaná volcano most likely started after 42 ka, and it was partly coeval with activity of Chichuj volcano. Tacaná (4,060 m) rises from an elevation of 1,700 m to the northwest nearby Tolquian in Guatemala, and from 1,500 m W of Union Juarez in Mexico (Fig. 6.1). The present cone of Tacaná has an approximate basal radius of 4 km N and S, but just 0.8 km E due to the buttressing influence of Chichuj, and 0.6 km to the west-southwest due to the partial burial by the younger San Antonio and Las Ardillas dome. Prior to the formation of the San Antonio edifice and Las Ardillas dome, the extension of the flanks of Tacaná to the SW was, in all probability, greater. To estimate the volume of the entire edifice, we consider a conservative basal diameter of 3 km and an average height of 2.6 km from its present base, which yields a volume estimation of ~25 km^3.

Around 29,510 year BP, Tacaná volcano began a period of explosive activity with the generation of plinian to subplinian plumes that were dispersed to the northeast by prevailing winds toward modern Guatemala. These events emplaced the Pumice fall 1 (29,510 +620/−575 year BP), the Pumice fall 2 (>24,650 +1,280/−1,100 year BP), the Ochre Pumice (undated), the Sibinal Pumice (23,540 +225/−245 year BP), and the Tacaná Pumice (14,435 +155/−150 year BP) (Fig. 6.8).

This period of plinian to subplinian eruptions of Tacaná volcano alternated with effusive eruptions, characterized by the generation of lava flows and/or by dome extrusions (Limón 2011). The partial destruction of summit domes has repeatedly generated pyroclastic density currents that deposited block-and-ash flow deposits, such as those of La Trinidad (42,000–38,630 +5,100/−3,100 year BP), Monte Perla (28,540 ± 260 year BP), and San Rafael (16,350 ± 50 year BP).

Sometime around ~15 ka, the collapse of the northwestern sector of Tacaná destroyed its conical morphology and was followed by the emplacement of a new summit dome. The volcano failed nearly parallel to the σ_{min} direction (Macías et al. 2010) generating the Agua Caliente debris avalanche towards the San Rafael River. This collapse event left a 600-m-wide summit amphitheater with a 30° opening to the northwest, and a >200 m thick deposit that blocked the San Rafael River. The flank failure weakened and destabilised the new summit dome, contributing to its collapse and the generation of a series of block-and-ash flows. The remobilization of this material produced debris flows that eroded the primary deposits and cascaded into the Coatán River.

The activity at Tacaná continued until ca. 10 ka with an eruption that produced a sustained pyroclastic fountain of ash and pumice that collapsed producing a widespread pyroclastic density current (Once de Abril ash flow). The event continued with hydromagmatic activity that dispersed pyroclastic surges. The pyroclastic density currents reached at least 8 km from the crater to the south and ~7 km to the northwest, ingesting large amounts of vegetation, as shown by charred logs and roots entrained in the deposit, suggesting the presence of a dense forest at that time. It is not clear if this eruption also caused the collapse of the northern flank of Tacaná emplacing the Tuimanj debris-avalanche deposit, locally overlain by deposits of the Once de Abril ash flow.

Sometime after the collapse of the Tacaná edifice to the northwest, which emplaced the Agua Caliente debris avalanche, volcanic activity continued with the production of 4–5 km long andesitic lava flows. These lava flows, including the Agua Zarca lava flow (10 ± 6 ka) were extruded from the summit area and flowed inside the collapse scar (Macías et al. 2010). During the early Holocene, the effusive activity of Tacaná alternated with small-volume explosive eruptions, apparently due to sustained pyroclastic fountains that dispersed pyroclastic density currents (e.g. the ~7,630 ka ash flow deposit, the ~6,910 ka pumiceous pyroclastic flow deposit, and the ~5,860 ka pyroclastic surge deposit). These explosive events alternated with extrusions of lava domes at the summit, and with lava flows that moved radially downslope (Limón 2011). Around ~2,660 ka, a hydromagmatic eruption dispersed pyroclastic surges that remained confined inside the crater area. Later, the activity became localized at the summit of Tacaná with the extrusion of three domes with different composition: andesitic to the east of the crater, olivine-bearing basaltic andesitic to the west, and dacitic to the northwest.

Another volcano was constructed 1.5 km southwest of Tacaná, forming a semi-conical edifice buttressed by the Tacaná edifice to the northeast but weak and unstable to the southwest. The inception of volcanism at San Antonio volcano is unknown, however, according to geologic field reconnaissance, stratigraphic relations, and $^{40}Ar/^{39}Ar$ dates (Table 6.1; Fig. 6.8), it probably started after the northwestern sector collapse of Tacaná (Agua Caliente debris avalanche), but prior to the emplacement of the Agua Zarca lava flow. San Antonio has an elongated base of 5–7 km, and rises from altitudes of 2,000 m from the northwest and 1,500 m from the southwest up to 3,500 m. Eventually, at its maximum growth stage, San Antonio reached a volume of ~16 km^3 and hosted a central andesitic lava dome.

Around 1950 year BP, San Antonio volcano resumed activity through phreatic-phreatomagmatic explosions, followed by the destabilization of the edifice. These explosions triggered a Peléan-style eruption that created a 30° wide scarp on the south-southwestern flank and the summit lava dome. The collapse of this dome generated pyroclastic density currents that emplaced block-and-ash flow deposits as far as 14 km from the source. This activity was followed by generation of lahars that isolated the Izapa ceremonial center from central Mexico, causing its abandonment. Later, San Antonio volcano resumed eruptive activity, with the effusion of andesitic and dacitic lava flows, which partially filled the horseshoe-shaped crater left by the collapse, and a summit dome (Macías et al. 2000). Las Ardillas dome was developed by the effusion of dacitic lava, between Tacaná and San Antonio edifices (Fig. 6.8).

Minor explosive eruptions occurred at Tacaná during the past 1,000 years, with the emplacement of thin pyroclastic density current deposits on the upper flanks of the volcano around 920 ± 35, 370 +80/−80, 280 ± 60, and 150 year BP (Table 6.2). These recent events were most likely generated by sustained low fountaining of pyroclasts, without the development of high eruption columns.

About ~850 year BP a low-altitude eruptive column dispersed fall deposits on the summit domes and upper flanks of Tacaná. This deposit is thickest southwest of Tacaná's present summit, and southeast of a green lake, which at the present is located at an elevation of 3,784 m (Fig. 6.1c). This lake resembles an explosion crater (60–80 m in diameter) that separates Las Ardillas dome from the southwestern base of Tacaná cone. Although further studies are needed, the distribution and thicknesses of the ~850 year BP fall deposit suggest that the area occupied by the greenish lake (Fig. 6.1c) represents the vent of this explosive event.

Both San Antonio volcano and Las Ardillas dome are overlain by a thin succession of paleosols, fine ash deposits, and pumice fall deposits (including the ~850 year fall) with a maximum age of 2,070 year BP, suggesting that younger events may have originated either from Tacaná or from a nearby vent.

The three youngest explosive eruptions of Tacaná occurred around 370 and 280, and 150 year BP, could have been originated somewhere around the summit area at small vents because the related pyroclastic deposits seem to have a dispersion limited to the flanks and the moat area. The 150 year BP deposit outcrops in the moat area at several locations. It was emplaced by dilute pyroclastic density currents that were able to carbonize trees in the moat area indicating a minimum temperature of ~343 °C. High-angle cross-beds, representing remobilisation by strong winds that are very common at those altitudes, attest to at least one eruption hiatus. Historical activity consisted of mild phreatic explosions in 1949 and 1986 that left no traceable deposits. Further mapping and stratigraphic studies are necessary to improve the stratigraphic record and document minor events of Tacaná as well as to precisely locate the vents for the eruptions recorded in the volcanic deposits of the region.

The Tacaná Volcanic Complex has been very active during geologic and historic times and produced both explosive and effusive eruptions, which ended with the extrusion of two summit lava domes that today occupy the horseshoe-shaped crater open to the northwest. The 1949 and 1986 phreatic eruptions occurred along the margin of the collapse scar produced during the Pleistocene. In addition to the extensive Holocene record, these historic events indicate that Tacaná can pose potential hazards to populations in the surrounding areas, especially those living within a 35 km radius. The city of Tapachula >350,000 inhabitants was in fact built on lahar deposits that originated from the volcano. Based on the stratigraphic record, Tacaná has experienced several shifts in its eruptive style, varying from low-magnitude effusive events to large-magnitude explosive events, such as the plinian eruptions that produced the Sibinal (2.3 km^3 DRE), and Tacaná (4 km^3 DRE) pumice fall deposits. Given the eruptive frequency of the TVC, future events of similar magnitude cannot be discounted.

Moreover, sector collapse represents a common phenomenon at the TVC. The occurrence of such activity, the presence of active fumarolic fields, hydrothermally altered areas at the summit, and the occasional intense seismic activity in the area, all combine to represent a significant debris flow hazard potential, even without the occurrence of concurrent volcanic activity, as documented elsewhere (Scott et al. 2001). More studies (geophysical, petrological, and stratigraphic) are necessary to understand the behavior of this active volcanic complex. The development and maintenance of permanent networks (seismic, geodetic, and geochemical) to monitor the TVC should also be seen as a high priority by national and regional governments to reduce the risk posed by this volcano.

References

Arce JL, Macías JL, Gardner J (2008) The ~14 ka Plinian-type eruption at Tacana Volcanic Complex, Mexico–Guatemala [Abstract]: AGU, Fall Meeting, San Francisco. 15–19 Nov

Arce JL, Macías JL, Gardner JE, Rangel E (2012) Reconstruction of the Sibinal Pumice, an andesitic Plinian eruption at Tacaná Volcanic Complex, México-Guatemala. J Volcanol Geoth Res 217–218:39–55

Arce JL, Walker J, Keppie JD (2014) Petrology of two contrasting Mexican volcanoes, the Chiapanecan (El Chichón) and Central American (Tacaná) volcanic belts: the result of rift- versus subduction-related volcanism. Int Geol Rev 56:501–524

Bergeat A (1894) Zur Kenntnis der jungen Eruptivgesteine der Republik Guatemala. Zeitschrift der Deutschen Zeitschrift der eutschen Geologischen Gesellschaft 46:131–157

Borjas HML (2006) Estratigrafía del sector NE del volcán Tacaná, Chiapas, México-Guatemala. B.S. Thesis, Instituto Politécnico Nacional, ESIA, México

Böse E (1902) Breve noticia sobre el estado actual del Volcán Tacaná, Chiapas. Memorias y Revista de la Sociedad Científica Antonio Alzate 18:266–270

Böse E (1903) Estado actual del volcán Tacaná. Inst. Geol. México, Bol. 20

Bourdier JL, Boudon G, Gourgaud A (1989) Stratigraphy of the 1902 and 1929 nuée-ardente deposits, Mt. Pelée, Martinique. J Volcanol Geotherm Res 38:77–96

Branney MJ, Kokelaar BP (2002) Pyroclastic density currents and the sedimentation of ignimbrites. Geol Soc London Memoirs 27:143 pp

Campa LE (2009) Depósito Pómez Blanca: Ejemplo de magmatismo riolítico en el Complejo Volcánico Tacaná, México-Guatemala. B.S. Thesis, Instituto Politécnico Nacional, ESIA, México, p 120

De la Cruz-Reyna S, Armienta MA, Zamora V, Juárez F (1989) Chemical changes in spring waters at Tacaná Volcano, Chiapas, México. J Volcanol Geoth Res 38:345–353. doi:10.1016/0377-0273(89)90047-4

De la Cruz V, Hernández R (1985) Estudio geológico a semidetalle de la zona geotérmica del Volcán Tacaná, Chiapas, México. Comisión Federal de Electricidad, Internal Report 41:28 pp

De Cserna Z, Aranda-Gómez JJ, Mitre-Salazar LM (1988) Mapa fotogeológico preliminar y secciones estructurales del volcán Tacaná. Universidad Nacional Autónoma de México, Instituto de Geología, Cartas Geológicas y Mineras 7, scale 1:50,000, 1 Sheet

Dollfus A, Monserrat E (1867) Arcive de la commission scientifique du Mexique Ministére de L´instruction Publique: El Renacimiento, 1, p. 451–457

Espíndola JM, Medina FM, De los Ríos M (1989) A C-14 age determination in the Tacaná volcano (Chiapas, México). Geofís Int 28:123–128

Espíndola JM, Macías JL, Sheridan MF (1993) El Volcán Tacaná: Un ejemplo de los problemas en la evaluación del Riesgo Volcánico. Proceedings, Simposio Internacional sobre Riesgos Naturales e Inducidos en los Grandes Centros Urbanos de América Latina: México D.F., Centro Nacional de Prevención de Desastres, pp 62–71

Fisher RV, Glicken HX, Hoblitt RP (1987) May 18, 1980, Mount St. Helens deposits in south coldwater creek, Washinghton. J Geophys Res 92(B10):10267–10283

García-Palomo A, Macías JL, Arce JL, Mora JC, Hughes S, Saucedo R, Espíndola JM, Escobar R, Layer P (2006) Geological evolution of the Tacaná Volcanic Complex, México-Guatemala. In: Rose WI, Bluth GJS, Carr MJ,

Ewert JW, Patino LC, Vallance JW (eds) Natural hazards in central America. Boulder, Colorado (Geol Soc Amer Spec Pap 412:39–57)

Humboldt A (1862) Cosmos: a sketch of the physical description of the universe, vol 4. H.G. Bohn, London, 575 pp

Limón C (2011) Estratigrafía y morfología de los flujos de lava y depósitos asociados a la actividad efusiva del volcán Tacaná, México-Guatemala. Master Thesis, Posgrado en Ciencias de la Tierra, UNAM, México

Macías JL (2007) Geology and eruptive history of some active volcanoes of México. In: Alaniz-Álvarez SA Nieto-Samaniego ÁF (eds) Geology of México: celebrating the centenary of the geological society of México. Geol Soc Am Spec Pap 422:183–232

Macías JL, Espíndola JM, García-Palomo A, Scott KM, Hughes S, Mora JC (2000) Late Holocene Peléan style eruption at Tacaná Volcano, Mexico-Guatemala: Past, present, and future hazards. Bull Geol Soc Am 112:1234–1249

Macías JL, Arce JL, García-Palomo A, Mora JC, Layer PW, Espíndola JM (2010) Late-Pleistocene flank collapse triggered by dome growth at Tacaná Volcano, México-Guatemala, and its relationship to the regional stress regime. Bull Volcanol 72:33–53

Macías JL, Espíndola JM, García-Palomo A, Arce JL, Mora JC, Saucedo R (2011) La actividad volcánica en el sur de México y su impacto en las poblaciones prehispánicas. In: Caballero M, Ortega-Guerrero B (eds) Escenarios de cambio climático: Registros del Cuaternario en América Latina II. Universidad Nacional Autónoma de México, México

Martíni M, Capaccioni B, Giannini L (1987) Ripresa dell'attivita sismica e fumarolica al Vulcano di Tacaná (Chiapas, Messico) dopo un quarantennio di quiescenza. Estrato da Bollettino del Grupo Nazionale per la Vulcanología, pp 467–470

Mercado R, Rose WI (1992) Reconocimiento geológico y evaluación preliminar de peligrosidad del Volcán Tacaná, Guatemala/México. Geofís Int 31:205–237

Mora JC, Macías JL, García-Palomo A, Espíndola JM, Manetti P, Vaselli O (2004) Petrology and geochemistry of the Tacaná Volcanic Complex, Mexico-Guatemala: evidence for the last 40,000 year of activity. Geofís Int 43:331–359

Mora JC, Gardner JE, Macías JL, Meriggi L (2013) Magmatic controls on eruption dynamics of the 1,950 yr B.P. eruption of San Antonio Volcano, Tacaná Volcanic Complex, Mexico–Guatemala. J Volcanol Geotherm Res 262:134–152

Mooser F, Meyer-Abich H, Mc Birney AR (1958) Catalogue of the active volcanoes of the world including Solfatara fields. Part VI, central America In: International volcanological association napoli, pp 26–30

Mujica MR (1987) Estudio petrogenético de las rocas ígneas y metamórficas en el Macizo de Chiapas: Instituto Mexicano del Petróleo, México, C-2009, 47 pp

Müllerried FKG (1951) La reciente actividad del Volcán de Tacaná, Estado de Chiapas, a fines de 1949 y principios de 1950. Informe del Instituto de Geología de la Universidad Nacional Autónoma de México, p 28

Murcia H (2008) Depósitos de lahar del Complejo Volcánico Tacaná y depósitos fluviales en el abanico de Tapachula, Chiapas, México. M.Sc. Thesis, Universidad Nacional Autónoma de México, México

Murcia H, Macías JL (2009a) Registro geológico de inundaciones recurrentes e inundación del 4 de octubre de 2005 en la ciudad de Tapachula, Chiapas, México. Rev Mex Cienc Geol 26:1–17

Murcia H, Macías JL (2009) Volcaniclastic sequences at the foot of Tacaná Volcano, southern México, implications for hazard assessment. Bull Volcanol, online

Newhall CG, Self S (1982) The volcanic explosivity index (VEI): An estimate of explosive magnitude for historical volcanism. J Geophys Res 87(C2):1231–1238

Ordaz MCA (2006) Reconstrucción de la historia eruptiva del Volcán Tacaná, Chiapas, México durante el Holoceno. B.S. Thesis, Instituto Politécnico Nacional, ESIA, México

Rangel E (2009) Estudio vulcanológico del depósito de caída Pómez Sibinal, Complejo Volcánico Tacaná, México-Guatemala. B.S. Thesis, Instituto Politécnico Nacional, ESIA, México

Rouwet D, Taran Y, Inguaggiato S, Varley N (2004) Hydrothermal activity at Tacaná volcano, Mexico–Guatemala. In: Wanty R, Seal R II (eds) WRI-11. Taylor and Francis Group, London, pp 173–176

Rouwet D, Inguaggiato S, Taran Y, Varley N, Santiago JA (2009) Chemical and isotopic compositions of thermal springs, fumaroles and bubbling gases at Tacaná Volcano (Mexico-Guatemala): implications for volcanic surveillance. Bull Volcanol 71:319–335. doi:10.1007/s00445-008-0226-x

Sapper K (1896) La geografía física y la geografía de la Península de Yucatán. Instituto Geológico de México 3:1–58

Sapper K (1897). Ueber die räumliche Anordnung der mittelamerikanischen Vulkane. Zeitschr Deum Geol Ges, Berlin, pp 672–682

Sapper K (1899) Ueper Gebirsbauund Boden des noerdlichen Mittelamerika. Petermanns Geogr Mitt 127:119

Saucedo GR, Esquivias H (1988) Evaluación del riesgo volcánico en el área del Volcán Tacaná, Chiapas. BS thesis, Instituto Politécnico Nacional, Escuela Superior de Ingenieria y Arquitectura, México

Scott KM, Macías JL, Naranjo JA, Rodriguez S, McGeehin JP (2001) Catastrophic debris flows transformed from landslides in volcanic terrains: mobility, hazard assessment, and mitigation strategies, vol 1630. USGS Professional Paper, 67 pp

Waibel L (1933) Die Sierra Madre de Chiapas. Mitteil d Geogr Gesell, Hamburg 43:13–162

Williams SN, Self S (1983) The October 1902 Plinian eruption of Santa Maria volcano, Guatemala. J Volcanol Geoth Res 16:33–56. doi:10.1016/0377-0273(83)90083-5

Fluid Geochemistry of Tacaná Volcano-Hydrothermal System

7

Dmitri Rouwet, Yuri Taran, and Salvatore Inguaggiato

Abstract

Tacaná hosts an active volcano-hydrothermal system, characterized by boiling temperature fumaroles, near the summit (3,600–3,800 m asl), and bubbling degassing thermal springs near its base (1,000–2,000 m asl). The magmatic signature of gases rising to the surface is attested by their high CO_2 contents ($\delta^{13}C_{CO2} = -3.6 \pm 1.3$ ‰), and relatively high $^3He/^4He$ ratios (6.0 ± 0.9 R_A), with a $CO_2/^3He$ ratio typical for the Central American Arc (2.3×10^{10}–6.9×10^{11}). Such magmatic signature is practically identical for the near-summit fumaroles, and the bubbling gases at the base of Tacaná edifice. Besides the HCO_3-enrichment in thermal spring waters, the springs (pH 5.8–6.7) show a SO_4-and minor Cl-enrichment: a CO_2 and H_2S + SO_2-rich magmatic steam condenses into a deeper geothermal aquifer, and the resulting hydrothermal fluid mixes with meteoric waters near the surface. The recharge area for the thermal springs is located at higher elevations (>400 m higher than spring outlet elevation), as inferred from the δD-$\delta^{18}O$ data for rivers, thermal and cold springs. These general insights of the Tacaná volcano-hydrothermal system serve as the baseline for future volcanic surveillance, and geothermal prospection. The main locus of hydrothermal activity is located inside the Tacaná horseshoe-shaped crater in the northwestern sector of the volcanic edifice. In terms of volcanic hazard, this sector can be considered the most probable site for future phreatic activity.

7.1 Introduction

Lining out a monitoring strategy for any volcano requires a general understanding of its volcano-hydrothermal system during periods of quiescence. Worldwide, about ten percent of the ~1,300 active volcanoes are well studied or monitored (Sigurdsson et al. 2000). Tacaná volcano, is one of those poorly studied, and densely populated volcanoes, in need of volcanic surveillance.

On 8 May 1986, a phreatic eruption breached through a major fault scarp at 3,600 m asl on the northwestern flank of Tacaná, renewing fumarolic activity (Chap. 6). Similar phreatic events occurred in

D. Rouwet (✉)
Istituto Nazionale di Geofisica e Vulcanologia, Sezione di Bologna, Via Donato Creti 12, 40128 Bologna, Italy
e-mail: dmitrirouwet@gmail.com

D. Rouwet · Y. Taran · S. Inguaggiato
Instituto de Geofísica, Universidad Nacional Autónoma de México, Ciudad Universitaria, Circuito Exterior s/n, Copilco, Del, Coyoacán, 04510, México DF, Mexico

D. Rouwet · S. Inguaggiato
Istituto Nazionale di Geofisica e Vulcanologia, Sezione di Palermo, Via Ugo La Malfa 153, 90146 Palermo, Italy

T. Scolamacchia and J.L. Macías (eds.), *Active Volcanoes of Chiapas (Mexico): El Chichón and Tacaná*,
Active Volcanoes of the World, DOI 10.1007/978-3-642-25890-9_7, © Springer-Verlag Berlin Heidelberg 2015

historic times (1855, 1878, 1949–1950) near the central Tacaná edifice, or in proximity of the southwestern San Antonio domes (Böse 1902; Mülleried 1951; Macías et al. 2000; García-Palomo et al. 2006). Phreatic eruptions in the past probably occurred after pressure build-up beneath sealed parts of the weathered Agua Caliente debris avalanche and block-and-ash flow deposits on the northwestern section of Tacaná (Macías et al. 2010). This process can be very shallow and local. These minor eruptions did not affect neither the population, nor the agricultural activity (mainly coffee plantations), in the surrounding areas.

It remains unclear why these phreatic eruptions occur, and relatively few knowledge exists on eventual precursory signals at Tacaná (Martini et al. 1987; De la Cruz-Reyna et al. 1989). Such as at many volcanoes, phreatic eruptions may anticipate phreatomagmatic or magmatic eruptive activity. Considering the mechanical instability of the volcanic edifice of Tacaná, as revealed by past volcanic deposits (Macías et al. 2010), minor phreatic eruptions, or even intense rain events, could trigger landslides.

Understanding the fluid geochemistry in a volcano-hydrothermal system can help in locating the more vulnerable sectors of the volcanic edifice. In addition it provides adequate means to estimate the geothermal potential of the volcano. The Comisión Federal de Electricidad published pioneering data obtained from geothermal prospection realized at Tacaná (Medina 1986). This chapter aims to give an overview of the fluid geochemistry of the Tacaná volcano-hydrothermal system, which can be considered a baseline for future volcano monitoring and geothermal prospection. The trace element geochemistry of thermal spring waters is discussed here for the first time.

7.2 Hydrothermal Manifestations

7.2.1 Fumaroles and Bubbling Degassing

The modern Tacaná volcano-hydrothermal system hosts two main fumarolic fields, both emitting vapors at boiling temperatures (89 °C at 3,600 m asl): (1) diffuse degassing occurs at the southwestern San Antonio dome (~3,600 m asl, Figs. 7.1, 7.2a), and (2) strong, noisy, fumaroles exhalations occur on the northwestern flank of Tacaná (Fig. 7.2b), at the base of a 40 m-high lava flow which remained ruptured after the Agua Caliente debris avalanche (Fig. 7.2c), and in two small explosion craters of ~15 m (3,600 m asl, Figs. 7.1 and 7.2d), reactivated during the 8 May 1986 phreatic eruption (Martini et al. 1987; Rouwet 2006; Rouwet et al. 2009). The northwestern fumaroles were sampled in 1986 (Medina 1986; Martini et al. 1987), and afterwards only twice in June 2003 and March 2005 (Rouwet et al. 2009). During the period of observation, the San Antonio fumaroles had a flux too low to prevent air contamination during sampling, so no geochemical data could be collected. In June 2003, an ephemeral crater lake was present in the northernmost explosion crater (Fig. 7.2e), this lake had disappeared in March 2005 (Fig. 7.2f).

Intense bubbling degassing takes place at thermal springs located at the contact between the crystalline basement and the volcanic deposits (1,000–2,000 m asl), despite the large vertical, and horizontal, distance among fumarolic fields that should be connected with the central magma conduit. The spring location suggests that degassing at Tacaná occurs over a broad area, and is structurally controlled. This observation already suggests some relationship between near-summit fumarolic and bubbling degassing at lower altitudes. With this regard, it is noteworthy that the geographical distribution of the thermal springs, and northwestern fumaroles, seems to be related to the same NW-SE trending Agua Caliente debris avalanche. Near the village of Agua Caliente, in the northwestern sector of Tacaná, four out of seven thermal springs manifest bubbling degassing: bubbling is most vigorous at the Agua Caliente spring (Fig. 7.3a, b), and apparently less intense at Toquián (Fig. 7.3c), Orlando, and Zarco. The three remaining northwestern springs (Agua Tibia, Barillas, and Zacarías) do not degas at the surface. A separate group of springs on the southwestern lower ends of the volcano (1,000 m asl), La Calera (Fig. 7.3d), shows the most vigorous bubbling of the entire volcano.

7.2.2 Thermal Springs and Hydrology

The dense river network around Tacaná drains eight thermal, and many cold, springs from the volcanic edifice towards the Pacific Ocean (Fig. 7.1). Several cold springs occur up to elevations of ~3,000 m asl (e.g. Santa María and Papales springs, Fig. 7.1). Two small lakes hosting cold waters of meteoric origin are located southwest of the central dome at an elevation of ~3,900 m asl.

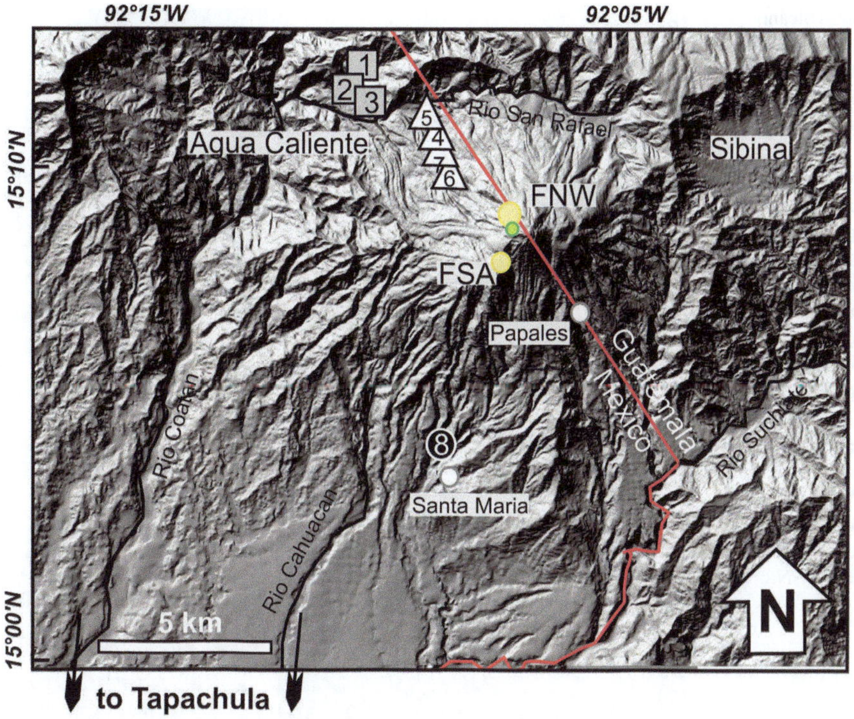

Fig. 7.1 Location map of the hydrothermal manifestations at Tacaná volcano. *Yellow circles* = fumarolic fields, *FNW* northwestern fumaroles, *FSA* San Antonio fumaroles. Thermal springs: *Grey squares* Group (1) springs: *1* Agua Caliente, *2* Toquián, *3* Orlando. *White triangles* Group (2) springs: *4* Agua Tibia, *5* Barillas, *6* Zarco, *7* Zacarías. *Black circle 8* La Calera spring. *Green circle* near-summit meteoric water lake, *white circle* = cold springs. *Orange dotted lines* indicate roads, *black lines* indicate major rivers, black marks indicate villages (modified from Rouwet et al. 2009)

Seven thermal springs discharge at the northwestern sector of Tacaná volcano. The strongly bubbling Agua Caliente spring (T = 50.1–51.1 °C, F = ~1 L/s, where T and F are spring temperature and discharge, respectively) discharges in a red pond next to a 70 m-high waterfall (Fig. 7.3a, b) at the base of the Agua Caliente debris avalanche deposit (Macías et al. 2010). The Toquián spring is the hottest of all (T = 63 °C, F = ~5 L/s) and discharges in a cave-like structure forming a "steamy cave" (Fig. 7.3c). The slightly bubbling Orlando spring (T = 44.2–45.5 °C, F = ~0.5 L/s) discharges from a muddy wall into a small red pond (~1 m diameter). The Zarco spring (T = 45–47.1 °C, F = ~2 L/s) ends up into a ~5 m large-1 m deep pool filled with turquoise water and reddish precipitates at its bottom. Weak bubbling degassing takes place near the edges of the pool. The Agua Tibia spring (T = 31.7–32.1 °C, F < 1 L/s) is captured into an artificial concrete reservoir, accessed through a metal cover. The Barillas spring (T = 38.9–40.3 °C, F > 40 L/s) shows a massive cascade-like discharge from a densely vegetated wall into a 5 m-wide pool of clear water. In 2003, the Zacarías spring (T = 39 °C, F < 0.5 L/s) discharged into a pond created by rocks. This spring disappeared in 2005. On the southwestern end of Tacaná volcano, a significantly cooler (T = 25.7–29.3 °C, F = ~3 L/s), although also strongly bubbling group of springs, La Calera, discharge into small ponds before ending up into cascading rivers (Fig. 7.3d). The dense forest, and big rivers, in the entire area testify the tropical-subtropical climate with an annual rainfall of >4,000 mm. It is clear that the high meteoric precipitation is the main source of water of the volcano-hydrothermal system at Tacaná, which is superimposed on a well-developed hydrologic network.

The thermal springs at Tacaná were sampled in 1985 (Fig. 7.4, Medina 1986), between 1986 and 1993 (only Agua Caliente spring, De la Cruz-Reyna et al. 1989; Armienta and De la Cruz-Reyna 1995), in April 1997 (only La Calera spring, Rouwet et al. 2009), 1998, February–March 2003, and in March 2005 (Rouwet et al. 2004, 2009; Rouwet 2006).

Fig. 7.2 **a** Tacaná volcano seen from Agua Caliente village, to the northwest, the *white square* indicates the near summit fumarolic field (3,600 m asl). **b** View of the San Antonio dome fumarolic field. The *white square* marks major diffuse degassing. **c** View of the northwestern fumarolic field, formed by the 1986 phreatic eruption, along the ~40 m high fault scarp. **d** Detail of the fumarolic field (person on the *left* for scale). **e** Ephemeral crater lake inside the northwestern fumarolic field in June 2003. **f** View of the same craterin (e), without the presence of the crater lake (Photographs a to f by N. Varley)

7.3 Hydrogeochemistry

7.3.1 Major Elements

The salinity and chloride contents of thermal spring waters show a positive correlation with temperature. The eight springs were subdivided in three groups (Fig. 7.4) based on the temperature and TDS (Total Dissolved Solids): (1) high temperature springs (>44 °C) with TDS > 2,750 mg/L, and a pH between 6.33 and 6.68 (Agua Caliente, Toquián and Orlando), (2) medium temperature springs (T = 30–45 °C) with TDS < 1,400 mg/L, and a pH ranging from 6.08 to 6.54 (Agua Tibia, Barillas, Zarco, and Zacarías), and (3) the La Calera the only low temperature (25.7–29.3 °C) spring, with a relatively high TDS (up to 1,870 mg/L) and the lowest pH (5.76–6.44) among all springs analyzed. Waters from Group (1) are relatively enriched in Cl with respect to those of Group (2), and La Calera; these last ones are relatively more HCO_3-rich compared to Group (1) and (2) springs (Table 7.1, Fig. 7.4).

Water-rock interaction processes determine the presence of different cations in thermal waters of geothermal aquifers. The waters of all springs are of Na–K-type, except those of La Calera spring, which are relatively more enriched in Na, Ca and Mg. All thermal spring waters probably result from the mixing between shallow groundwater, and a SO_4–HCO_3-enriched thermal water (Rouwet et al. 2009). Similar SO_4-HCO_3-rich waters are common at lower elevations of large volcano-hydrothermal systems (Sturchio

Fig. 7.3 **a** Waterfall (~70 m high) next to the Agua Caliente spring (*white square*). **b** Sampling of bubbling gas at the Agua Caliente spring. **c** Toquián spring manifested as a "steamy cave". **d** La Calera bubbling spring

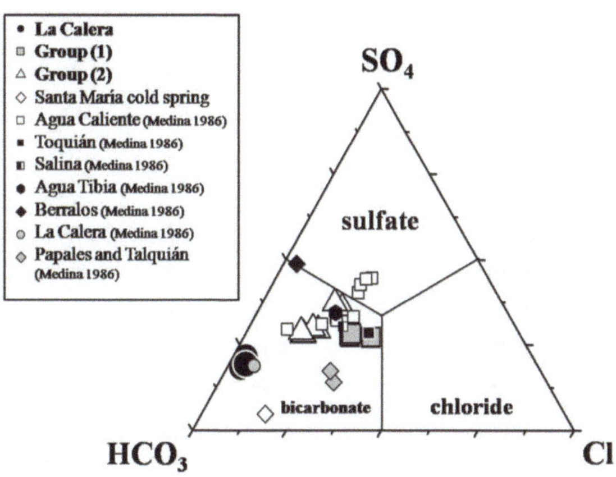

Fig. 7.4 SO_4-HCO_3-Cl triangular diagram for the Tacaná thermal spring waters. Data for La Calera, Group (1) and (2) springs are taken from Rouwet et al. (2009). The remaining, pre-1986 data, are compiled from Medina (1986)

et al. 1988; Giggenbach et al. 1990; Giggenbach and Corrales Soto 1992; Delfín et al. 1996; Fischer et al. 1997). The relatively high Cl contents, in all springs, in particular for Group (1), probably derive from a deep, geothermal aquifer hosting Na–Cl-type waters. The waters of all springs are characterized by a very low (<1 ppm) content in fluorine.

7.3.2 Trace Elements

The first data set on trace element composition for Tacaná thermal spring waters is presented in Table 7.1, together with major components. The most abundant trace elements are Sr, Fe, Mn and Al. The Orlando and Zarco spring waters are Fe-dominant; Agua Caliente, Toquián, and Agua Tibia are Sr-dominant; Barillas is Al-dominant, while La Calera is Mn-dominant (Table 7.1). For Group (1) and (2), the As content ranges from 25 to 133 ppb; La Calera water has considerably lower As contents (2.4 ppb, Table 7.1). Log-Log correlation plots of large data sets of major and trace element concentrations can be used for a qualitative classification of different groups of springs (Taran et al. 2008). The water composition of the same group of springs (e.g. Agua Caliente, Orlando and Toquián), show almost no scatter, while the composition of other groups may scatter by orders of magnitude (Fig. 7.5). The relative abundances of major and trace elements in thermal waters normalized to Na, a rock-forming element with a maximum mobility in the rock-water interaction, and to the same element/Na ratio in the Tacaná rocks (Mora et al. 2004; Macías et al. 2010) is a way to express the

Table 7.1 Major cations and trace metals for Tacaná thermal spring waters

	Agua Caliente	Toquian	Orlando	Agua Tibia	Barillas	Zarco	Zacarías*	La Calera	Rock AC	Rock SA
Na (ppm)	405	536	399	81	126	177	107	106	10,940	10,925
K (ppm)	57	52	59	14	19	29	18	15	7,470	6,815
Ca (ppm)	301	245	287	41	67	82	46	131	32,634	34,619
Mg (ppm)	154	128	144	30	42	69	32	74	10,890	12,073
HCO_3 (ppm)	1,000	903	930	232	311	354	271	683	–	–
SO_4 (ppm)	977	995	948	181	276	476	217	272	–	–
Cl (ppm)	354	443	341	37	57	89	40	23	–	–
F (ppm)	–	–	–	0.24	0.41	0.47	0.25	–	–	–
Li	235	588	262	39	50	90	–	67	na	na
Be	0.002	0.015	0.004	0.005	0.003	0.003	–	0.032	na	na
Al	–	16	833	–	1,168	–	–	479	31,021	31,164
V	0.76	0.76	1.26	7.09	8.67	0.08	–	1.84	113	129
Cr	0.06	0.09	0.06	–	–	0.01	–	0.01	6.8	18
Mn	1,626	476	717	0.86	2.07	671	–	697	714	714
Fe	1,839	1,833	2,928	9.4	21.4	5,296	–	–	10,120	22,442
Co	2.12	0.95	0.98	0.02	0.03	1.49	–	2.02	14	15
Ni	0.81	0.53	0.37	0.04	0.15	0.07	–	0.47	2.7	5.6
Cu	0.09	0.28	0.01	0.15	0.13	–	–	0.13	12	23
Zn	–	5.9	–	–	–	–	–	3.8	77	71
Ga	2.99	6.36	2.77	1.88	2.40	3.42	–	0.73	na	na
As	81	121	133	25	30	118	–	2.4	na	na
Se	0.02	0.02	0.03	0.03	0.03	0.02	–	0.04	na	na
Rb	134	172	130	53	67	104	–	42	52	51
Sr	2,168	2,921	1,927	292	576	436	–	554	514	514
Mo	4.82	1.26	3.07	1.56	4.02	5.75	–	1.05	na	na
Ag	0.06	0.05	0.07	–	–	0.13	–	–	na	na
Cd	0.06	0.05	0.09	0.24	0.12	0.03	–	0.24	na	na
Sn	0.09	0.10	0.05	0.09	0.03	0.18	–	0.04	na	na
Sb	0.10	19.07	0.07	0.09	0.09	0.05	–	0.05	0.65	na
Cs	69	109	100	5.0	7.2	20	–	8.4	2.2	2.2
Ba	53	96	52	21	29	41	–	8.9	724	685
Tl	0.18	0.43	0.23	0.03	0.04	0.09	–	0.01	na	na
U	0.59	0.22	0.22	0.13	0.28	0.04	–	0.38	1.2	1.2

Major elements (February–March 2005) are taken from Rouwet et al. (2009). Trace element content (in ppb) are unpublished data. *Rock AC* average rock composition for the Agua Caliente deposit (in ppm, 13 samples, Macías et al. 2010), *Rock SA* average rock composition for the San Antonio dome-La Calera area (in ppm, 18 samples, Mora et al. 2004).—: no data available, na: not analyzed. *Zacarías spring water content refer to sampling in March 2003 as the spring disappeared in 2005

relative mobility of trace elements during water-rock interaction and weathering processes (Aiuppa et al. 2000; Rouwet 2006; Taran et al. 2008):

$$RM = (RFE/Na)_w/(RFE/Na)_r \qquad (7.1)$$

where RM is the relative mobility of an element within the fluid-rock system, with respect to the reference element Na, and RFE is each of the rock-forming elements analyzed. The relative mobility pattern for different elements is similar for all thermal springs of

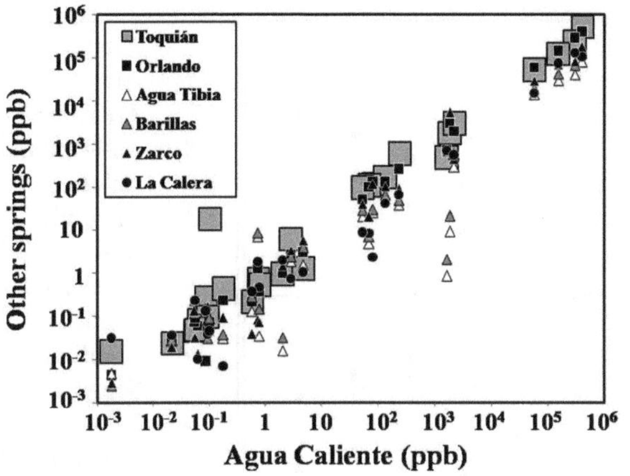

Fig. 7.5 Log-log plot of major and minor elements for Agua Caliente versus the other springs. Toquián and Orlando show a good correlation, while scattering is larger for Group (2) and La Calera spring

Tacaná (Fig. 7.6). The most scattered points are those of the Fe-group elements (Mn, Fe, Co, Ni). Almost all thermal springs at Tacaná precipitate colored Fe-oxides, and probably siderite, therefore the variability in Fe and Fe-group elements, up to almost two orders of magnitude, may be related to different degrees of super-saturation of sampled waters with respect to Fe-minerals.

7.3.3 Solute Geothermometry

The bicarbonate-rich Tacaná thermal spring waters can be considered "immature", as the presence of bicarbonate suggests a continuous, and widespread, input of deep magmatic CO_2. Such condition implies that the classic solute geothermometer approach of Giggenbach (1988) is not straightforward (e.g. Chap. 4), proven by the fact that Na/K temperatures range from 230 to 280 °C, contrasting the significantly lower K^2/Mg temperatures (54–77 °C), or near spring outlet temperatures. Amorphous silica is near equilibrium at spring outlet temperatures (Rouwet 2006). The saturation indices, SI = log (Q/K), (where Q is the activity product, and K the equilibrium constant for each mineral) for a wide temperature range (25–300 °C), using the major and trace element composition of spring waters, allows to determine if water is in equilibrium with certain minerals at specific temperatures (Reed 1982; Pang and Reed 1998; Palandri and Reed 2001). The SI was modeled with PHREEQC Interactive Version 2 (Parkhurst and Appelo 1999), and represents an alternative method to estimate geothermometric temperatures at depth (e.g. Taran et al. 2008). Based on the silicate assemblage for La Calera, the most HCO_3-rich spring, most clay minerals are in

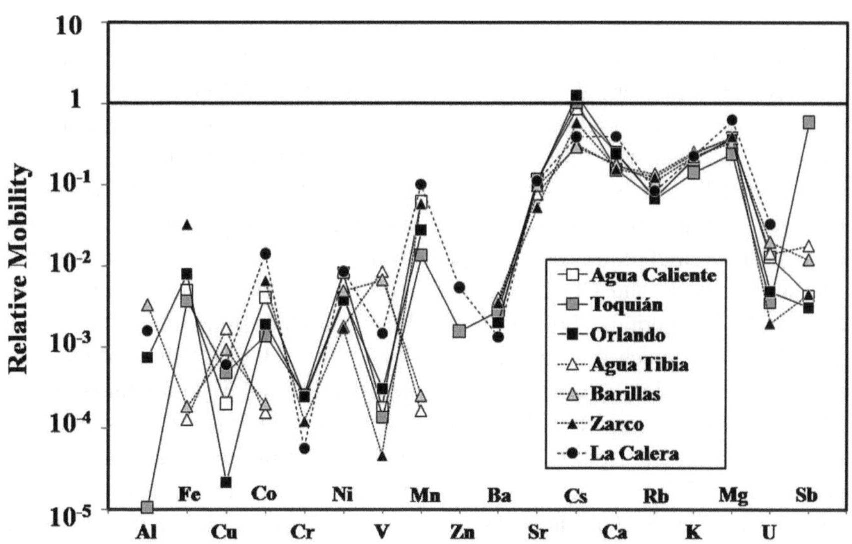

Fig. 7.6 Spider diagram for the Agua Caliente, Barillas, Zarco and La Calera spring waters based on the RM (Eq. 7.1) of trace elements. The values for $(X/Na)_r$ (Eq. 7.1) are average values calculated from 13 samples of the Agua Caliente block-and-ash flow (Macías et al. 2010) (for Group (1) and (2) thermal springs), and from 18 samples of the San Antonio dome (Mora et al. 2004) (for the La Calera spring) (see Table 7.1)

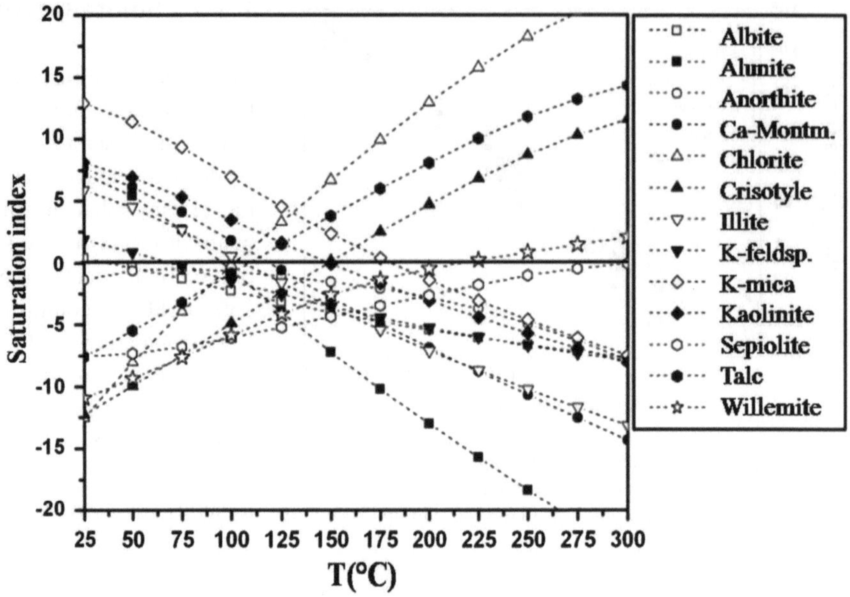

Fig. 7.7 Saturation index (SI) modeling for silicate mineral assemblage of the La Calera spring waters: T (°C) versus SI (saturation index)

equilibrium for the temperature range of 100–180 °C (Fig. 7.7). Lower geothermometric temperatures are deduced (50–125 °C) for Toquián, the hottest spring. Equilibrium temperatures of carbonate and sulfate assemblages (not shown) are in accordance with these silicate geothermometric temperatures (Rouwet 2006).

7.3.4 Isotopic Composition: δD and $\delta^{18}O$

The δD versus $\delta^{18}O$ graph (Fig. 7.8a) of all analyzed fluids clearly suggests that both thermal waters, and fumarole condensates have a meteoric origin. Data points for La Calera waters are slightly shifted towards lower $\delta^{18}O$ values (dotted line in Fig. 7.8a); taking into account a high gas discharge from these springs, such data could indicate a higher oxygen isotope exchange between spring waters and CO_2 (Chiodini et al. 2001). Almost all other springs show a slight positive $\delta^{18}O$-shift (Fig. 7.8a), which is interpreted as the result of water-rock interaction processes in the feeding aquifer. Toquián, and to lesser extent Orlando, spring waters possibly underwent isotopic fractionation upon evaporation (Fig. 7.8a).

The correlation between Cl contents and δD for thermal springs (Fig. 7.8b) is an effective method to deduce water origin. A positive correlation suggests the occurrence of mixing between local meteoric water, and an isotopically heavier end member, which can be either magmatic vapor, or partially evaporated hydrothermal water (Taran et al. 1989; Giggenbach 1992; Ohba et al. 2000); a negative correlation suggests that the main hydrothermal aquifer has a recharge area located at higher elevations of the springs themselves. The Tacaná thermal spring waters (Fig. 7.8b) follow the second scenario. Again, two trends can be observed: Group (1) waters show the lowest δD for higher Cl contents, while Group (2), and La Calera waters, plot along the same trend of higher δD for lower Cl contents (Fig. 7.8b). An isotope survey of meteoric waters at various altitudes of the Tacaná edifice (Rouwet et al. 2009), indicate a gradient of 1.6 ‰/100 m height difference. Such result implies that the recharge area of Group (1) waters occurs at highest elevations of the volcano (~3,000 m asl), while for Group (2) and (3) springs meteoric waters infiltrate from only ~400 to ~950 m higher than the spring outlet (Rouwet et al. 2009). This meteoric recharge mechanism is a common feature at similar large volcanoes or mountain ranges (Giggenbach et al. 1983, 1990). It is highly probable that the extremely negative δD and $\delta^{18}O$ for fumaroles condensates (δD ranges from −110 to −128 ‰, and $\delta^{18}O$ from −19.9 to −15.5 ‰) result from fractionation between water, and its steam, of an already isotopically light boiling meteoric aquifer in the summit part of Tacaná (Rouwet et al. 2009) (Fig. 7.8a). The June 2003 ephemeral crater lake in the northern section of the northwestern fumarolic field (δD = −109 ‰ and $\delta^{18}O$ = −13.8 ‰) was mainly filled with fumarolic condensate (≈steam-heated pool) (Fig. 7.8a).

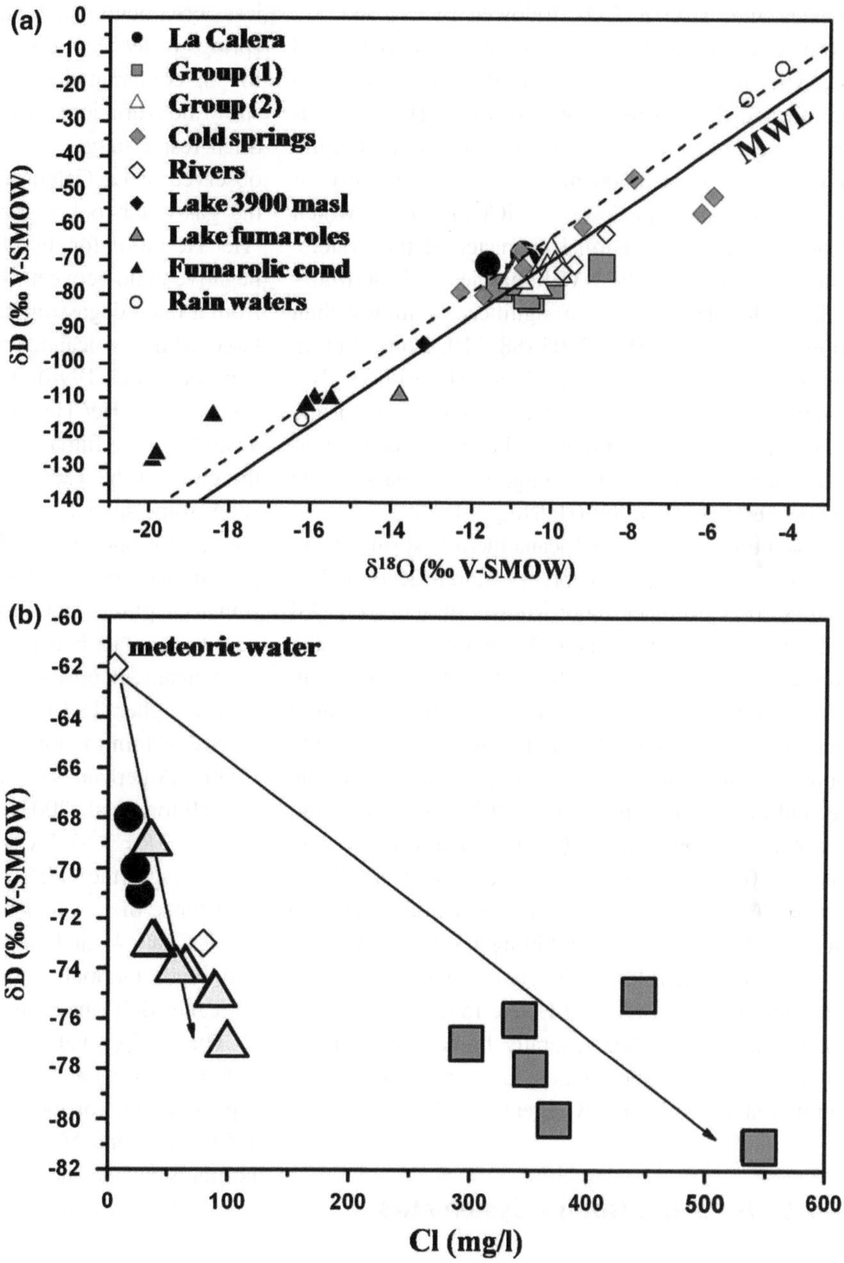

Fig. 7.8 a $\delta^{18}O$ versus δD diagram of the Tacaná thermal and cold spring waters, rivers, meteoric water lakes, the lake at the June 2003 NW fumarolic field, fumarole condensates and rain waters. *MWL* Meteoric Water Line. **b** Cl versus δD diagram for the Tacaná thermal spring waters

7.4 Gas Geochemistry

7.4.1 Chemical Composition of Fumarole Exhalations and Spring Bubbling Gases

Medina (1986), Martini et al. (1987), Rouwet (2006), and Rouwet et al. (2009) analyzed the chemical composition of fumarolic gases at Tacaná, in a time lapse of roughly 20 years. A comparison between the composition of fumarolic gases collected in 1986, with those collected in June 2003, shows that the fumaroles were more vapor-rich in 1986 (938–987 mmol/mol H_2O) with respect to 2003 (880 mmol/mol H_2O), had lower N_2, and slightly higher H_2 contents (Martini et al. 1987; Rouwet et al. 2009). As it is generally the case for fumarolic gases, the second most abundant species after H_2O is CO_2, followed by N_2 and S_{tot} (Rouwet et al. 2009). Even if all analyses considered S_{tot}, with no distinction between SO_2 and H_2S, it can be inferred that sulfur occurs mainly as H_2S, considering the temperatures of the fumaroles (at boiling point of 89 °C at 3,600 m asl), and the absence of other acidic gas species (i.e. HCl and HF), which determine the hydrothermal character of the system. The N_2/Ar ratios of the NW fumaroles in 1986 (920–1,320; Medina 1986) were significantly higher than those measured in 2003–2005 (88–110; Rouwet et al. 2009), indicating an additional non-atmospheric N_2-source throughout the sampling period of fumarolic emissions. The He/Ne ratios of the 2003–2005 fumaroles vary from 0.6 to 1.0, suggesting a considerable atmospheric contribution ($He/Ne_{air} = 0.288$).

The bubbling gases of Tacaná thermal springs were sampled only twice: in February-March 2003 and in March 2005 (Rouwet 2006; Rouwet et al. 2009). All these gases show the same CO_2-dominance as for the fumarolic gases (90.7–99.9 mol%; Rouwet et al. 2009). Noteworthy is the lower CO_2-content in Zarco bubbling gases (74.0–75.4 mol%) with respect to the bubbling gases at the other springs. Nitrogen is the second most abundant gas: 0.07–9.2 mol% for all but the Zarco bubbling gases (24.0–25.5 mol%; Rouwet et al. 2009). Sulfur species are absent as being absorbed in the spring water by oxidation of H_2S to SO_4. Methane contents in bubbling gases are similar to those in the fumaroles, while H_2 is practically absent. The N_2/Ar (up to 306) and He/Ne ratios (up to 20.3) for bubbling gases are generally higher than for the fumarolic gases. Such ratios are typical for gases emitted at volcanic arcs (Giggenbach 1992).

7.4.2 He and C Isotope Systematics

The He and C isotopic composition of gases emitted from fumaroles and as bubbling gases in springs were first compiled and discussed by Rouwet (2006) and Rouwet et al. (2009). The $^3He/^4He$ ratios for fumarolic gases, uncorrected for air contamination, vary from 3.5 to 3.8 R_A (where R_A is the $^3He/^4He$ ratio in air = 1.40×10^{-6}), and from 2.2 to 6.0 R_A for bubbling gases at thermal springs. Despite the low He/Ne ratios in fumarolic gases, the values of $^3He/^4He$ of 5.1–6.6 R_A corrected for air contamination can be accepted, considering that similar high values were measured in less air-contaminated bubbling gases at the thermal springs ($^3He/^4He$ varies from 2.6 to 6.9 R_A, corrected for air contamination). All springs except La Calera show no variations with time in terms of $^3He/^4He$. Such result suggests that the high bubbling intensity observed at La Calera might cause discontinuities in the gas-water-rock regime, also shown by the lower $^3He/^4He$ ratios for dissolved gases (1.3–1.8 R_A). The negative trend between $^3He/^4He$ ratios and the distance from a main degassing spot, observed at some similar large edifice volcanoes as Tacaná (Sano et al. 1990; van Soest et al. 1998), is not clear for Tacaná gases: at Tacaná the $^3He/^4He$ ratio remains high even 5 km away from the fumarolic field. The $^3He/^4He$ ratios for the gases at the Tacaná volcano-hydrothermal system show some spatial and temporal heterogeneity among the various springs and fumaroles, although, with the present data set, no clear distinction exists between the fumarolic gases (3,600 m asl) and those at the springs, near the contact between the crystalline basement and the volcanic deposits (1,000–2,000 m asl).

The higher $^3He/^4He$ ratios range from 5.1 to 6.9 R_A fit well within the range for gases at volcanoes of the Central American Volcanic Arc (CAVA, 5.97 ± 1.44 R_A; Hilton et al. 2002), and of volcanic arcs worldwide (5.37 ± 1.87 R_A; Hilton et al. 2002) (Fig. 7.9). On average, the $^3He/^4He$ ratios at Tacaná are lower than those of El Chichón (up to 8.1 R_A, Mazot et al. 2011; Chap. 4), and those measured at other volcanoes and hydrothermal systems of the Trans-Mexican Volcanic Belt (maximum > 7.5 R_A; Polyak 2000).

The $\delta^{13}C_{CO2}$ values of all free gases (fumarolic and bubbling gases) at Tacaná vary from −3.1 to −4.9 ‰, typical values for gases in volcanic arcs (e.g., Taylor 1986; Sano and Marty 1995; Shaw et al. 2003), and normally heavier than $\delta^{13}C_{CO2}$ for MORB-type gases (~−5 ‰, Pineau and Javoy 1983). The large variation in $^3He/^4He$ and fractionation effects in the CO_2-dominated system will inevitably lead to variations in $CO_2/^3He$ ratios (Fig. 7.9). Besides the "fitting" of $CO_2/^3He$ ratios of a volcano inside its volcanic arc, in the Tacaná case, $CO_2/^3He$ ratios can give insights into near-surface physical degassing processes in a single volcano-hydrothermal system (Rouwet et al. 2009). A general trend becomes obvious: bubbling gases at thermal springs with lowest gas/water ratios (i.e. less intensely bubbling) show lower $CO_2/^3He$ ratios, due to a loss of CO_2 by dissolution, and/or to calcite precipitation (e.g. Zarco spring). Besides these

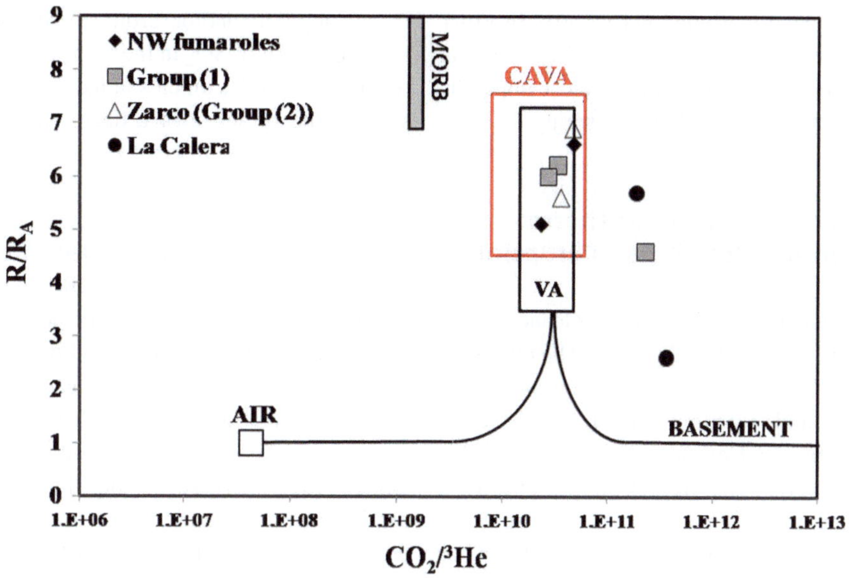

Fig. 7.9 $CO_2/^3He$ vs R/R_A ($^3He/^4He$) diagram for Tacaná fumarolic and bubbling gases. *VA* average values for Arc Volcanoes, *MORB* values for Mid Ocean Ridge Basalts, *CAVA* average values for Central American Volcanic Arc volcanoes. The values for northwestern fumaroles, and Group (1) and (2) bubbling gases generally fall within the CAVA range. La Calera and Toquián bubbling gases are CO_2-enriched with respect to the other gases

heterogeneities in the near-surface degassing regime of the aqueous hydrothermal system, the "corrected" $CO_2/^3He$ ratios for Tacaná generally fit well within established ranges found in volcanoes/hydrothermal systems of the Central American Volcanic Arc and volcanic arcs worldwide ($3.1 \pm 1.6 \times 10^{10}$, Sano and Williams 1996; Goff and McMurtry 2000; Shaw et al. 2003; Snyder et al. 2003; Rouwet et al. 2009) (Fig. 7.9). As demonstrated by the relative HCO_3-enrichment of La Calera spring waters, also the bubbling gases show a CO_2-enrichment with respect to the other springs, leading to higher $CO_2/^3He$ ratios (Fig. 7.9).

7.5 Magmatic Volatile Output and Geothermal Potential

As discussed before, Tacaná fumaroles are apparently SO_2-free, and their total steam discharge is low, not manifesting a clear plume degassing. These findings suggest the absence of significant magmatic volatile output (SO_2, HCl, HF), except for the low reactive species He and CO_2. Nevertheless, several major streams of the northwestern sector of the volcano drain the thermal springs. These rivers have high discharges (up to 1,000 L/s) and contain high amounts of SO_4 (e.g. Río Toquián $SO_4 > 400$ mg/L, Rouwet et al. 2004; Rouwet 2006). Assuming that all SO_4 in thermal spring waters originates from the scrubbing (dissolution) of SO_2–H_2S into the geothermal aquifers, an "SO_2 output" of 20–30 t/d can be calculated for the single Toquian river (Rouwet et al. 2004; Rouwet 2006). Considering that at least two more similar rivers drain the thermal springs, we estimate a total SO_2 output for Tacaná of 60–90 t/d. Such SO_2 fluxes are typical for quiescent degassing volcanoes (Andres et al. 1993; Taran et al. 2002; Rodríguez et al. 2004; Aiuppa et al. 2005; Inguaggiato et al. 2012).

Besides dissolved magmatic volatiles, the thermal springs efficiently transport heat from the volcano-hydrothermal system towards the surface (springs, rivers). Estimating this heat output gives an insight into the geothermal potential of the Tacaná aquifers. A total measured discharge of the eight thermal springs of 53 L/s with an average water temperature of 42 °C ($\approx$176 kJ/kg), results in a geothermal potential of >9 MW (Rouwet 2006; Rouwet et al. 2009). For a minimum steam output from the northwestern fumarolic field of 1 kg/s (at 89 °C, $\approx$2,600 kJ/kg) and additional $\sim$2.6 MW heat output is accounted for by the fumarole emissions, totaling $\sim$12 MW for the entire Tacaná volcano-hydrothermal system. In order to better estimate the thermal output of the Tacaná volcano-hydrothermal system, future field campaigns should focus on discovering "hidden" springs.

7.6 Tacaná Volcano-Hydrothermal System and Its Phreatic Eruptions

A conceptual model of the Tacaná volcano-hydrothermal system is presented in Fig. 7.10. The relatively high discharge of thermal waters and degassing features (fumaroles and bubbling gases), with a clear magmatic signature ($^3He/^4He$ = 6.0 ± 0.9 R_A, $\delta^{13}C_{CO2} = -3.6 \pm 1.3$ ‰), suggest the presence of a cooling and degassing magma body at unknown depth (Chaps. 2 and 6). Thermal spring waters are of a rather uncommon HCO_3–SO_4-Cl-type. Bicarbonate originates from the absorption of a CO_2-rich vapor into an aquifer. Sulfate possibly originates from the oxidation of H_2S, emitted from a shallow hydrothermal system, in an aquifer and/or from direct absorption and recombination of SO_2 in the hydrothermal system. The relatively high Cl contents in the thermal spring waters suggest the presence of a deep, "more mature" geothermal aquifer, with Na–Cl waters. As shown by the δD and $\delta^{18}O$ composition, thermal spring waters have a meteoric origin. Moreover, these waters are isotopically lighter than what expected for meteoric waters at the elevation of the spring outlets. Such occurrence implies that (1) spring waters are meteoric waters infiltrated at higher altitudes, and/or (2) an isotopically lighter vapor condensates in a local meteoric aquifer at the elevation of the spring outlets. Group (1) springs show lower δD values compared to the Group (2) waters (Fig. 7.10), suggesting that Group (1) waters originate from meteoric waters infiltrating from higher elevations. The relative HCO_3-CO_2-enrichment of La Calera waters could be attributed to the degassing of the apparently most recent magmatic intrusion (near the San Antonio dome; 1950 year BP, Macías et al. 2000). The diffuse San Antonio fumaroles are

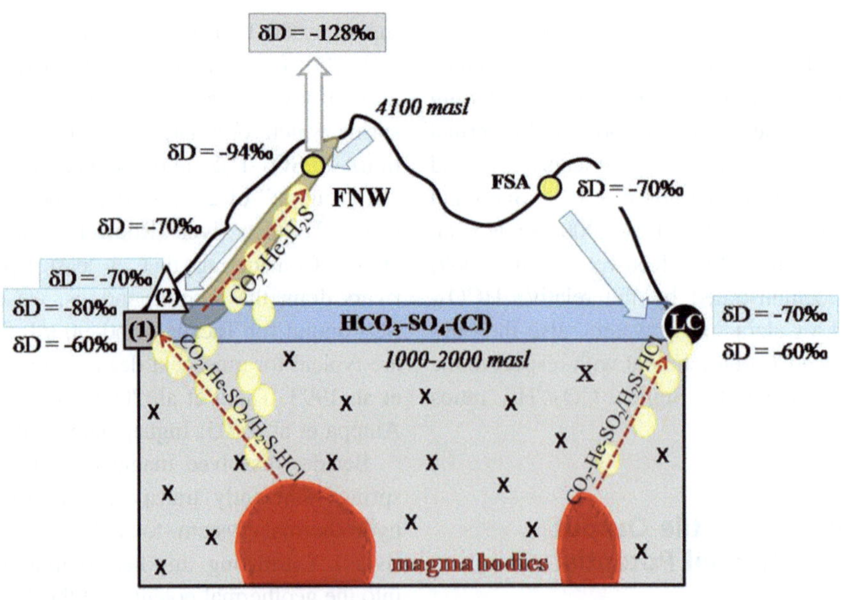

Fig. 7.10 Conceptual model of the Tacaná volcano-hydrothermal system (modified from Rouwet et al. 2009). *FNW* northwestern fumaroles, *FSA* San Antonio fumaroles. The *blue arrows* indicate the infiltration pathways of meteoric water. The *white arrow* indicates the fumarolic steam, and its respective δD (*grey square*). The *blue box* shows the geothermal aquifer at the contact between the crystalline basement (X's) and the volcanic edifice. The *black numbers* indicate the δD of the thermal springs (Group (1), (2) and *LC* La Calera), the numbers in the *light blue boxes* indicate the δD of the meteoric water at various altitudes. *Yellow circles* and *red dotted arrows* delineate degassing pathways. The *brown* area is the Agua Caliente block-and-ash flow deposit (Macías et al. 2010). Magma bodies, at uncertain depth, are colored in *red*. The acidic magmatic gas species (HCl and SO_2/H_2S) are probably partly "scrubbed" into a deeper aquifer, which not necessarily reaches the surface as a spring (and thus not shown). Deep H_2S can be also lost by sulfidation at depth, before re-equilibration inside a shallower hydrothermal system

probably the remnant surface manifestation of this relatively recent magmatic activity. Moreover, the distance between La Calera springs and the other seven springs, supports the hypothesis of a separate aquifer feeding the La Calera springs. The extremely low δD and $\delta^{18}O$ for vapors of the northwestern fumaroles result from steam separation from a meteoric aquifer in the upper 1,000 m of Tacaná volcano. It can be concluded that the Tacaná volcano-hydrothermal system is water-dominated. All fluids (i.e. spring waters and fumarolic vapors) have a meteoric origin. Such meteoric overprint leads to a strong dilution, and cooling, of a boiling hydrothermal system, eventually masking (buffering) possible changes in the volcanic gas input. Therefore, solute geochemistry of the thermal spring waters does reflect near-surface processes, rather than possible changes in the degassing regime of the volcano. The volatile output mainly occurs in the water phase, rather than in a gas/steam phase.

The most peculiar phenomenon of the water-dominated Tacaná volcano-hydrothermal system is the identical CO_2- and He-rich degassing observed in the near-summit as fumaroles and as bubbling gases at thermal springs, despite a difference in elevation of 2,000 m. The isotopic composition of near-summit fumarolic and bubbling degassing at thermal springs at low elevations is clearly magmatic ($^3He/^4He = 6.0 \pm 0.9$ R_A, $\delta^{13}C_{CO2} = -3.6 \pm 1.3$ ‰). There is a clear structural control on the distribution of thermal spring outlets and degassing features, as attested by the location of the bubbling thermal springs (except La Calera) in the northwestern sector of the volcano at the base of the Agua Caliente debris avalanche deposits (DAD) and escarpment. This particular distribution also allows an easier monitoring of the changes in volcanic degassing from more accessible thermal springs, rather than from the near-summit, hardly accessible fumarolic field.

A flank collapse occurred in the northwestern sector of the Tacaná edifice during Late Holocene (Macias et al. 2010, Chap. 6). The Agua Caliente debris avalanche, and the following block-and-ash flows are highly permeable deposits to infiltrate meteoric waters and rising gases, and represent the perfect locus to develop a shallow hydrothermal system, as demonstrated by the presence of thermal springs and boiling temperature vapors. Gas-water-rock interaction in this avalanche deposit (0.8 ± 0.5 km^3, Macías et al. 2010) possibly causes rock alteration, and secondary mineral precipitation, leading to the sealing of the shallow hydrothermal system. The most recent phreatic eruptions in 1949 and 1986, occurred in correspondence of the same DAD to the northwest, could be explained by a pressure build-up, following rock alteration. Future phreatic activity could follow similar scenarios. Despite the lack of evidence of a magmatic system at shallow depths, as suggested by the absence of highly acidic SO_4–Cl type springs, and high temperature fumaroles, such minor phreatic eruptions could also trigger avalanches, inducing depressurization, and the upward migration of a deeper magma. Unfortunately, with the present insights on fluid geochemistry and circulation, such phreatic eruptions are difficult to predict, as they largely depend on local variations in vapor pressure along the NW-SE trending fault and related Agua Caliente deposit.

7.7 Tacaná Versus Other Volcano-Hydrothermal Systems

Oftentimes, acid sulfate or sulfate-chlorine waters discharge near the central conduit, while bicarbonate-rich waters discharge towards the base of active volcanoes. Neutral-chlorine waters originating from deep geothermal aquifers might discharge at the surface at large distances from the eruption center. Other volcanoes lack completely such thermal springs. It remains a question why some volcanoes develop volcano-hydrothermal systems and others don't, despite having similar volcano-tectonic settings, and volcanic activity. Three major controlling factors seem to exist: (1) an abundant meteoric recharge of volcanic aquifers, (2) a stratigraphic and/or structural control on fluid migration inside these thermal aquifers, and (3) the age, and thus the evolution stage, of a volcano-hydrothermal system, which is often difficult to quantify.

The Tacaná volcano-hydrothermal system covers an area of ~ 25 km^2 between the southernmost La Calera spring, and the near-summit fumarolic field spring groups (Group (1) and (2), Fig. 7.1). El Chichón volcano-hydrothermal system is only slightly smaller (~ 10 km^2; Chap. 4), despite the significant smaller size of the volcanic edifice. Volcanoes with a similar size as Tacaná hosting volcano-hydrothermal systems of similar dimension are Cumbal, Colombia (~ 20 km^2; Lewicki et al. 2000), and Irazú-Turrialba volcanic complex, Costa Rica (~ 50 km^2; Rouwet et al. 2010). Some smaller size volcanoes have volcano-hydrothermal systems with similar dimension of Tacaná (e.g. Kusatsu-Shirane, Japan, ~15 km^2; Kakihana et al. 1987; Zāo, Japan, ~50 km^2; Ishikawa et al. 2007; Pinatubo, Philippines, ~70 km^2; Stimac et al. 2004). More extended volcano-hydrothermal system, seem more common in large volcanic edifices, such as Mt. Rainier (USA, ~2×10^2 km^2; Frank 1995), Rincón de la Vieja (Costa Rica, ~10^2 km^2; Tassi et al. 2005), Miravalles volcanic complex (Costa Rica, ~10^3 km^2; Giggenbach and Corrales Soto 1992), Poás (Costa Rica, ~10^2 km^2; Rowe et al. 1995), Nevado del Ruíz (Colombia, ~10^3 km^2; Giggenbach et al. 1990; Sturchio and Williams 1990), and Galeras (Colombia, ~10^2 km^2; Fischer et al. 1997). Nevertheless, other active large volcanic edifices do not manifest hydrothermal activity on their flanks, despite the abundant meteoric recharge of possible thermal aquifers; examples of such type in Mexico are Popocatépetl, and Volcán de Fuego de Colima. The absence of a volcano-hydrothermal system in these cases is probably due to the presence of a shallow magmatic plumbing system, which determines high temperature degassing along a central conduit. Considering the physical characters of these volcanoes, volcano-hydrothermal systems seem to develop mostly in volcanic dome complexes, rather than in cone-shaped stratovolcanoes. Apparently, magma rise and fluid migration in volcanic complexes are controlled by local and/or regional faults and structures, as for Tacaná volcano, lying within the Motagua-Polochic fault system near the Caribbean-North America-Cocos Plate boundaries.

References

Aiuppa A, Allard P, D'Alessandro W, Michel A, Parello F, Treuil M, Valenza M (2000) Mobility and fluxes of major, minor and trace metals during basalt weathering and groundwater transport at Mt. Etna volcano (Sicily). Geochim Cosmochim Acta 64(11):1827–2184

Aiuppa A, Inguaggiato S, McGonigle AJS, O'Dwyer MO, Oppenheimer C, Padgett MJ, Rouwet D, Valenza M (2005) H_2S fluxes from Mt. Etna, Stromboli, and Vulcano (Italy) and implications for sulfur budget at volcanoes. Geochim Cosmochim Acta 69:1861–1871

Andres RJ, Rose WI, Stoiber RE, Williams SN, Matías A, Morales R (1993) A summary of sulfur dioxide emission rate measurements from Guatemalan volcanoes. Bull Volcanol 55:379–388

Armienta MA, De la Cruz-Reyna S (1995) Some hydrogeochemical fluctuations observed in Mexico related to volcanic activity. Appl Geochem 10:215–227

Böse E (1902) Breve noticia sobre el estado actual del Volcán Tacaná, Chiapas. Memorias de la Sociedad Científica Antonio Alzate 18:266–270

Chiodini G, Marini L, Russo M (2001) Geochemical evidence for the existence of high-temperature hydrothermal brines at Vesuvio volcano, Italy. Geochim Cosmochim Acta 65 (13):2129–2147

De la Cruz-Reyna S, Armienta MA, Zamora V, Juárez F (1989) Chemical changes in spring waters at Tacaná volcano, Chiapas, Mexico: a possible precursor of the May 1986 seismic crisis and phreatic explosion. J Volcanol Geoth Res 38:45–353

Delfín FG Jr, Villarosa HG, Layugan DB, Clemente V, Candelaria MR, Ruaya JR (1996) Geothermal exploration of the pre-1991 Pinatubo hydrothermal system. In: Newhal CG, Punongbayan S (eds) Fire and mud. University of Washington Press, Seattle, pp 197–212

Fischer TP, Sturchio NC, Stix J, Arehart GB, Counce D, Williams SN (1997) The chemical and isotopic composition of fumarolic gases and spring discharges from Galeras Volcano, Colombia. J Volcanol Geoth Res 77:229–253

Frank D (1995) Surficial extent and conceptual model of hydrothermal system at Mount Rainier, Washington. J Volcanol Geoth Res 65:51–80

García-Palomo A, Macías JL, Arce JL, Mora JC, Hughes S, Saucedo R, Espíndola JM, Escobar R, Layer P (2006) Geological evolution of the Tacaná Volcanic Complex, Mexico-Guatemala. GSA Special Paper 412 Natural Hazards in Central America, pp 39–57

Giggenbach WF (1988) Geothermal solute equilibria. Derivation of Na-K-Mg-Ca geoindicators. Geochim Cosmochim Acta 52:2749–2765

Giggenbach WF (1992) The composition of gases in geothermal and volcanic systems as a function of tectonic setting. Proc Int Symp Water-Rock Interaction. Balkema 7:873–878

Giggenbach WF, Corrales Soto R (1992) Isotopic and chemical composition of water and steam discharges from volcanic-magmatic–hydrothermal systems of the guanacaste geothermal province, Costa Rica. Appl Geochem 7:309–332

Giggenbach WF, Gonfiantini R, Jangi BL, Truesdell AH (1983) Isotopic and chemical composition of Parbati Valley geothermal discharges, NW Himalaya, India. Geothermics 12:199–222

Giggenbach WF, García N, Londoño A, Rodríguez L, Rojas N, Calvache ML (1990) The chemistry of fumarolic vapor and thermal-spring-discharges from the Nevado del Ruíz volcanic–magmatic–hydrothermal system, Colombia. J Volcanol Geoth Res 42:13–39

Goff F, McMurtry GM (2000) Tritium and stable isotopes of magmatic waters. J Volcanol Geoth Res 97:347–396

Hilton DR, Fischer TP, Marty B (2002) Noble gases and volatile recycling at subduction zones. Rev Mineral Geochem 47:319–370

Inguaggiato S, Mazot A, Diliberto IS, Inguaggiato C, Madonia P, Rouwet D, Vita F (2012) Total CO_2 output from Vulcano

Island (Aeolian Islands, Italy). Geochem Geophys Geosyst 13:Q02012. doi:10.1029/2011GC003920

Ishikawa H, Ohba T, Fujimaki H (2007) Sr isotope diversity of hot spring and volcanic lake waters from Zão volcano, Japan. J Volcanol Geoth Res 166:7–16

Kakihana H, Ossaka T, Oi T, Musashi M, Okamoto M, Nomura M (1987) Boron isotopic ratios of some hot spring waters in the Kusatsu-shirane area, Japan. Geochem J 21:133–137

Lewicki JL, Fischer T, Williams SN (2000) Chemical and isotopic compositions of fluids at Cumbal Volcano, Colombia: evidence for magmatic contribution. Bull Volcanol 62:347–361

Macías JL, Espíndola JM, García-Palomo A, Scott KM, Hughes S, Mora JC (2000) Late Holocene Peléan-style eruption at Tacaná volcano, Mexico and Guatemala: past, present and future hazards. GSA Bull 112:1234–1249

Macías JL, Arce JL, García-Palomo A, Mora JC, Layer PW, Espíndola JM (2010) Late-Pleistocene flank collapse triggered by dome growth at Tacaná volcano, México-Guatemala, and its relationship to the regional stress regime. Bull Volcanol 72:33–53

Martini M, Capaccioni B, Giannini L (1987) Ripresa dell'attività sismica e fumarolica al Vulcano di Tacaná (Chiapas, Messico) dopo un quarantennio di quiescenza. Estrato da Bolletino del Gruppo Nazionale per la Vulcanologia, pp 467–470

Mazot A, Rouwet D, Taran Y, Inguaggiato S, Varley N (2011) CO_2 and He degassing at El Chichón volcano, Chiapas, Mexico: gas flux, origin and relationship with local and regional tectonics. Bull Volcanol 73:423–441

Medina A (1986) Geoquímica de aguas y gases del volcán Tacaná. Rev Geotermia Geoenergía 2:95–110

Mora JC, Macías JL, García-Palomo A, Arce JL, Espíndola JM, Manetti P, Vaselli O, Sánchez JM (2004) Petrology and geochemistry of the Tacaná Volcanic Complex, Mexico-Guatemala: evidence for the last 40,000 years of activity. Geofis Int 43:331–359

Müllerried FKG (1951) La reciente actividad del volcán Tacaná, Estado de Chiapas, a fines de 1949 y principios de 1950. Informe del Instituto de Geología de la UNAM, 28

Ohba T, Hirabayashi J, Nogami K (2000) D/H and $^{18}O/^{16}O$ ratios of water in the crater lake at Kusatsu-Shirane volcano, Japan. J Volcanol Geoth Res 97:329–346

Palandri JL, Reed M (2001) Reconstruction in situ composition of sedimentary formation waters. Geochim Cosmochim Acta 65:1741–1767

Pang ZH, Reed M (1998) Theoretical chemical thermometry on geothermal waters: problems and methods. Geochim Cosmochim Acta 62(6):1083–1091

Parkhurst DL, Appelo CAJ (1999) User's guide to PHREEQC (version 2)-a computer program for speciation, batch-reaction, one-dimensional transport, and inverse geochemical calculations. US Geological Survey Water-Resources Investigations Report 99–4259 p 312

Pineau F, Javoy M (1983) Carbon isotopes in mid ocean ridge basalts. Earth Planet Sci Lett 62:239–257

Polyak B (2000) Helium isotope tracer of geothermal activity. In: Proceedings of World Geothermal Congress, Kyushu-Tohoku, Japan, May 28–June 10, pp 1581–1586

Reed M (1982) Calculation of multicomponent chemical equilibria and reaction processes in systems involving minerals, gases, and an aqueous phase. Geochim Cosmochim Acta 46:513–528

Rodríguez LA, Watson IM, Rose WI, Branan YK, Bluth GJS, Chigna G, Matías O, Escobar D, Carn SA, Fischer TP (2004) SO_2 emissions to the atmosphere from active volcanoes in Guatemala and El Salvador, 1999-2002. J Volcanol Geoth Res 138:325–344

Rouwet D (2006) Estudio geoquímico comparativo de los sistemas hidrotermales de los volcanes activos en Chiapas: El Chichón y Tacaná. PhD thesis, Universidad Nacional Autónoma de México, México DF, p 216

Rouwet D, Taran Y, Inguaggiato S, Varley N (2004) Hydrothermal activity at Tacaná volcano, Mexico-Guatemala. In: Wanty R, Seal R II (eds) WRI-11. Taylor and Francis Group, London, pp 173–176

Rouwet D, Inguaggiato S, Taran Y, Varley N, Santiago SJA (2009) Chemical and isotopic compositions of thermal springs, fumaroles and bubbling gases at Tacaná Volcano (Mexico-Guatemala): implications for volcanic surveillance. Bull Volcanol 71:319–335

Rouwet D, Mora-Amador R, Ramírez-Umaña C, González G (2010) Hydrogeochemical model of the Irazú and Turrialba "twin volcanoes" (Costa Rica). In: Abstracts of AGU SF fall meeting, V23A, p 2392

Rowe GL Jr, Brantley SL, Fernández JF, Borgia A (1995) The chemical and hydrologic structure of Poás Volcano, Costa Rica. J Volcanol Geoth Res 64:233–267

Sano Y, Marty B (1995) Origin of carbon in fumarolic gas from island arcs. Chem Geol 19:265–274

Sano Y, Williams SN (1996) Fluxes of mantle and subducted carbon along convergent plate boundaries. Geophys Res Lett 23:2749–2752

Sano Y, Wakita H, Williams SN (1990) Helium-isotope systematics at Nevadodel Ruíz volcano, Colombia: implications for the volcanic hydrothermal system. J Volcanol Geoth Res 42:41–52

Shaw AM, Hilton DR, Fischer TP, Walker JA, Alvarado GE (2003) Contrasting He–C relationships in Nicaragua and Costa Rica: insights into C cycling through subduction zones. Earth Planet Sci Lett 214:499–513

Sigurdsson H, Houghton B, McNutt SR, Rymer H, Stix J (2000) Encyclopedia of volcanoes. Academic Press, San Diego, p 1442

Snyder G, Poreda R, Fehn U, Hunt A (2003) Sources of nitrogen and methane in central American geothermal settings: noble gas and ^{129}I evidence for crustal and magmatic volatile components. Geochem Geophys Geosyst 4:9001. doi:10.1029/2002GC000363

Stimac JA, Goff F, Counce D, Larocque ACL, Hilton DR, Morgenstern U (2004) The crater lake and hydrothermal system of Mount Pinatubo, Philippines: evolution in the decade after the eruption. Bull Volcanol 66:149–167

Sturchio NC, Williams SN (1990) Variations in chemistry of acid-sulfate-chloride springs at Nevado del Ruíz volcano, Colombia: November 1985 through December 1988. J Volcanol Geoth Res 42:203–210

Sturchio NC, Williams SN, Garcia NP, Londoño AC (1988) The hydrothermal system of Nevado del Ruíz volcano, Colombia. Bull Volcanol 50:399–412

Taran YA, Pokrovsky BG, Dubik YM (1989) Isotopic composition and origin of water from andesitic magmas (trans: Dokl). Ac Sci USSR 304:440–443

Taran Y, Gavilanes JC, Cortés A (2002) Chemical and isotopic composition of fumarolic gases and the SO_2 flux from Volcán de Colima, México, between the 1994 and 1998 eruptions. J Volcanol Geoth Res 117:105–119

Taran Y, Rouwet D, Inguaggiato S, Aiuppa A (2008) Major and trace element geochemistry of neutral and acidic thermal springs at El Chichón volcano, Mexico. Implications for monitoring of volcanic activity. J Volcanol Geoth Res 178:224–236

Tassi F, Vaselli O, Capaccioni B, Giolito C, Duarte E, Fernández E, Minissale A, Magro G (2005) The hydrothermal-volcanic system of Rincón de la Vieja volcano (Costa Rica): a combined (inorganic and organic) geochemical approach to understanding the origin of the fluid discharges and its possible application to volcanic surveillance. J Volcanol Geoth Res 148:315–333

Taylor BE (1986) Magmatic volatiles: isotopic variation of C, H, and S. In: Valley JW et al (ed) Stable isotopes in high temperature geological processes. Mineralogical Society of America, Washington, DC, pp 185–225 (Rev Miner 16)

van Soest MC, Hilton DR, Kreulen R (1998) Tracing crustal and slab contributions to arc magmatism in the Lesser Antilles island arc using helium and carbon relationships in geothermal fluids. Geochim Cosmochim Acta 62:3323–3335

Risk Management of El Chichón and Tacaná Volcanoes: Lessons Learned from Past Volcanic Crises

8

Servando De la Cruz-Reyna and Robert I. Tilling

Abstract

Before 1985, Mexico lacked civil-protection agencies with a mission to prevent and respond to natural and human-caused disasters; thus, the government was unprepared for the sudden eruption of El Chichón Volcano in March–April 1982, which produced the deadliest volcanic disaster in the country's recorded history (~2,000 fatalities). With the sobering lessons of El Chichón still fresh, scientists and governmental officials had a higher awareness of possible disastrous outcome when Tacaná Volcano began to exhibit unrest in late 1985. Seismic and geochemical studies were quickly initiated to monitor activity. At the same time, scientists worked actively with officials of the Federal and local agencies to develop the "Plan Operativo" (Operational Plan)—expressly designed to effectively communicate hazards information and reduce confusion and panic among the affected population. Even though the volcano-monitoring data obtained during the Tacaná crisis were limited, when used in conjunction with protocols of the Operational Plan, they proved useful in mitigating risk and easing public anxiety. While comprehensive monitoring is not yet available, both El Chichón and Tacaná volcanoes are currently monitored—seismically and geochemically—within the scientific and economic resources available. Numerous post-eruption studies have generated new insights into the volcanic systems that have been factored into subsequent volcano monitoring and hazards assessments. The State of Chiapas is now much better positioned to deal with any future unrest or eruptive activity at El Chichón or Tacaná, both of which at the moment are quiescent as of 2014. Perhaps more importantly, the protocols first tested in 1986 at Tacaná have served as the basis for the development of risk-management practices for hazards from other active and potentially active volcanoes in Mexico. These practices have been most notably employed since 1994 at Volcán Popocatépetl since a major eruption under unfavorable prevailing winds may constitute a substantial threat to densely populated metropolitan Mexico City.

S. De la Cruz-Reyna (✉)
Instituto de Geofísica, Universidad Nacional Autónoma de México, Ciudad de México, 04510 D.F., México
e-mail: sdelacruzr@gmail.com

R.I. Tilling
Volcano Science Center, U.S. Geological Survey, Menlo Park, California 94025, USA

T. Scolamacchia and J.L. Macías (eds.), *Active Volcanoes of Chiapas (Mexico): El Chichón and Tacaná*, Active Volcanoes of the World, DOI 10.1007/978-3-642-25890-9_8, © Springer-Verlag Berlin Heidelberg 2015

While the 1982 El Chichón disaster was a national tragedy, it greatly accelerated volcanic emergency preparedness and multidisciplinary scientific studies of eruptive processes and products, not only at El Chichón but also at other explosive volcanoes in Mexico and elsewhere in the world.

8.1 Introduction

The State of Chiapas is located in southeastern Mexico in a region populated mainly by the indigenous Zoque Maya-related ethnic group. The region hosts two volcanoes—El Chichón and Tacaná (Fig. 8.1)—that have erupted historically (Siebert et al. 2010). El Chichón is the youngest of the Quaternary volcanoes forming the Chiapanecan Volcanic Arc, and Tacaná is the northernmost active volcano of the Central America Volcanic Arc (Damon and Montesinos 1978; García-Palomo et al. 2006; Macías 2007; and references therein; Fig. 1). Their geologic and tectonic settings are discussed in detail in other chapters of this volume.

In March–April 1982, El Chichón volcano suddenly gained worldwide notoriety with an unexpected weeklong eruption that produced the worst volcanic disaster in the recorded history of Mexico. Two explosive outbursts—each lasting about 2 h—during 3–4 April killed nearly 2,000 people and caused severe economic loss and human misery in southern Chiapas. Detailed accounts of the impacts of this disastrous eruption may be found in De la Cruz-Reyna and Martin Del Pozzo (2009), and in Tilling (2009a). With the memory of the 1982 El Chichón disaster still relatively fresh, Tacaná, located only 280 km SSE of El Chichón, began to show seismic unrest, including felt earthquakes. The activity quickly attracted scientific attention and produced public anxiety in December 1985. Seismicity increased irregularly and ultimately culminated in a phreatic explosion on 8 May 1986 that, fortunately, had relatively minor impacts and no fatalities (De la Cruz Reyna et al. 1989).

For both the El Chichón and Tacaná volcanic crises, whose eruption intensities and adverse impacts differed by orders of magnitude, scientists, government officials, and the local inhabitants were ill prepared to respond effectively to the volcanic unrest. However, in the Tacaná case, the level of awareness of volcano hazards was much higher. Lessons learned from these two volcanic eruptions, especially the El Chichón tragedy, were applied to the development of volcanic-crisis management procedures currently used in Mexico (De la Cruz-Reyna et al. 2000; De la Cruz-Reyna and Tilling 2008). The employment of these procedures was tested during the reactivation of Volcán Popocatépetl in December 1994 (see Delgado-Grandos et al. 2008 and references therein). In this paper, we review some aspects of the El Chichón and Tacaná volcanic crises as they bear on current mitigation of volcanic risk in Mexico.

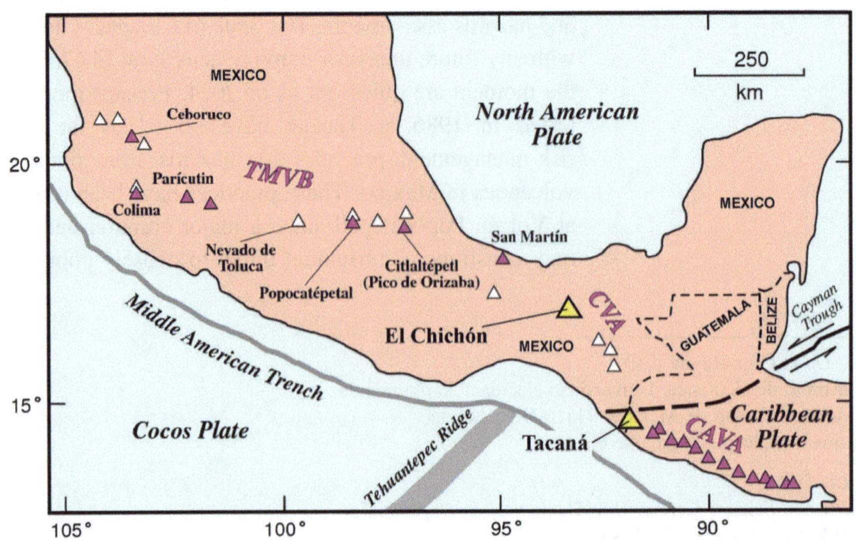

Fig. 8.1 Location of El Chichón and Tacaná Volcanoes (large yellow triangles). TMVB = Trans-Mexican Volcanic Belt, CVA = Chiapanecan Volcanic Arc, and CAVA = Central America Volcanic Arc. Purple triangles show other historically active volcanoes, and white triangles indicate extinct or dormant volcanoes. (Modified from Macías et al. 1997a, Fig. 2)

8.2 1982 El Chichón Eruption

8.2.1 Background and Pre-1982 Situation

El Chichón was only "discovered" as a volcano in 1928 during geologic reconnaissance by Frederich Müllerried, a geologist of the Instituto Geológico of the University of Mexico; his reconnaissance study was in part spurred by reports of felt earthquakes in the Chiapas region at the time. Müllerried described the volcano as a large, geomorphically youthful central dome within a summit crater (Müllerried 1932, 1933). However, the seismicity waned and did not culminate in eruption, leaving El Chichón as an obscure, little-studied volcano for about the next half-century. Beginning in the 1970s, the proximity of the volcano to an oil-producing region and potential for geothermal energy exploitation motivated geological studies of the El Chichón region (González-Salazar 1973; Molina-Berbeyer 1974; Canul and Rocha 1981). In these studies, El Chichón was considered to be an active or potentially active volcano. However, this fact was not public knowledge, because these studies were internal reports of the Comisión Federal de Electricidad (CFE) and were never published. By the early 1980s, there was still no definitive geological evidence of historical eruptive activity, although some anecdotal accounts referred to an eruption about 100 years before.

Thus, before 1982, with the limited knowledge then available, the local inhabitants and officials did not think of El Chichón as potentially dangerous. It is also important to remember that, at the time, most people, including many volcanologists, regarded volcanoes with no record of historical activity as posing little or no threat to life and property. Accordingly, such volcanoes—in México and elsewhere—typically received less attention from scientists and generally were of little concern to emergency-management entities. What is considered as "historical" time varies from country to country; for México, "historical" generally refers to the era after the arrival of the Spanish (in 1520 AD). However, in areas where pre-hispanic calendars can be coupled with European calendars, as is the case of the Aztec region in Central Mexico, "historical" may extend back to about 1300 AD. In any case, it is now known that 10 of the largest eruptions (≥VEI 5) in the world since 1800 AD were the first historical eruptions for the respective volcanoes (Siebert et al. 2010, Table 11). Initially, the 1982 El Chichón eruption was believed to be its first in historical time, but post-1982 radiocarbon dating studies have documented several historical (though not recorded) eruptive episodes, the youngest at 550 B.P. (Rose et al. 1984; Tilling et al. 1984; Espíndola et al. 2000).

In March 1982, another serious extenuating circumstance was that Mexico did not have a designated national- or regional-level governmental entity officially responsible for responding to and managing natural-hazard emergencies. Such an official body—Sistema Nacional de Protección Civil (National System of Civil Protection)—was not created until 1986, after the disastrous magnitude-8.1 Michoacán earthquake in September 1985 that caused fatalities and severe damage in Mexico City and other cities (UNAM Seismology Group 1986; Anderson et al. 1986; Campillo et al. 1989). Thus, the absence of awareness of potential volcanic hazards, combined with the no volcano monitoring or emergency-response planning, meant the volcano's reawakening and destructive eruptions came as a nearly complete surprise to scientists, civil authorities, and the local populace. Therein lies the tragedy of the El Chichón disaster.

8.2.2 Summary of Activity and Impacts

The brief summary of the 1982 El Chichón eruption and its impacts given below is distilled from many publications (and references therein), to which we refer interested readers wanting further detail (e.g., Weintraub 1982; Alcayde 1983; Sigurdsson et al. 1984; Macias 2007; Macias et al. 1997a, b; De la Cruz-Reyna and Martin del Pozzo 2009; Tilling 2009a; see Chap. 3). Scientific and emergency responses undertaken during the eruption are discussed in the following section ("Lessons learned").

Following as many as 26 months of irregular precursory seismicity, as determined in retrospect by Jiménez et al. (1999), including some felt earthquakes, a powerful explosive eruption suddenly began at 23:15 (local time) on 28 March 1982. Although lasting only about 6 h, this explosion produced extensive ashfalls that affected areas more than 200 km from the volcano. Heavy ashfalls in the region made many roads impassable (Fig. 8.2) and forced the closure of major airports at Villahermosa and Tuxtla Gutiérrez. As

Fig. 8.2 After the 28 March explosion, many roads in the region were blanketed and made nearly impassable by heavy ashfalls, causing traffic jams by people evacuating or returning home after being evacuated. (Photograph by Servando de la Cruz-Reyna on 1 April 1982; De la Cruz-Reyna and Martin del Pozzo 2009, Fig. 5)

much as 30 cm of ash were deposited in Pichucalco, about 20 km to the NNE of the volcano (Figure 2 in Macías et al. 2008). Dozens of people were killed from collapse of ash-laden roofs, and thousands began to flee ("self-evacuate") the area. For the next 6 days, eruptive activity was weak and intermittent, consisting of occasional small ash explosions accompanied by nearly continuous seismicity. Then, the two most energetic explosions of the eruption occurred at ~19:35 on 3 April and ~05:20 on 4 April. Each lasted only about 2 h, but both were much more powerful and lethal than the 28 March explosion. In addition to producing voluminous ashfalls, they also generated pyroclastic flows and surges that devastated everything in their paths, including nine villages within a radius of about 8 km from the volcano summit (Fig. 8.3). Some inhabited areas as far as 15 km from the volcano also were heavily damaged by ashfall (Sigurdsson et al. 1984, 1987; Scolamacchia and Macías 2005; Macias 2007; Macías et al. 2008, 2011). The pyroclastic surges of April 3 and 4 resulted in the overwhelming majority of the more than 2,000 deaths, including dozens of soldiers dispatched to assist people of the villages isolated after the onset of the eruption on March 28.

The March–April explosions blasted away nearly all of El Chichón's summit dome, creating a new crater about 1-km-wide and 230-m deep. Thereafter, except for minor ash emission from phreatic explosions in September 1982, eruptive activity essentially ceased. However, with the beginning of the rainy season, several secondary lahars were triggered in valleys draining the volcano. The largest of such rain-induced lahars occurred on 26 May on the Río Magdalena, generated by the catastrophic failure of a natural dam of still-hot pyroclastic debris (Riva Palacios-Chiang 1983; Macias et al. 2004). The resulting near-boiling lahar and flood killed 1 worker, badly burned 3 others, and destroyed a bridge. The drifting stratospheric volcanic clouds from the April eruptions affected global climate, lowering surface temperature by 0.2–0.5 °C in the northern hemisphere (Simarski 1992). In addition to the deaths and property damage directly related to the activity, "...economic loss and social damage was also high as thousands of people lost their belongings and had to be resettled or remained on the surrounding areas for years working in whatever occupation was available, ...[mostly] menial jobs with scarce reward." (Macías et al. 1997a, p. 19).

8.2.3 Lessons Learned

As emphasized earlier, the deadly 1982 El Chichón eruption came as a deadly surprise to local inhabitants and officials alike. Interestingly, after the eruption, involved volcanologists learned of an ancient Chiapas Maya oral tradition that tells of a "boiling flood" that

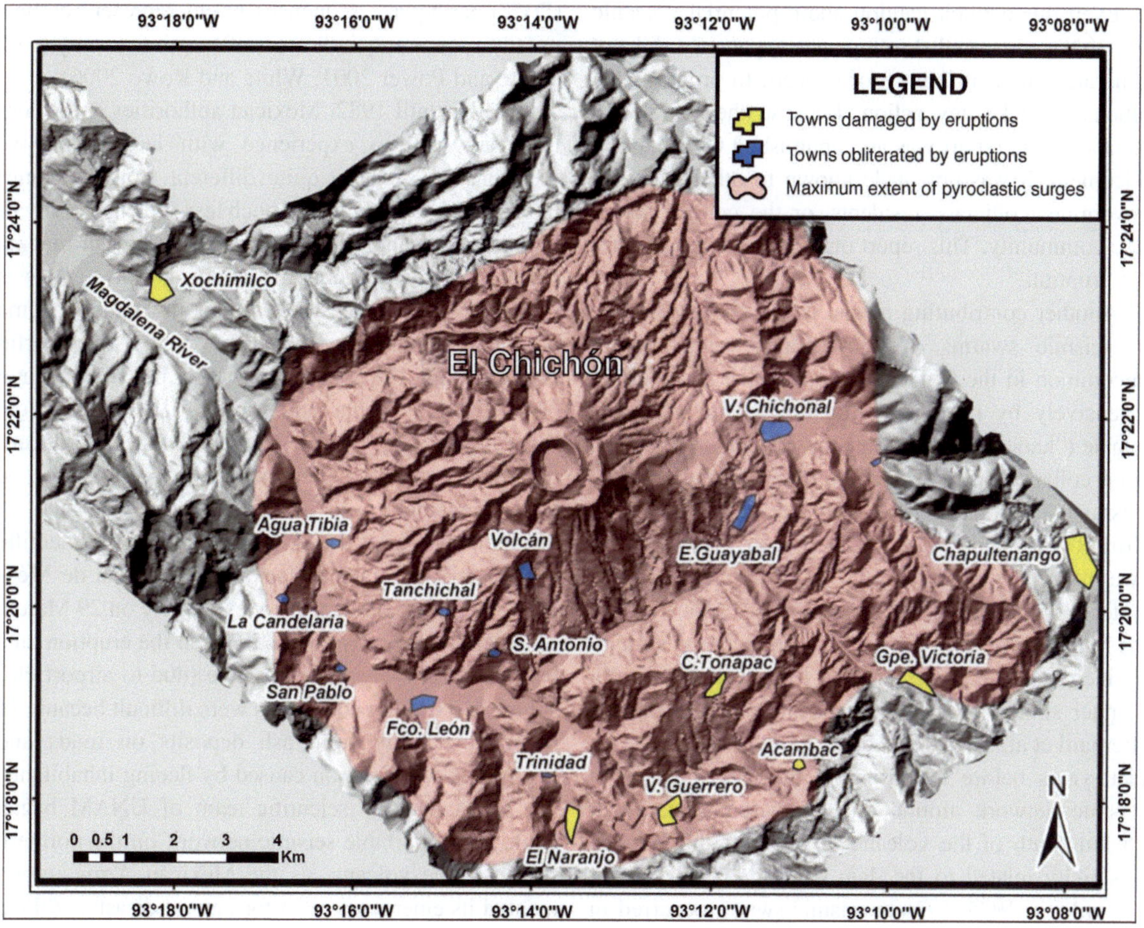

Fig. 8.3 Digital elevation model showing the area around El Chichón Volcano and the towns that were obliterated or damaged by ashfalls and pyroclastic flows and surges of the 1982 eruptions. (Modified from Limón-Hernández and Macías 2009, Fig. 3a)

almost completely destroyed their civilization in the past (e.g., Duffield 2007). With the occurrence of the above-described failed pyroclastic dam and hot flood on 26 May, the basis for this previously enigmatic legend became clear: there were similar catastrophic events during the prehistoric eruptive history of El Chichón. Nonetheless, the "…inhabitants of the region in the months preceding the 1982 eruption were not really aware that El Chichón represented an active volcano, let alone a threat to their communities." (Limón-Hernández and Macías 2009, p. 115). Admittedly, scientific information about the volcano was scarce before the eruption, but why did not the increasing occurrence of felt earthquakes in the area draw scientific and public attention? The answer mostly lies in the fact that, at the time, El Chichón was not known to have erupted in recorded history and its prehistoric eruptive behavior was poorly known. Yet, the heightened pre-eruption seismicity had made a profound impression on two geologists of the Comisión Federal de Electricidad (CFE) doing fieldwork around the volcano during 1980–1981. Feeling many earthquakes and hearing frequent rumbling noises prompted them to express their concern about the volcano in an internal report (Canul and Rocha 1981) submitted to the Geothermal Department of the CFE in September 1981, 6 months before the eruption:

> …during December/80–January/81, strong noises and small earthquakes were heard from the subsurface, being stronger and more frequent in the crater…Possibly, they are related to some subsurface magmatic activity and/or tectonic movements. *It is thus concluded that in this zone exists a high volcanic risk that must be considered if one wishes to develop a geothermal field* (italics added; translated from Canul and Rocha 1981, pp. 26–27).

Even this explicit caution about potential volcanic risk to a geothermal development apparently did not generate sufficient interest or concern to prompt CFE officials to take any action. Because the Canul and Rocha (1981) report was an unpublished CFE internal document, it was not made known to other Mexican government officials, residents, or the general scientific community. This report only became known after the eruption.

Another contributing reason is that the occurrence of seismic swarms of non-volcanic origin is not uncommon in the Chiapas region, which is underlain extensively by intensively fractured, water-rich carbonate ("karstic") formations; these strata are prone to karst collapses and associated ground water-triggered seismicity (Figueroa 1973; Mota et al. 1984; De la Cruz-Reyna and Martin Del Pozzo 2009). Such swarms apparently are caused by the rapid loading of water—by intense rainfall or other mechanisms—on the highly permeable karstic rocks, inducing seismicity by increasing pore pressure and fracture at depth in a manner similar to that observed in dam impounding (Bernard et al. 2006; Kraft et al. 2006; Miller 2008). In fact, years before the eruption, the CFE installed a seismic network around the Chicoasen dam, about 25 km south of the volcano, to monitor the induced seismicity related to the dam (Rodriguez 1977). For example, in 1975–1976, a seismic swarm occurred in the area of Cerro Brujo, near the town of Chiapa de Corzo, located about 75 km SSE of El Chichón. This swarm lasted for about 1 year and produced earthquakes of magnitude up to 3, causing cumulative damage to buildings and houses. Other non-volcanic swarms occurred in October 1983 in the area of Chavarría and Garrido-Ganabal and in February 1984 in the area of Agua Blanca (Palenque), at a similar distance to the east of the volcano. It is difficult for the small Chicoasen seismic network—dedicated to register the locally induced dam-induced seismicity—to discriminate between seismic swarms of volcanic-tectonic origin from those related to karst collapse or pore-elastic media water- induced seismicity in an area much larger than its coverage. Karstic swarms occurred in a range of 20–100 km from the reference, and most sensitive, station of the Chicoasen network (CH3), which was located too far (62 km SSE) from El Chichón to be useful for volcano monitoring (Havskov et al. 1983). Moreover, in the mid-1980s, it was not then known that occurrences of distal volcanic-tectonic (DVT) earthquake sequences might represent earliest possible precursory indicators of eruptive activity (e.g., White and Power 2001; White and Rowe 2006).

Finally, until 1982, Mexican authorities and scientists had had no experience with large explosive eruptions—a scenario quite different from the long-lived, slowly evolving and much less violent activity of Parícutin Volcano 39 years earlier. The mean magma production rate of Parícutin (about 45,000 kg/s) was orders of magnitude smaller than the mean magma production rates of El Chichón's most powerful explosions in 1982, estimated to be between 40 and 60×10^6 kg/s (Luhr and Simkin 1993; Carey and Sigurdsson 1986). Hence, in hindsight, lack of concern regarding pre-1982 precursory seismicity may be understandable.

Upon hearing of the 28 March eruption, a scientific team from Universidad Nacional Autónoma de México (UNAM) departed from Mexico City on 29 March. The team took nearly 2 days to reach the eruption area because air travel was impossible (due to airport closures), and driving conditions were difficult because of reduced visibility from ash deposits on roads and heavy traffic congestion caused by fleeing inhabitants. On 30 March, the scientific team of UNAM began deploying a portable seismic network on the northern sector of the volcano, as the Mexican Army implemented its emergency plan for natural disasters ("Plan DNIII"). Over several days following the initial eruption, the Army evacuated about 20,000 people from affected areas and set up camps to house displaced residents. Areas within a 12-km radius of the volcano, were unreachable because of impassable roads; thus, with little or no communications, the fate of many hundreds of people stranded in towns and villages close to the volcano was unknown. The Army attempted to reach people trapped in these eruption-isolated settlements, which were only occasionally accessible via helicopter under hazardous flying conditions. Several days into the eruption, the death toll was estimated to be less than 100.

By April 1, a network of 5 smoke-drum seismographs was operating. Because the data from these instruments were not acquired and processed in real-time, they were of little practical use in anticipating the course of the eruption (De la Cruz-Reyna and Martin del Pozzo 2009; Tilling 2009a). However, post-eruption processing and analysis of data from this temporary network, combined with those from the CFE

Chicoasen network later provided useful information on seismicity before, during, and after the eruption (e.g., Havskov et al. 1983; Medina et al. 1992; Yokoyama et al. 1992; Jiménez et al. 1999).

Despite addition of a seismic monitoring network, during the lull in activity after 28 March, confusion and uncertainty among the military, civil authorities, and the affected populations did not diminish. The evacuations were not strictly enforced, and contradictory opinions and articles in newspapers prompted many people to return to their homes in the high-risk areas (Macías et al. 1997a). These problems were compounded by the paucity of real-time information about activity at the volcano and a diversity of scientific opinions about the status of the volcano. During the chaotic days of 29 March–3 April, military authorities decided to rescind the evacuation order, thereby allowing many more evacuees to return home. The Army's decision was based on the opinion and advice of a prominent scientist, who "...predicted that the volcano would not erupt again. It has passed the critical point of activity. So, with that information, we told the people they could go home." (as quoted in the news media). Then in the evening of 3 April, the first of the two most powerful explosions occurred—tragically *after* many evacuees had returned home. Almost certainly, had the evacuation remained in place and carried out efficiently, the death toll from the El Chichón eruption would have been substantially lower than 2,000.

Not surprisingly, the reawakening of El Chichón Volcano and its tragic aftermath prompted many scientific investigations, not only of the processes, products, and environmental impacts (geologic and atmospheric) of the 1982 eruption as well as its prehistoric eruptive behavior and magmatic system. Before 1982, only a few reconnaissance studies of the volcano existed in the published scientific literature; since then, hundreds of papers (in Spanish and English) have been published in scientific journals, books, and monographs. It is beyond the scope of this paper to review the abundant literature on El Chichón; instead we outline key insights and lessons learned from these post-eruption studies, especially within the context of risk management.

Many investigators have discussed the causes of the disastrous outcome of the 1982 El Chichón eruption (e.g., Macías et al. 1997a; Espíndola et al. 2000, 2002; Macias 2007; De la Cruz-Reyna and Martin del Pozzo 2009; Limón-Hernández and Macías 2009; Tilling 2009a; and references therein). From these detailed post-eruption studies, we emphasize three important lessons for reducing volcano risk:

(1) Detailed geological studies supported by the best available dating methods are necessary to fully understand the hazards at all young volcanoes. The "discovery" of the morphologically youthful volcano in 1928 (Müllerried 1933) should have spurred geologic mapping, geochronology, and other geoscience studies during the following decades. Unfortunately, such studies were not begun until after the volcanic disaster. For example, post-1982 petrologic, stratigraphic and geochemical studies (e.g., Luhr et al. 1984; Rose et al. 1984; McGee et al. 1987; Tilling et al. 1987; Macías et al. 1997b; Layer et al. 2009) show that El Chichón's magmatic system has produced eruptive products of remarkably uniform trachyandesitic, generally sulfur-rich, composition over the past 300,000 years. Most of the pre-1982 eruptions produced pyroclastic and surge deposits that were significantly more voluminous than the 1982 deposits. Most importantly from a volcano-hazards perspective, however, geologic mapping and dating studies (e.g., Tilling et al. 1984; Macías et al. 1997b; Espíndola et al. 2000; Nooren et al. 2009) indicate that the prehistoric eruptive history of El Chichón is characterized by frequent and violent activity during the Holocene (i.e., past 10,000 years). Including the 1982 outburst, at least 11 eruptions have occurred in the past 4,000 years, with repose intervals ranging between 100 and 600 years. Post-eruption studies thus clearly emphasize that El Chichón is a highly dangerous volcano, capable of repeated large explosive eruptions that can adversely impact human activity (Espíndola et al. 2000; Mendoza-Rosas and De la Cruz-Reyna 2010). Ironically, applying the criteria proposed at several UNESCO-sponsored workshops (e.g., Yokoyama and others 1984, Appendix 1, Annex 3) to identify the world's potentially high-risk volcanoes, El Chichón would not have considered high-risk before 1982, simply because it had no eruptions in recorded history and its eruptive history during the very recent geologic past was unknown.

In hindsight, had El Chichón's violent past been known when it first began to exhibit unrest in 1980, an effective scientific-response and emergency-management strategy might have been developed long before the 1982 eruption. Would having such a strategy in place before the initiation of volcanic unrest at El Chichón helped to contribute to a less disastrous outcome? We will never know the answer. In any case, the compelling lesson here is: to reconstruct its eruptive history and to characterize its eruptive style, geologic mapping and dating studies should be conducted for any volcano having youthful morphology and intense fumarolic activity, which was the state of El Chichón when first visited by Müllerried in 1928. An assessment then can be made from such data of the probabilities of future eruptions and of the potential hazards posed by future activity (De la Cruz-Reyna 1996; Mendoza-Rosas and De la Cruz-Reyna 2008). Such assessments can also help guide the design of monitoring networks and the preparation of contingency plans preferably before volcanic unrest begins.

(2) Real-time geophysical monitoring of potentially active volcanoes is required to reliably detect signs of unrest and support early warning. Without a long-term seismic monitoring network, the frequently occurring felt earthquakes in 1980–1981 could not be evaluated by comparison with long-term baseline-monitoring data. Post-eruption studies of seismic activity at El Chichón (e.g., Havskov et al. 1983; Medina et al. 1992; Jiménez et al. 1999) demonstrate that the 1982 activity indeed was preceded by premonitory seismic activity associated with an inferred shallow magmatic source (Jiménez et al. 1999, Fig. 10). Unfortunately however, the 1980–1982 seismic data collected by the Chicoasen dam-monitoring network were only analyzed after the volcanic disaster. Would having the results of near-real time analysis of the seismic data during the course of the pre-1982 volcanic unrest enabled a less disastrous outcome at El Chichón? Again, we cannot answer this rhetorical question. The lesson here is also clear: once a volcano has been identified as being high-risk, it must be monitored on a regular—preferably continuous—basis at whatever level scientific and economic resources permit. Data from long-term volcano monitoring provide the scientific basis for early detection of deviations from the volcano's "normal" level of activity or inactivity.

(3) Guidance to authorities during volcanic crises must be based upon the best available scientific understanding of eruption scenarios. Perhaps the most important lesson from the El Chichón disaster relates to the imperative need for sound scientific judgment. The "lull" in eruptive activity during 29 March–2 April was incorrectly interpreted to augur the end of the eruption. Tragically, the Army, acting on this scientific advice, abandoned the evacuation and allowed many evacuees to return to their villages and homes in highly hazardous zones. Of all the causative factors, this decision alone greatly contributed to the high number of fatalities. Even before 1982, available worldwide data suggested that most eruptions generally last much longer than a week (Simkin et al. 1981). This observation has since been confirmed by the now more abundant data for durations for Holocene eruptions (Siebert et al. 2010; Tilling 2009a, Fig. 12). In hindsight, the El Chichón tragedy emphasizes the critical importance of quickly obtaining the best possible scientific input in assessing the full range of possible scenarios during the course of a crisis or an eruption. In 1982, the Army and the scientists on site at El Chichón doubtless would have benefited from the advice of more experienced volcanologists regarding evacuation decisions. Fortunately nowadays, Mexico has many knowledgeable and experienced volcanologists in responding to volcanic crises and eruptions, who can cooperate with colleagues worldwide via rapid communications and mutual-assistance programs, such as the U.S. Geological Survey's Volcano Disaster Assistance Program (Ewert et al. 1998).

8.3 1985–1986 Tacaná Seismic Crisis and Phreatic Eruption

Because published reports about the 1985–1986 Tacaná crisis and eruption are relatively meager (e.g., SEAN 1986a, b; Rose and Mercado 1986; De la Cruz-Reyna et al. 1989), the following discussion draws heavily from unpublished sources, mostly personal observations, impressions, and experiences of one of the authors (De la Cruz-Reyna), who was on site and directly involved in the response to the crisis.

8.3.1 Background and Pre-1985 Situation

Tacaná Volcano—the highest edifice (4,060 m asl) of the Tacaná Volcanic Complex (Macias 2007, Fig. 30)—straddles the Mexico-Guatemala border (Figs. 8.1 and 8.4) and marks the northern end of the Central American Volcanic Arc (García-Palomo et al. 2006). It is located in the Soconusco region of Chiapas State and neighboring San Marcos Department (Guatemala), an area with fertile soils that supports an important agricultural economy. Over time, the populations around the volcano have become highly ethnically diverse, consisting of a mixture of indigenous Mayan peoples and immigrants from various European countries, Japan, China, and the United States (Orozco-Zuarth 1994; Zebadua 1999). The population density is relatively high, especially during times of crop harvest when influxes of migrant workers augment the permanent populations. It is during the harvest seasons that the exposure to volcanic hazards posed by Tacaná is at its highest.

In contrast to El Chichón, Tacaná was described in detail by Böse (1902, 1903, 1905) and later recognized as a historically active volcano, with a small eruption in September 1949 (Espíndola et al. 1989; Macias 2007, and references therein; Siebert et al. 2010). The 1949 eruptive activity was studied by Müllerried (1951). As with El Chichón, in the early 1980s, CFE also conducted studies to make the first geological map of the Tacaná volcanic complex and to evaluate its geothermal potential (Medina 1985; De la Cruz and Hernández 1985).

Before 1985, while the scientific community had some knowledge about Tacaná Volcano, inhabitants in the region had long forgotten about the small eruption in 1949 which produced only minor ashfalls on a nearby town and resulted in no fatalities or significant damage. Nonetheless, when the Tacaná crisis began in December 1985 (SEAN 1986a, b), the populations and public officials around Tacaná were more aware of natural hazards than those around El Chichón before 1982 (Fig. 8.5). This relatively heightened awareness was due in large part to having some familiarity with three recent geological disasters in Latin America: the catastrophic 1982 eruption of El Chichón (280 km to the NNW); the September 1985 Michoacán earthquake; and the volcanic disaster in November 1985 at Nevado del Ruiz, Colombia (Williams 1990a, b), which killed more than 23,000 people, received worldwide coverage in the news media.

Fig. 8.4 Panoramic view looking northerly at the Tacaná volcanic complex. The international border, which runs through the summit of Tacaná Volcano (highest point on skyline), is clearly seen as the sharp vertical feature in the picture, with Guatemala to the right and Mexico to the left. (Photograph by Servando De la Cruz-Reyna, 7 February 1986)

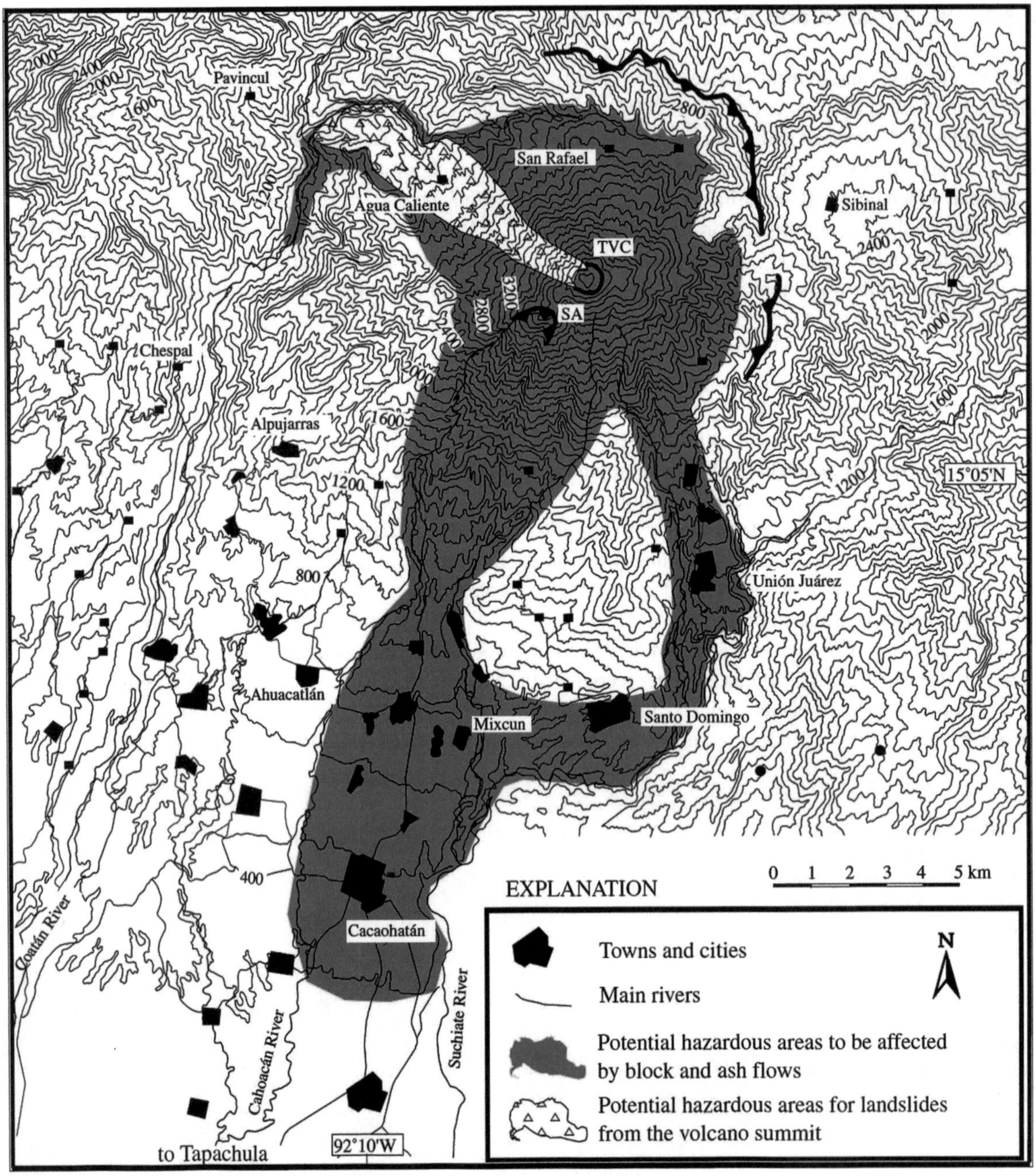

Fig. 8.5 Hazards map of pyroclastic flows and small-size landslides produced by flank instability, based on geologic and dating studies since the 1985-1986 volcanic crisis at Tacaná (From Macias and others 2000, Fig. 14A.) This map, which does not consider hazardous zones for ashfall, shows the many population centers that could be impacted by a potential major explosive eruption, such as that at El Chichón in 1982. The largest city in the region, Tapachula, is located about 30 km to the SSE

8.3.2 Summary of Activity

Starting in mid-December 1985, felt earthquakes accompanied by strong rumbling were reported over a wide area around the Mexico-Guatemala border. The regional seismic network at the time, while adequate for determination of locations of the strongest earthquakes, it was inadequate for determination of lower magnitudes. This noticeable earthquake activity, although not accompanied by any surface signs of volcanic activity, caused grave concern among the population, not only because of the circumstances mentioned above but also some of the people now living there had been affected by the 1982 activity of El Chichón volcano. The largest single earthquake during this swarm, recorded on February 3, 1986 with magnitude 4.7, damaged adobe constructions in the town of Ixchiguán, Department of San Marcos in Guatemala, about 25 km east-northeast of the Tacaná summit.

The seismic activity then declined but persisted at weaker levels through early April. During this period, the epicenters were confined to an area between 15 and 25 km east and east-northeast of the volcano. However, around April 20, 1986, stronger earthquakes began to be felt and heard in the immediate area of the volcano, and the seismic network located them in a volume below the west and southwest flanks of the mountain. This new activity steadily increased until May 7, when an energetic earthquake swarm again caused alarm among the population.

On May 8, when earthquakes were felt at a rate of two or more per minute, a small phreatic explosion opened a small crater on the northwestern flank of the volcano at 3,600-m elevation, almost exactly on the Mexico-Guatemala border (Fig. 8.6). The approximate dimensions of this crater were 20 × 10 m, and a steady, white steam column rose approximately 1,000 m when unperturbed by winds. Two days later, seismicity declined steadily, returning to the pre-May levels 2 days later. However, the fumarole persisted with little change for over 5 years. Afterwards, fumarolic emissions waned slowly until 2002 when it became imperceptible (Martini et al. 1987; De la Cruz-Reyna et al. 1989; Armienta and De la Cruz-Reyna 1995; see Chap. 7).

Intermittent fumarolic activity at or near the new crater and occasional seismic activity were reported through early 1988. Since 2002, Tacaná has exhibited little or no sign of volcanic activity. Geological studies since the 1985–1986 seismic crisis have shown that the volcano in the recent geologic past has produced powerful eruptions, some triggering massive lahars and flooding (Mercado and Rose 1992; Macias et al. 2000; García-Palomo et al. 2006). These new data and hazards assessments show that Tacaná is capable of producing an eruption much more powerful than that in 1986 (Fig. 8.5).

8.3.3 Operational Planning and Scientific Response During the Crisis

During precursory seismicity at Tacana in early 1986, both residents and authorities were extremely apprehensive because the evolving crisis was strikingly similar to the pre-eruptive conditions that prevailed previously at El Chichón (anxiety, confusion, disorganization, miscommunication), and panic slowly began to build among the affected communities (e.g., La Republica 1986). However, the incipient and largely improvised Chiapas State Civil Protection System

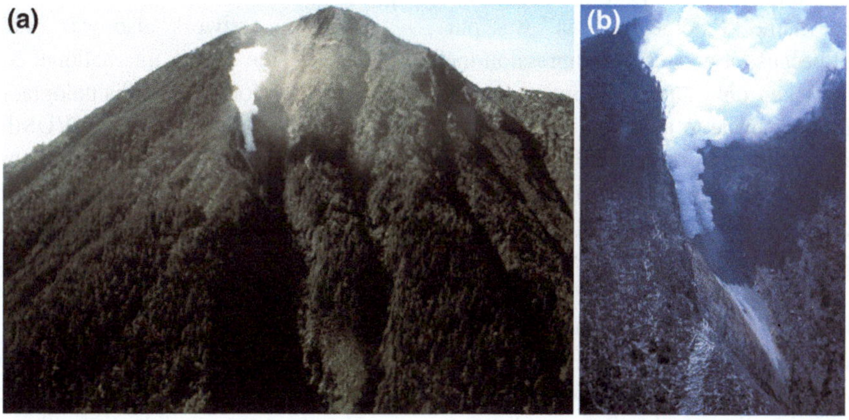

Fig. 8.6 Views in June 1986 of the steam cloud rising from the new active fumarole at the explosion crater that formed on 8 May: (a), distant view (photograph by W.I. Rose in June 1986); and (b), closer view (photograph by S. De la Cruz-Reyna on 11 May, 1986). During May 1986, the steam cloud on occasion reached maximum heights of about 1,000 m above the summit

worked together with scientists of UNAM and CFE to analyze the situation and to begin a coordinated effort to prepare for a possible eruption. They aimed to identify the main causes of the El Chichón disaster, and construct potential scenarios to develop the basic framework of an Operational Plan ("Plan Operativo") that would help guide responses to changes in activity level for Tacaná during the crisis. The cooperating government officials and scientists became convinced that, if authorities, populations and scientists had held a common perception of the risk, and if the authorities had developed in advance the directives of what to do at each stage of the escalating threat, the El Chichón disaster could probably have been prevented, or at least minimized in magnitude. Given the disastrous incorrect interpretation of the "lull" during the El Chichón eruption, the issue of uncertainty in outcomes of volcano unrest received considerable attention in the meetings between on-site scientists and government officials. In hindsight, it should be emphasized that, even with the advances in volcanology since the early 1980s, there is still no reliable capability for predicting the highly variable outcomes of every case of volcano unrest (Tilling 2009b).

The basic elements of the plan developed at for Tacaná eventually evolved into preventive general methodology used in the preparedness plans for the management of volcanic risk throughout Mexico (De la Cruz-Reyna et al. 2000; De la Cruz-Reyna and Tilling 2008), specifically into the current Plan Operativo (Sistema Estatal de Protección Civil Chiapas 2011) for the volcanoes of Chiapas. At Tacaná, one of the first decisions was to emphasize the "prevention" element of the plan. The name "Emergency Plan" was purposely avoided, because it creates the impression that actions are to be taken only after an eruption begins. Another decision was made to avoid, or at least minimize the confusion, lack of criteria, and communication difficulties among the numerous authorities involved in decision-making. Several groups of governmental authorities, plus cultural and ethnic groups, who normally did not communicate, were faced with an unexpected new situation in which their respective roles and responsibilities were not clear.

To effectively manage the developing crisis at Tacaná, it was imperative to develop coordination, procedures, and protocols among the many governmental and non-governmental entities at the Federal, State, Municipal, and Town levels. The organizations involved included: the military authorities of the Ministry of Defense (including Air Force and Navy), Ministry of the Interior (including Civil Protection), Ministry of Transportation and Communications, Civilian aviation (including Airports and Airspace Control), Ministries of Health, Urban Development; Social Development; Agriculture; Education; Water Management; Electric Power; Oil and Gas; Telephone; Foreign Affairs; and other agencies. Also, the Plan considered the roles and responsibilities of scientists from universities, including the Universidad Nacional Autónoma de México (UNAM) and the Universidad de Ciencias y Artes de Chiapas (UNICACH). The structure and characteristics of the Operational Plan that became the model for similar plans developed for use at Popocatépetl, Colima and other volcanoes are described in the Appendix.

As the Tacaná Operational Plan and related protocols were being developed, earthquakes continued to be felt and a small portable seismic network of five smoked-paper seismographs was installed around the volcano. Changing of the records from these instruments and other maintenance were done manually by scientists at the stations. The first stations were located at sites with relatively good access during the last days of January 1986, and other stations were deployed near the volcano at more remote sites in early February. These more remote stations operated irregularly until mid-June 1986, when, because of access and logistical difficulties (e.g., high-energy consumption requiring heavy batteries, need for daily smoking and fixing of the records in the field), the remote stations ceased to operate.

A seismic network also was established on the Guatemala side of the international boundary by the Instituto Nacional de Sismología, Vulcanología, Meterología e Hidrología (INSIVUMEH). Data from one of the Guatemala seismic stations (mostly from SGB, see Fig. 8.7), together with those from the Guatemala national seismic network, were shared with UNAM scientists and were useful in locating the larger earthquakes. During the crisis, there were several meetings between the Mexican and INSIVUMEH scientists, and the Mexican Civil Protection officials met with their Guatemalan counterparts to reach agreements about possible evacuation procedures and other mitigative measures.

To monitor possible ground deformation, a bubble tiltmeter was installed during January–February 6 km

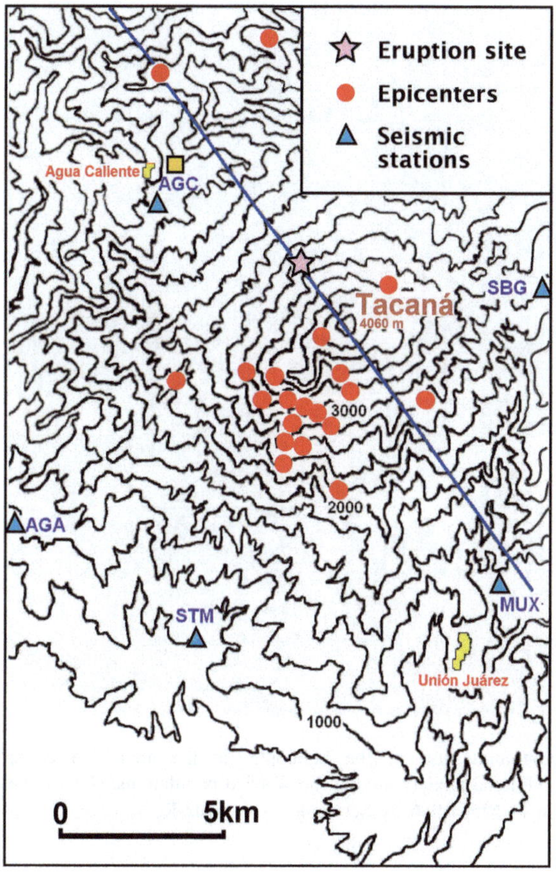

Fig. 8.7 The summit area of Tacaná Volcano and the distribution of the well-located epicenters of the May 1986 seismic swarm (*filled red circles*); *blue triangles* mark the locations of the smoke-drum seismic monitoring stations used at the time. *Orange square* represents the Agua Caliente thermal spring, site of geochemical monitoring (see text). Star indicates the site of the phreatic eruption on 8 May. *Yellow polygons* mark the population centers of Unión Juárez and Agua Caliente, and the *blue line* is the boundary between Mexico and Guatemala. (After De la Cruz-Reyna and others 1989, Fig. 1)

south of the crater, at an elevation of 1,800 m. Extending 1 km from that point towards the volcano, and 4 km away, a leveling line was established and reoccupied periodically. A dry-tilt (spirit-level) array was also set at one of the leveling points. The bubble tiltmeter showed erratic behavior, probably because of unstable local ground settling, and neither the leveling nor the tilt data indicated evidence of any significant deformation on the south flank of the volcano. Water sampling at Agua Caliente, the only known thermal spring related to Tacaná Volcano (Fig. 8.7), began at the same time (Martini et al. 1987; De la Cruz-Reyna et al. 1989). The sampling of thermal waters was logistically arduous, involving several hours of hiking to reach the spring.

Interpretations of the available monitoring data were mainly limited to analysis of seismic data (Fig. 8.7) recorded at the five portable stations and results of chemical analyses of hot spring waters at Agua Caliente. The seismic stations became fully operational by February 1986, and the first water sample from Agua Caliente was collected on 26 January 1986. To obtain analytical results for the water samples, the entire process—sample collection, chemical analysis, and interpretation—involved a maximum of 2 weeks but generally less, around one week. Thus, the results on the four samples collected during January-February (De la Cruz-Reyna 1989, Table 1) were available to indicate the sharp increase in sulfate concentrations that was interpreted to be of volcanic origin. Post-crisis analysis and interpretation of these data (De la Cruz-Reyna et al. 1989) showed that the seismicity was characterized by swarms of very high-frequency earthquakes with very sharp P and S phases. It lacked persistent LP events and harmonic tremor, suggesting that it was generated in the crystalline granitic basement underlying the volcanic structure and was not directly caused by magma movement. Furthermore, the time evolution of the hypocenter locations, which first appeared at about 15–25 km to the ENE of the volcano and later migrated near to it, made it unclear whether the seismicity was cause or the effect of the phreatic activity. On the other hand, 2 months prior to the seismic swarm crisis and phreatic explosion, a marked increase in SO_4^{2-} concentration in the spring water confirmed a relationship between the seismicity and the volcano.

Throughout the monitoring studies, the scientific team improved their understanding of the volcanic crisis and associated risks. At the same time, they worked in parallel with the authorities and the Civil Protection team to develop the Operational Plan (summarized in the Appendix).

Development of the plan was based on the following tenets:

1. Scientists, authorities, and the public must have common understanding of the hazards and risks related to the phenomenon.
2. The best way to communicate hazard and risk information is through the use of simple scenarios.

Fig. 8.8 People near the village of Unión Juárez, curious about the Government of Chiapas helicopter that was dispatched to assist in the response to the volcanic crisis at Tacaná. However, continuous use of the helicopter in the area also caused additional anxiety among the worried populations. (Photograph on 11 May 1986 by Servando De la Cruz-Reyna)

3. Specific actions must be defined for each scenario and such actions must be preventive.
4. Preventive actions require a significant scenario-forecasting capability, and long-term instrumental monitoring is an essential tool to acquire such capability.
5. Each authority and stakeholder should know their role and responsibilities for each possible scenario.
6. The public must be informed as clearly as possible about the scenarios and the reasons why subsequent actions are taken or recommended.
7. The Operational Plan for volcanic crises must complement the existing DN-III military-led response plan, the only one then in existence in the country to manage a disaster situation.

Although the Tacaná Operational Plan was in development during the 1985–1986 unrest, these ideas were tested informally during the seismic crisis of February 1986. The plan received its first rigorous test when the phreatic explosion occurred on 8 May. Following the explosion, Mexican civil authorities were prepared to evacuate all residents within 15 km of the volcano summit, including the migrant workers throughout the high-risk area. The authorities also planned to evacuate part of the Guatemalan population through Mexican territory. However, the low intensity of the explosion and the rapid decay of seismicity in the following days caused the authorities, in consultation with the scientists, to delay evacuation plans and wait for further developments. The authorities requested from the scientists a forecast of possible future activity to help them consider long-term mitigative measures, including the relocation of the most exposed towns. In these forecasts, the scientists never dismissed the possibility of an evolution into a magmatic phase. Seismicity decreased significantly after 8 May. With this development, together with interpretation of analytical data on water samples collected after the explosion, the determination was made that magma had not ascended to shallow levels beneath the volcano. The scientists concluded that further escalation of activity was unlikely, and, based on this belief, the authorities

Fig. 8.9 The "Viejo Volcán" seismic station, located on the NE flank of El Chichón volcano, at about 400 m from the new crater rim (skyline). The lower left inset shows the newly built Centro de Investigación en Gestión de Riesgos y Cambio Climático on the campus of the Universidad de Ciencias y Artes de Chiapas in Tuxtla Gutiérrez, Chiapas, which hosts the "Centro de Monitoreo Sísmico y Volcánico" (CMVS, *upper left* inset), where the seismic monitoring information is received in parallel with CENAPRED and the National Seismological Service at UNAM. This state-of-the-art facility is part of the efforts of the Chiapas State government to be better prepared for the next natural hazards crisis. (Photographs courtesy of Dr. Silvia Ramos H.)

abandoned evacuation and relocation plans. During the height of the crisis, emergency shelters were prepared and food supplies were stocked. However, although some people moved out of the region by their own decision, to our knowledge, no people were evacuated by government order, as the activity never exceeded a moderate phreatic phase (Fig. 8.8).

8.4 Current Management of Volcanic Hazards in Chiapas

The successful application of the—at the time—informal Operational Plan (Plan Operativo Tacaná) during the Tacaná seismic swarm and phreatic explosion crisis in 1986 led to official adoption of the plan (See Appendix) by the Chiapas government in December 1986. This document (the Chiapas Operational Plan) later served as a model for similar operational plans at other volcanoes such as Colima and Popocatépetl, plans that have been later integrated to each state civil protection system (see for example, http://qacontent.edomex.gob.mx/dgproteccion_civil/planoperativopopocatepetl/index.htm).

The Chiapas Operational Plan (applicable to both Tacaná and El Chichón) is periodically reviewed and updated. The 2011 version of the plan (Sistema Estatal de Protección Civil Chiapas 2011) incorporates a structured alert system, called the Volcanic Traffic Light Alert System (VTLAS) (De la Cruz-Reyna and Tilling 2008). In addition to improved planning, scientific infrastructure in support of volcano monitoring and crisis response has also improved since the mid-1980s. In Chiapas, monitoring is carried out by the Centro de Monitoreo Sísmico y Volcánico (CMVS) at the recently created Centro de Investigación en Gestión de Riesgos y Cambio Climático (Fig. 8.9; http://www.cmvs.chiapas.gob.mx/) of the Universidad de Ciencias y Artes de Chiapas. CMVS is also in charge of the State of Chiapas seismic and hydro-meteorological monitoring and offers to the university students a bachelor program in Earth sciences. Presently, the CMVS monitors El

Chichón and Tacaná with three telemetric seismic stations at each volcano. Two of the Tacaná stations are broadband seismometers; data from these instruments are processed by the Mexican Seismological Service at UNAM. Data are also telemetered to the National Center of Disaster Prevention (CENAPRED) in Mexico City where the data are systematically processed and interpreted to complement preliminary analysis by the CMVS.

Two campaign Global Positioning System (GPS) stations (one on each volcano), web-hosted video cameras, and periodic water sampling for geochemical analysis are also part of the routine monitoring for both El Chichón and Tacaná. Since the time of their respective episodes of activity, water from El Chichón crater lake has been sampled and analyzed about three times a year, and water from Tacaná water springs is analyzed once a year on average. (Casadevall et al. 1984; Armienta and De la Cruz-Reyna 1995; Armienta et al. 2000, 2008, 2014; Rouwet et al. 2009a, b). At present, there is no permanent geodetic monitoring at El Chichón and Tacaná; however, such monitoring networks are contemplated by the Centro de Investigación en Gestión de Riesgos should additional resources become available. Any anomalous activity is communicated immediately to the State of Chiapas Civil Protection where discussions with scientists will determine the appropriate course of action. Currently, an electronic Volcanic Traffic Light Alert System (green–yellow–red with increasing severity of hazard) is being implemented in the small towns around the volcanoes.

Admittedly, the current monitoring program clearly would benefit from denser networks and more frequent geochemical sampling, but the State of Chiapas is now much better prepared to cope with a volcanic crisis compared to the situation in the early 1980s. In addition to these new monitoring networks, volcanic hazards assessments and maps are now available for Tacaná (Lahar zonation, Mercado and Rose 1992; Pyroclastic flow zonation, Macías et al. 2000) and El Chichón (Hazards map, Macias et al. 2011). Recent detailed geologic studies of volcaniclastic sequences of the Tacaná Volcanic Complex emphasized the vulnerability of Tapachula City ($\sim$200,000 population) to potential hazards from future lahars (Murcia and Macías 2014).

On 4 April 2012, the latest version of the El Chichón Hazards map was officially presented to the Governor of Chiapas; the Chiapas State government is now making decisions and plans to distribute the map and to implement an associated public education program. Moreover, a study was recently made to evaluate the awareness and perception of volcanic risk among the inhabitants living around El Chichón (Limón-Hernández and Macías 2009); the results suggest that a large percentage of the population—especially the older residents—in the high-risk area is still not well prepared for future activity. Thus, it is necessary to implement long-term governmental programs to increase public awareness of volcano hazards and to develop hazards-mitigation strategies. Nonetheless, a lasting legacy of both the 1982 El Chichón eruption and the 1985–1986 Tacaná crisis is the immense amount of new scientific data generated by post-eruption investigations at these two volcanoes. The increase in the quantity and quality of data should serve well to enhance volcano-monitoring, early detection and interpretation of volcanic unrest, and public warning process. In time, such improved data will in turn enable authorities to more effectively manage—and for the local populace to gain greater awareness of—the risk posed by potential future eruptions affecting the Chiapas region.

8.5 Appendix: Outline of the Plan Operativo (Operational Plan) Developed in 1986 During the Tacaná Volcanic Crisis

At the time of its development, protocols within this Operational Plan were limited to a series of simple instructions for participating authorities and scientific groups. The instructions outlined roles and responsibilities in place for each of 5 possible conditions of volcano status, which were defined as:

(A) "Latent". The volcano has a potential for activity, but no precursory signals are detected.

(B) "May happen". The volcano is showing some sign of unrest without evidence of a timescale for further activity.

(C) "High probability of occurrence". The volcano is displaying increasing signs of unrest suggesting that an eruption may occur in hours to days.

(D) "Ongoing eruptive activity". Any eruptive manifestation is detected.

(E) "Eruptive activity ended". No evidence that the activity may resume in the near term.

The Plan Operativo Tacaná, was a 153-page document outlining operative procedures—consisting of 2–7 short and precise instructions for each of the above conditions to officials of three executive coordination offices:

1. The General Coordination at the Center of Operations, including the top State Government executive officials (the governor, the secretary of government, and all the state secretaries), and the military.
2. The Coordination of Public Information including two sections:
   - Location of persons, and keeping families together in case of evacuation.
   - Public education about the nature of volcanic activity, the associated hazards, and the current status of the volcano.
3. The Operations Coordination consisting of:
   - An operations Center, where all available information is received and stored.
   - A communications and Transportation coordination center where the information on the availability and conditions of roads, airspace and vehicles was concentrated.
   - Health coordination. A Human health information and mitigation of impacts center, including the Red Cross and other emergency services.
   - The Supply and Commerce center, in charge of the management of food and water resources in the affected area.
   - The Urban Development and Ecology coordination, in charge of the management of housing that may be affected by the activity, and possible relocation of population.
   - The Shelter administration, in charge of the temporary shelters in case of evacuation.

This plan included 4 additional documents with further detail on key elements of the emergency response:

a. Evacuation procedures
b. Emergency Communication procedures
c. Population and resources distribution in the affected area
d. Operation of shelters in case of evacuation.

These were the conditions mentioned in the original plan of 1986, which was far to be perfect, but nevertheless marked an abyssal difference with the management of the El Chichón 1982 crisis.

Later, and mostly as a consequence of the Popocatépetl volcano activity, a new operational plan known as the Traffic Light Alert System (TLAS) was developed (De la Cruz-Reyna 1995; De la Cruz-Reyna and Tilling 2008; http://www.cenapred.unam.mx/es/Instrumentacion/InstVolcanica/MVolcan/Semaforo/).

Although the TLAS maintains some of the operational structure of the Tacaná original Operational Plan, it has three main differences:

1. The colors define the state of alert of the exposed population and not in a direct way the current state of the volcano activity.
2. The phases within the colors define the state of alert of the Civil Protection System, very much in the way the Tacaná Operative Plan did.
3. The color and phases are set by the authorities of the Civil Protection System (the General Coordinator, with the level of an undersecretary of state of the Ministry of the Interior) according to the consensual recommendations of a Scientific Committee about the most likely scenarios, and not only based on a list of precursors (which seems to be a problem with other volcano alert codes when unexpected signals or lulls appear). The members of the Scientific Committees (as stated by the National Civil Protection Organization and Operation Manual, Diario Oficial 1995, 2006) should be recognized experts in the field, and it is assumed that such group of experts is well aware about the significance and relevance of unexpected signals or lulls.

References

Alcayde M (ed) (1983) El Volcán Chichonal, Ponencias presentadas en el simposio sobre el Volcán Chichonal, VI Convención Geológica Nacional de la Sociedad Geológica Mexicana. Instituto de Geología, UNAM, México, D.F., 120 pp (Collection of 9 papers)

Anderson JG, Bodin P, Brune JN, Prince J, Singh SK, Quaas R, Onate M (1986) Strong ground motion from the Michoacan, Mexico, earthquake. Science 233(4768):1043–1049. doi:10.1126/science.233.4768.1043

Armienta MA, De la Cruz-Reyna S (1995) Some hydrogeochemical fluctuations observed in México related to volcanic activity. Appl Geochem 10:215–227

Armienta MA, De la Cruz-Reyna S, Macías JL (2000) Chemical characteristics of the crater lakes of Popocatépetl, El Chichón and Nevado de Toluca Volcanoes, México. J Volcanol Geoth Res 97:105–125

Armienta MA, Vilaclara G, De la Cruz-Reyna S, Ramos S, Ceniceros N, Cruz O, Aguayo A, Arcega-Cabrera F (2008) Water chemistry of lakes related to active and inactive Mexican volcanoes. J Volcanol Geoth Res 178(2):249–258

Armienta MA, De la Cruz-Reyna S, Ramos S, Ceniceros N, Cruz O, Aguayo A, Arcega-Cabrera F (2014) Hydrogeochemical surveillance at El Chichón volcano crater lake, Chiapas, Mexico. J Volcanol Geoth Res 285:118-128 doi: 10.1016/j.jvolgeores.2014.08.011

Bernard P, Lyon-Caen H, Briole P, Boudin F, Makropulus K, Papadimitriou P, Lemeille F, Patau G, Billiris H, Paradissis D, Papazissi K, Castarede H, Charade O, Nercessian A, Avallone A, Pacchiani F, Zahradnik J, Sacks S, Linde A (2006) Seismicity, deformation and seismic hazard in the western rift of Corinth: new insights from the Corinth Rift laboratory (CRL). Tectonophysics 426:7–30

Böse E (1902) Breve noticia sobre el estado actual del Volcán Tacaná, Chiapas. Memorias y Revista de la Sociedad Científica "Antonio Alzate" 18:266–270

Böse E (1903) Los temblores de Zanatepec, Oaxaca a fines de septiembre de 1902 y el estado actual del Volcán de Tacaná. Parergones del Instituto Geológico de México 1(1):25 pp

Böse E (1905) Reseña acerca de la geología de Chiapas y Tabasco. Boletin del Instituto Geológico de México 20:116 pp

Campillo M, Gariel JC, Aki K, Sánchez-Sesma FJ (1989) Destructive strong ground motion in Mexico city: source, path, and site effects during great 1985 Michoacan earthquake. Bull Seismol Soc Am 79:1718–1735

Canul RF, Rocha VS (1981) Informe geológico de la zona geotérmica de "El Chichónal", Chiapas: informe 32–81, unpublished report of the Geothermal Department of the Comisión Federal de Electricidad, Morelia, Michoacán, México, completed September 1981, 30 pp, 5 figs, and 9 plates

Carey S, Sigurdsson H (1986) The 1982 eruptions of El Chichon volcano, Mexico (2): Observations and numerical modelling of tephra-fall distribution. Bull Volcanol 48:127–141

Casadevall TJ, De la Cruz-Reyna S, Rose WI, Bagley S, Finnegan DL, Zoller WH (1984) Crater lake and post-eruption hydrothermal activity, El Chichon Volcano, Mexico. J Volcanol Geoth Res 23:169–191

Damon P, Montesinos E (1978) Late Cenozoic volcanism and metallogenesis over an active Benioff Zone in Chiapas, México. Arizona Geol Soc Dig 11:155–168

De la Cruz-Reyna S (1995) Un código de alerta para el manejo de emergencias antes y durante potenciales erupciones del Volcán Popocatépetl. En: Volcán popocatépetl estudios realizados durante la crisis de 1994–1995. Coedición del Sistema Nacional De Protección Civil, el Centro Nacional De Prevención De Desastres y la Unam, pp 327–333 (ISBN: 970-628-127-4)

De la Cruz-Reyna S (1996) Long-term probabilistic analysis of future explosive eruptions. In: Tilling RI, Scarpa R (eds) Monitoring and mitigation of volcanic hazards. A IAVCEI/UNESCO volume. Springer, Berlin, pp 599–629

De la Cruz V, Hernández R (1985) Estudio geológico a semidetalle de la zona geotérmica del Volcán Tacaná, Chiapas. 41/85, Comisión Federal de Electricidad, México D.F.

De la Cruz-Reyna S, Tilling RI (2008) Scientific and public responses to the ongoing volcanic crisis at Popocatépetl Volcano, Mexico: importance of an effective hazards-warning system. J Volcanol Geoth Res 170:111–120

De la Cruz-Reyna S, Armienta MA, Zamora V, Juárez F (1989) Chemical changes in spring waters at Tacaná Volcano, Chiapas, México. J Volcanol Geoth Res 38:345–353. doi:10.1016/0377-0273(89)90047-4

De la Cruz-Reyna S, Meli R, Quaas R (2000) Volcanic crisis management. In: Sigurdsson H, Houghton B, McNutt S, Rymer H, Stix J (eds) Encyclopedia of volcanoes. Academic Press, San Diego, pp 1199–1214

De la Cruz-Reyna S, Martin Del Pozzo AL (2009) The 1982 eruption of El Chichón Volcano: eyewitness of the disaster. Geofísica Internacional 48(1):21–31

Delgado Grandos H, De la Cruz-Reyna S, Tilling RI (eds) (2008) The 1994-present eruption of Popocatépetl Volcano: background, current activity, and impacts: special issue. J Volcanol Geoth Res 170:134 pp (Collection of 11 papers)

Diario Oficial (1995) Primera sección, Secretaría de Gobernación, Acuerdo mediante el cual se crean los Comités Científicos Asesores del Sistema Nacional de Protección Civil, 6 June 1995

Diario Oficial (2006) Segunda sección, Secretaría de Gobernación, Manual de Organización y Operación del Sistema Nacional de Protección Civil, 23 Oct 2006

Duffield WA (2007) Oral tradition, El Chichón, and beyond. In: Espíndola JM, Arce JL, Macías JL (eds) Proceedings of the commemorative conference El Chichón Volcano: twenty-five years later. San Cristóbal de las Casas, Chiapas, Mexico 2007, Publicación Especial 6, Instituto de Geología, UNAM, p 26

Espíndola JM, Medina FM, De los Rios M (1989) A C-14 age determination in the Tacaná Volcano (Chiapas, Mexico). Geofísica Internacional 28:123–128

Espíndola JM, Macías JL, Sheridan MF, Tilling RI (2000) Eruptive history of El Chichón Volcano (Chiapas, Mexico) and its impact on human activity. Bull Volcanol 62:90–104

Espíndola JM, Macías JL, Godínez L, Jiménez Z (2002) La erupción de 1982 de Volcán Chichonal, Chiapas, México. In: Lugo HJ, Inbar M (eds) Desatres naturales en América Latina. Fondo de Cultura Económica, México, D.F., pp 37–65

Ewert JW, Miller CD, Hendley JW II, Stauffer PH (1998) Mobile response team save lives in volcano crises. U.S. Geological Survey Fact Sheet 064-97, 2 pp (revised June 1998)

Figueroa J (1973) Sismicidad en Chiapas. Series del Instituto de Ingeniería UNAM, Sismología e Instrumentación Sísmica, vol 316. Instituto de Ingeniería, México D.F., 50 pp

García-Palomo A, Macías JL, Arce JL, Mora JC, Hughes S, Saucedo R, Espíndola JM, Escobar R, Layer P (2006) Geological evolution of the Tacaná Volcanic complex, México-Guatemala. In: Rose WI, Bluth GJS, Carr MJ, Ewert JW, Patino LC, Vallance JW (eds) Natural hazards in Central America. Geol Soc Am Spec Pap 412:39–57

Gonzalez-Salazar A (1973) Informe preliminar de la zona geotérmica del Volcán Chichonal, Chiapas. Comision Federal de Electricidad. Internal report (unpublished) 4 pp

Havskov J, De la Cruz-Reyna S, Singh SK, Medina F, Gutiérrez C (1983) Seismic activity related to the March-April, 1982 eruptions of El Chichón Volcano, Chiapas, Mexico. Geophys Res Lett 10(4):293–296

Jiménez Z, Espíndola VH, Espíndola JM (1999) Evolution of the seismic activity from the 1982 eruption of El Chichon Volcano, Chiapas, Mexico. Bull Volcanol 61:411–422

Kraft T, Wassermann J, Schmedes E, Igel H (2006) Meteorological triggering of earthquake swarms at Mt. Hochstaufen, SE-Germany. Tectonophysics 424:245–258

La Republica (1986) Existe aparente tranquilidad pero algunos ya han abandonado el lugar: headline of article on 11 May 1986 (Periodico La Republica en Chiapas 11 May 1986)

Layer PW, García-Palomo A, Jones D, Macías JL, Arce JL, Mora JC (2009) El Chichón Volcanic complex, Chiapas, México: stages of evolution based on field mapping and $^{40}Ar/^{39}Ar$ geochronology. Geofísica Internacional 48(1):33–54

Limón-Hernández C, Macías JL (2009) Volcanic hazards and risk perception at the "Zoque" community of Chapultenango: El Chichón Volcano, Chiapas, Mexico. Geofísica Internacional 48(1):113–132

Luhr JF, Simkin T (eds) (1993) Parícutin: the volcano born in a Mexican cornfield. Geoscience Press Inc., Phoenix, 427 pp

Luhr JF, Carmichael ISE, Varekamp JC (1984) The 1982 eruptions of El Chichón Volcano, Chiapas, Mexico: mineralogy and petrology of the anhydrite-bearing pumices. J Volcanol Geoth Res 23:69–108

Macías JL (2007) Geology and eruptive history of some active volcanoes of México. In: Alaniz-Álvarez SA, Nieto-Samaniego ÁF (eds) Geology of México: celebrating the centenary of the Geological Society of México. Geol Soc Am Spec Pap 422:183–232. doi:10.1130/2007.2422(06)

Macías JL, Espíndola JM, Taran Y, Sheridan MF, García A (1997a) Explosive volcanic activity during the last 3,500 years at El Chichón Volcano, México. In: Excursion no. 6 field guide, general assembly of the international association of volcanology and chemistry of the earth's interior (IAVCEI), 12–18 Jan 1997

Macías JL, Sheridan MF, Espíndola JM (1997b) Reappraisal of the 1982 eruptions of El Chichón Volcano, Chiapas, Mexico: new data from proximal deposits. Bull Volcanol 58:459–471

Macías JL, Espíndola JM, García-Palomo A, Scott KM, Hughes S, Mora JC (2000) Late Holocene Peléan style eruption at Tacaná Volcano, Mexico-Guatemala: past, present, and future hazards. Geol Soc Am Bull 112(8):1234–1249

Macías JL, Capra L, Scott KM, Espíndola JM, García-Palomo A, Costa JE (2004) The 26 May 1982 breakout flows derived from failure of volcanic dam at El Chichón, Chiapas, Mexico. Geol Soc Am Bull 116:233–246

Macias JL, Capra L, Arce JL, Espíndola JM, García-Palomo A, Sheridan MF (2008) Hazard map of El Chichón Volcano, Chiapas, Mexico: constraints posed by eruptive history and computer simulations. J Volcanol Geoth Res 175:444–468

Macías JL, Capra L, Arce JL, Espíndola JM, García-Palomo A, Sheridan MF (2011) Mapa de Peligros del Volcán Chichonal. CIGA-UNAM and Instituto de Geofísica-UNAM, México D.F.

Martíni M, Capaccioni B, Giannini, L (1987) Ripresa dell'attivita sismica e fumarolica al Vulcano di Tacana (Chiapas, Messico) dopo un quarantennio di quiescenza. Bollettino del Grupo Nazionale per la Vulcanologia:467–470

McGee JJ, Tilling RI, Duffield WA (1987) Petrologic characteristics of the 1982 and pre-1982 eruptive products of El Chichón Volcano, Chiapas, Mexico. Geofísica Internacional 26:85–108

Medina HA (1985) Geoquímica de agues y gases del Volcán Tacaná. Chiapas, Geotermia: Revista Mexicana de Geoenergía 2:95–110

Medina F, Gonzalez L, Gutierrez C, Aguilera R, Espíndola JM (1992) Analysis of the seismic activity related to the 1982 eruptions of El Chichon Volcano, Mexico. In: Gasparini P, Scarpa R, Aki (eds) Volcanic seismology IAVCEI proceedings in volcanology, vol 3, pp 97–108

Mendoza-Rosas AT, De la Cruz-Reyna S (2008) A statistical method linking geological and historical eruption time series for volcanic hazard estimations: applications to active polygenetic volcanoes. J Volcanol Geoth Res 23:147–167. doi:10.1016/j.jvolgeores.2008.04.005

Mendoza-Rosas AT, De la Cruz-Reyna S (2010) Hazard estimates for El Chichón Volcano, Chiapas, Mexico: a statistical approach for complex eruptive histories. Nat Hazards Earth Syst Sci 10:1159–1170. doi:10.5194/nhess-10-1159-2010

Mercado R, Rose WI (1992) Reconocimiento geológico y evaluación preliminar de peligrosidad del Volcán Tacaná, Guatemala/México. Geofísica Internacional 31(3):205–237

Miller SA (2008) Note on rain-triggered earthquakes and their dependence on karst geology. Geophys J Int 173:334–338

Molina-Berbeyer R (1974) Informe preliminar geoquímico de los fluidos geotérmicos del Volcan Chichonal, Chiapas. Comisión Federal de Electricidad. Internal report (unpublished), 23 pp

Mota R, De la Cruz-Reyna S, Mena M (1984) Enjambres sísmicos en Chiapas: un fenómeno frecuente. GEOS (Boletin de la Unión Geofísica Mexicana) 2:B-17

Müllerried FKG (1932) Der Chichón, ein bischer unbegannter tätiger Vulkan im nordlichen Chiapas, Mexiko: Zeit. Vulknologie 14:191–209

Müllerried FKG (1933) El Chichón. Unico volcán en actividad en el sureste de México. Rev Inst Geol Mex 33:156–170

Müllerried FKG (1951) La reciente actividad del Volcán de Tacaná, Estado de Chiapas, a fines de 1949 y principes de 1950. Tuxtla Gutiérrez, Departomento de Prensa yTurismo, Sección Autográfica, Universidad Nacional Autónoma de México, Instituto de Geología, México, D.F. 25 pp

Murcia H, Macías JL (2014) Volcaniclastic sequences at the foot of Tacaná Volcano, southern México: implications for hazard assessment. Bull Volcanol 76:27 pp. doi:10.1007/s00445-014-0835-5

Nooren CAM, Hoek WZ, Tebbens LA, Martin del Pozzo AL (2009) Tephrochronological evidence for the late Holocene eruption history of El Chichón Volcano, Mexico. Geofísica Internacional 48(1):97–112

Orozco-Zuarth MA (1994) Sintesis de Chiapas. Edysis, Tuxla Gutierrez Chiapas, 128 pp

Riva Palacios-Chiang R (1983) Informe y comentarios acerca del Volcán Chichonal, Chiapas. In: Alcayde M (ed) El Volcán Chichonal, Ponencias presentadas en el simposio sobre el Volcán Chichonal, VI Convención Geológica Nacional de la Sociedad Geológica Mexicana. Instituto de Geología, UNAM, México, D.F., pp 49–56

Rodriguez R (1977) Enjambre de temblores ocurrido en Chiapa de Corzo, Chiapas, 1975. Tesis Lic. Fisica, Fac. de Ciencias, UNAM, Resultados Sismologicos, 55 pp

Rose WI, Mercado R (1986) Report on UNDRO/OFDA mission to Tacaná Volcano Guatemala/México. INSIVUMEH, Guatemala, 30 pp + 3 maps (in English or Spanish)

Rose WI, Bornhorst TJ, Halsor SP, Capaul WA, Plumley PS, De la cruz-Reyna S, Mena M, Mota M (1984) Volcán El Chichon, México: pre-1982 S-rich eruptive activity. J Volcanol Geoth Res 23:147–167

Rouwet R, Bellomo S, Brusca L, Inguaggiato S, Jetzeler M, Mora R, Mazot A, Bernard R, Cassidy M, Taran Y (2009a) Major and trace elements of El Chichón Volcano-hydrothermal system (Chiapas, México) in 2006–2007: implications for future geochemical monitoring. Geofísica Internacional 48(1):55–72

Rouwet D, Inguaggiato S, Taran Y, Nicholas Varley N, José A, Santiago SJA (2009b) Chemical and isotopic compositions of thermal springs, fumaroles and bubbling gases at Tacaná Volcano (Mexico–Guatemala): implications for volcanic surveillance. Bull Volcanol 71:319–335. doi:10.1007/s00445-008-0226-x

Scolamacchia T, Macías JL (2005) Distribution and stratigraphic of deposits produced by diluted pyroclastic density currents of the 1982 eruptions of El Chichón Volcano, Chiapas, México. Revista Mexicana de Ciencias Geológicas 22:159–180

SEAN (1986a) Tacaná (México/Guatemala): 2 months of increased seismicity. Bull Sci Event Alert Netw 1(1):11–12

SEAN (1986b) Tacaná (México/Guatemala): local seismicity continues. Bull Sci Event Alert Netw 11(2):4–5

Siebert L, Simkin T, Kimberly P (2010) Volcanoes of the world, 3rd edn. Smithsonian Institution and University of California Press, Berkeley, 551 pp

Sigurdsson H, Carey SN, Espíndola JM (1984) The 1982 eruptions of El Chichón Volcano: stratigraphy of pyroclastic deposits. J Volcanol Geoth Res 23:11–37

Sigurdsson H, Carey SN, Fisher RV (1987) The 1982 eruptions of El Chichón Volcano: physical properties of pyroclastic surges. Bull Volcanol 49:467–488

Simarski LT (1992) Volcanism and climate change: special report. American Geophysical Union, Washington, D.C., 27 pp

Simkin T, Siebert L, McClelland L, Bridge D, Newhall C, Latter JH (1981) Volcanoes of the world: a regional directory, gazetteer., and chronology of volcanism during the last 10,000 years. Smithsonian Institution and Hutchinson Ross Publishing Company, Stroudsburg, 232 p

Sistema Estatal de Protección Civil Chiapas (2011) Plan operativo Volcan Tacaná. Instituto de Protección Civil, UNICACH. Centro de Investigación en Gestión de Riesgos y Cambio Climático, Centro de Monitoreo Volcanológico, Sismológico. Gobierno del Estado de Chiapas, 106 pp

Tilling RI (2009a) El Chichón's surprise eruption in 1982: lessons for reducing volcanic risk. Geofísica Internacional 48 (1):3–19

Tilling RI (2009b) Volcano hazards and early warning. In: Meyers RA (ed) Encycloped complexity and systems science, vol 9. Springer, New York, pp 9861–9872

Tilling RI, Rubin M, Sigurdsson H, Carey S, Duffield WA (1984) Holocene eruptive activity of El Chichón Volcano, Chiapas, Mexico. Science 224(4650):747–749

Tilling RI, Bornhorst TJ, Taggart JE Jr, Rose WI, McGee JJ (1987) Inter-laboratory comparison of X-ray fluorescence analyses of eruptive products of El Chichón Volcano, Chiapas, Mexico. Appl Geochem 2(3):337–345

UNAM Seismology Group (1986) The September 1985 Michoacán earthquakes: aftershocks distribution and history of rupture. Geophys Res Lett 13(6):573–576

Weintraub B (1982) The disaster of El Chichón: fire and ash, darkness at noon. Natl Geogr Mag 162(5):655–684

White RA, Power JA (2001) Distal volcano-tectonic earthquakes: diagnosis and use in eruption forecasting. EOS transactions-AGU 82 (47) (Abstract #U32A-0001)

White R, Rowe C (2006) Volcano-tectonic earthquake sequences near active volcanoes and their use in eruption forecasting. Seismol Res Lett 77:240 pp

Williams SN (ed) (1990a) Special issue on Nevado del Ruiz, Colombia I. J Volcanol Geoth Res 41:1–377

Williams SN (ed) (1990b) Special issue on Nevado del Ruiz, Colombia I. J Volcanol Geoth Res 42:1–224

Yokoyama I, Tilling RI, Scarpa R (1984) International mobile early-warning system(s) for volcanic eruptions and related seismic activities: report FP/2106-82-01 (2286), UNESCO, Paris, 105 pp

Yokoyama I, De la Cruz-Reyna S, Espíndola JM (1992) Energy partition in the 1982 eruption of El Chichón Volcano, Chiapas, Mexico. J Volcanol Geoth Res 51:1–21

Zebadúa E (1999) Breve historia de chiapas. Fondo de Cultura Economica, Mexico D.F., 187 pp

Outlook

9

Juan Manuel Espindola

The activity of El Chichón and Tacaná volcanoes described in Chaps. 3 and 6, and further commented from the point of view of risk analysis in Chap. 8, illustrate the typical circumstances during volcanic crisis in little studied volcanoes. The fact that those crises had a very different outcome also illustrates the difficulties encountered by volcanologists when facing critical situations in active volcanoes about which little background information exists, and no monitoring is implemented. The outcome in the worst of cases can be a large number of causalities in one instance, or a social and economic slump besides loss of confidence in scientists in another.

The occurrence of the crises at El Chichón and Tacaná volcanoes had as a consequence a greater awareness of the hazardous nature of these volcanoes, which spurned their study and resulted in a greater knowledge of their nature. The chapters of this volume summarize the most recent research on the volcanoes and in the references one can follow the course of such research, most of it carried out after the crises. The question now is, what else is needed so that we get a sound understanding of these volcanoes? The answer to this question is manifold and depends on our focus. It is, indeed, a matter of scientific interest to continue with the study of these volcanoes yet this subject, as many others in earth science, is particularly important from the point of view of its impact on society. Therefore, research should also be directed to provide the basis for undertaking appropriate measures to reduce risk. In both cases, though, it would be pretentious to give a definitive list of actions to be carried out since scientific research, theoretical and applied, is multifaceted and continuously evolving. Thus, in what follows only a few aspects that are lacking and seem basic are pointed out.

9.1 Chichón Volcano

One of the most puzzling aspect of El Chichón is its post-eruptive behavior, in particular the total absence of dome formation and relatively low seismicity, as compared to that in other active volcanoes. Jimenez et al. (1999) suggested a depth of 8 km to the top of a seismic zone of quiescence that could be interpreted as the magma chamber, this is in agreement with the inferences made by Luhr et al. (1984) on the basis of geochemical data. A few years before the eruption a gravity survey had been carried out on El Chichón region but not a sign of an anomaly that could be related to a magma chamber was identified (Medina et al. 1990). After the eruption a the largest number of shocks occurred at a depth of 15 km under the volcano but the seismicity did not show a zone of quiescence as in the case of the 1980 Mt Saint Helens and 1991 Pinatubo eruptions (Espindola et al. 2006). The gap observed in the later eruptions is congruent with the fact that only a small part of the stored magma is ejected during eruptions (e.g. Bower and Woods 1997); therefore the behavior at El Chichón calls for further research to provide a model of its behavior. Probably part of the answer to these questions can be derived from further geochemical studies. Davidson

J.M. Espindola (✉)
Instituto de Geofísica, Universidad Nacional Autónoma de México, Ciudad Universitaria, 04510 México D.F, Mexico
e-mail: jmec@unam.mx

T. Scolamacchia and J.L. Macías (eds.), *Active Volcanoes of Chiapas (Mexico): El Chichón and Tacaná*,
Active Volcanoes of the World, DOI 10.1007/978-3-642-25890-9_9, © Springer-Verlag Berlin Heidelberg 2015

et al. (2001) have provided evidence on the episodes of replenishment of the magma chamber through studies of the radial variation of $^{87}Sr/^{86}Sr$ in feldspars. Therefore further studies of melt inclusions in phenocrystals and variations of Li in plagioclase and inclusions as well as abundance of ^{210}Pb from whole rock samples as studied for Mt St Helens by Berlo et al. (2004) would shed light on the evolution of the magma before the eruption. This would help to clarify the evolution of the magma transfer related to the 1982 El Chichón.

In Chap. 4 Peiffer and coworkers summarized the results of several years of study of the hydrothermal system, and this study provides an excellent basis for the continuous geochemical monitoring of the volcano. Although the same cannot be said of the geophysical methods, at least the seismicity is being continuously monitored with a basic array of three telemetric seismic stations. As stated in Chap. 8 the monitoring is carried out by the Centro de Monitoreo Sísmico y Volcánico (CMVS) of the Universidad de Ciencias y Artes de Chiapas. This organism is advised by Mexico's Servicio Sismológico Nacional (SSN) in the interpretation and analysis of the data. At present arrangements are being made to send the signals to the central station of the SSN at the Universidad Nacional Autónoma de México (UNAM) in Mexico City (C. Valdéz, communicacion personal). Fortunately the existence of an institution in the Sate of Chiapas taking charge of the monitoring of the volcanoes is a good starting point and it is desired that they count with the adequate support, financial and academic, to play a leading role in this pursuit. A few other geophysical measurements have been carried out over the years but none of them has yielded fundamental information about the structure of the volcano or provided the basis for a different type of monitoring (see Chap. 8). As pointed out in Chap. 8 it would be desired to continue with the deformation monitoring. Several methods are available such as the traditional methods of wet and dry inclinometry, gravimetry, and nowadays GPS (see Chap. 8). Although continuous monitoring of the parameters related to deformation is relatively costly, periodic systematic measurements good yield basic information in case of a new crisis, which although would not necessarily be related to an eruption could be related to a period of resurgence, which could instill concern among the population. In addition, seismic tomography would yield fundamental information on the structure underneath the volcano, which could be completed with a new gravity and magnetic survey of the area. This would provide a good velocity model, and therefore better location of hypocenters, as well as the possible location of a magma reservoir. These are basic measurements but of course the geophysical toolbox includes a large number of methods that can be employed to the study and monitoring of El Chichón.

9.2 Tacaná Volcano

Much has been learned about Tacaná volcano since its crisis in 1986; Chaps. 6 and 7 summarize our present state of understanding on this volcanic complex and in Chap. 8 the monitoring actions carried out nowadays were reviewed. The results of the research carried out since 1986 has presented Tacaná volcano in a new light making scientists and authorities conscious of the hazard it poses. Archaeological research has found evidences that secondary effects of the activity at Tacaná has had an intense impact on human settlements in its surrounding such as the nearby ancient city of Izapa (Rodríguez Vázquez 2010). From a consideration to the time passed since its last large eruption, the population settled around it and the height of its summit, Tacaná is riskier than El Chichón. This is further increased by its proximity to the Motagua-Polochic fault systems origin of many strong regional earthquakes. The relationships between earthquakes and volcanic activity are illustrated by the crisis of 1986 in the Tacaná itself. At present basic surveillance steps have been taken in this volcano, namely: an array of 4 seismic stations from the SSN are in operation (see www.ssn.unam.mx for details), and geochemical monitoring of the hydrothermal fluids are performed periodically (see Chap. 8). This system provides the basic information on the activity of volcano on the Mexican side; yet Tacaná is a volcano shared by two nations and for a better location of hypocenters seismic stations should also be operating on the southern side. According to the information provided by INSIV-UMEH, the Guatemalan Institution in charge of monitoring geophysical hazards in that country, the closest seismic station of their permanent network is El Palmar, Quetzaltenango, some 80 km SE of the volcano, but during the 1986 events several stations were operating at distance less than 10 km from the volcano

(Quevec Robles and Molina Cruz 1984). This is an unsatisfactory situation introduced by the political boundaries, but it can be circumvented by taking steps to pursue a joint surveillance of the volcano; during the 1986 crisis a good communication between Guatemalan and Mexican scientists was achieved. The same should be made in regard to the construction of a hazards map of the volcano, which is still lacking. The hazard map of Mercado and Rose (1992) was opportune after the 1986 events, but it was made on the basis of photogeological interpretation with little field control. Given the present state of knowledge on the volcano and the development of numerous codes for the computer simulation of volcanic phenomena, a new map is mandatory; and if is to be effective it must include the surroundings of the volcano on both sides of the Guatemala-Mexico border.

With respect to the geochemical monitoring the general insights of the Tacaná volcano-hydrothermal system are described in Chap. 7 and they provide the reference frame for its geochemical surveillance, which should be continued regularly. However because interpretation of geochemical data is often complicated by several factors it should be supplemented with other techniques, for instance radon monitoring to mention one.

In regard to geophysical studies and monitoring what was said of El Chichón applies here, even with more emphasis because the volcano has been less studied than former.

The events of 1982 at El Chichón and 1986 at Tacaná drew attention to these volcanoes and spurned volcanological research about their eruptive history, the nature of their eruptions and schemes to diminish their risk through surveillance and preparedness. It is clear that the conditions are now much better than at the time of those crises, namely: there exists is a clear conscience of their danger, a baseline exists to interpret geochemical and geophysical warning events, basic monitoring is being carried out systematically or at least frequently. However, we are still far from having a complete understanding of these volcanoes and the surveillance and studies should be continued and complemented for one thing is sure: that sooner or latter this volcanoes will reactivate while the population in its surroundings is continuously growing. Therefore it would be wise for the relevant authorities to provide research funds specifically tagged to further study El Chichón and Tacaná volcanoes; the aim should be to count with a volcano observatory run by a team of volcanologists with experience in eruptions and management of volcanic risk.

References

Berlo K, Blundy J, Turner S, Cashman K, Hawkesworth C, Black S (2004) Geochemical precursors to volcanic activity at Mount St. Helens, USA. Science 306:1167. doi:10.1126/science.1103869

Bower SM, Woods AW (1997) Control of magma volatiles content and chamber depth on the mass erupted during explosive volcanic eruptions. J Geophys Res 102 (B5):10273–10290

Davidson JP, Tepley FJ, Palacz Z, Meffan-Main S (2001) Magma recharge, contamination and residence times revealed by in situ laser ablation isotopic analysis of feldspar in volcanic rocks. Earth Planet Sci Lett 184:427–442

Espindola JM, Zamora-Camacho A, Jiménez Z (2006) Some aspects of the seismicity associated with the 1982 eruption of El Chichon Volcano, Chiapas, Mexico. J Volcanol Geoth Res 157:367–374

Jimenez Z, Espíndola VH, Espíndola JM (1999) Evolution of the seismic activity from the 1982 eruption of El Chichón Volcano, Chiapas, Mexico. Bull Volcanol 61:411–422

Luhr JF, Carmichael ISE, Varekamp JC (1984) The 1982 eruption of El Chichón volcano, 1347 Chiapas, Mexico: mineralogy and petrology of the anhydrite bearing pumices. J Volcanol Geoth Res 23:69–108

Medina F, Gonzalez-Moran T, Gonzalez L (1990) Gravity and seismicity analyses of the El Chichon Volcano, Chiapas, Mexico. PAGEOPH 133(1):149–165

Mercado R, Rose WI (1992) Reconocimiento geológico y evaluación preliminar de peligrosidad del Volcán Tacaná. Guatemala/México Geof Intern 31(3):205–237

Quevec Robles ER, Molina Cruz JE (1984) Actividad sísmica en los alrededores del Tacaná (Reporte Preliminar). Departamento de Sistemas Geofísicos, Sección Sismologia, INSIV-UMEH, Guatemala

Rodríguez Vázquez E (2010) Impacto de las erupciones volcánicas del Tacaná durante el Preclásico. In: Arroyo B, Linares y A, Paiz L (eds) En XXIII Simposio de Investigaciones Arqueológicas en Guatemala, 2009. Museo Nacional de Arqueología y Etnología, Guatemala, pp 658–664 (digital version, in Spanish)

Index

T. Scolamacchia and J.L. Macías (eds.), *Active Volcanoes of Chiapas (Mexico): El Chichón and Tacaná*,
Active Volcanoes of the World, DOI 10.1007/978-3-642-25890-9, © Springer-Verlag Berlin Heidelberg 2015

MIX
Papier aus verantwortungsvollen Quellen
Paper from responsible sources
FSC® C105338

If you have any concerns about our products,
you can contact us on
ProductSafety@springernature.com

In case Publisher is established outside the EU,
the EU authorized representative is:
Springer Nature Customer Service Center GmbH
Europaplatz 3, 69115 Heidelberg, Germany

Printed by Libri Plureos GmbH
in Hamburg, Germany